'This is an incredibly important book and I wan~ ~ ~ world to read it.'

author of *Please Don't* ~ ng

OS

'I underlined so many things ~ ~end a long time considering the implicatio~ ~rnt. Clear, concise, measured but urgent. Essential r~ ~ ~or all.'

JAN CARSON
author of *The Fire Starters*

'A powerful, up-to-date, and sometimes terrifying primer on the stupendous global problems we face today. By alternating good news and bad news chapters, Paul Behrens gives us a vivid, stereoscopic insight into the wicked challenges we humans face in coming decades.'

DAVID CHRISTIAN
Professor of Russian and European History at Macquarie University
and author of *Origin Story: A Big History of Everything*

'Behrens provides a wealth of critically important facts, accessibly and insightfully related by presentations alternately slanted to pessimistic and optimistic attitudes – an ideal structure for stimulating thoughtful discussion both in the classroom and public forum.'

HERMAN DALY
Emeritus Professor at the School of Public Policy, University of Maryland
and author of *Beyond Growth: The Economics of Sustainable Development*

'Highly readable, passionate, nuanced.'

HENRY MANCE
Chief Features Writer for the *Financial Times*

'An extraordinary distillation of science, policy, and common sense without being tedious or dismal . . . the author's grasp of the complexity and urgency of our predicament is masterful. Deserves a large global readership.'

DAVID ORR
Distinguished Professor of Environmental Studies and Politics at Oberlin College Emeritus and author of *Dangerous Years: Climate Change, the Long Emergency, and the Way Forward*

'Paul Behrens's nod to Charles Dickens in his title – *The Best of Times, The Worst of Times* – is a fitting one, for Behrens writes with the verve of a novelist, and the story he tells – how our environmental future is entangled in issues of equality, employment, housing, food, energy, and much else – a page-turner. While there are many books out there on the impact of climate change, I know of no other book that weighs the evidence so even-handedly, from both optimistic and pessimistic perspectives, enabling readers to evaluate the scientific data in an informed way. This is an illuminating and deeply researched book, one that deserves a wide readership.'

JAMES SHAPIRO
Professor of English at Columbia University and author of *Shakespeare in a Divided America: What His Plays Tell Us About Our Past and Future*

'Scientists have warned that tipping points could drive the Earth System past a fork in the road to two different futures. This book – beautifully written with a powerful format – vividly describes what these futures might look like, and how we might steer society towards a liveable future.'

WILL STEFFEN
Emeritus Professor at the Fenner School of Environment and Society, Australian National University

'Paul's book is truly amazing and I think everyone should read it. You'll learn so much, I promise. It's like a thought manual for the future. It's just plain great.'

JULIA STEINBERGER
Professor of Ecological Economics at the University of Lausanne

'Behrens is the friend that gives it to you straight: unflinching on the bad stuff, but he won't crush you with despair. This book is an excellent assessment of where we are and how we might proceed, as we navigate the uncharted terrain of the Anthropocene. Rich in complexity, deeply researched and, importantly, seeded with hope.'

GAIA VINCE
author of Adventures in the Anthropocene: A Journey to the Heart
of the Planet We Made and Transcendence: How Humans
Evolved through Fire, Language, Beauty, and Time

'The Best of Times, The Worst of Times is written in a style that brings all the data but is clear, concise, and at times poetic. Buy this book for your friends. Make them read it. It will change the way you think about the future and live your life in the present.'

STUART VYSE
behavioural scientist and author of
The Uses of Delusion: Why It's Not Always Rational to be Rational

THE
INDIGO
PRESS

THE BEST OF TIMES,
THE WORST OF TIMES

Futures from the Frontiers of Climate Science

PAUL BEHRENS

THE

INDIGO

PRESS

THE INDIGO PRESS
50 Albemarle Street
London W I S 4BD
www.theindigopress.com

The Indigo Press Publishing Limited Reg. No. 10995574
Registered Office: Wellesley House, Duke of Wellington Avenue,
Royal Arsenal, London SE I 8 6SS

This edition published in Great Britain in 2022 by The Indigo Press
Paul Behrens asserts the moral right to be identified as the author of this work in accordance with
the Copyright, Designs and Patents Act 1988
First published in Great Britain in 2020 by The Indigo Press

A CIP catalogue record for this book is available from the British Library

ISBN: 978-1-911648-09-3
eBook ISBN: 978-1-911648-10-9

Design by House of Thought
Author photo © Arash Nikkhah
Typeset in Goudy Old Style by Tetragon, London

Extract from *The Principle of Hope*, Ernst Bloch, 1954 reprinted courtesy of The MIT Press

For Caoilinn

CONTENTS

Prologue: 'Do you think we're going to be okay?'   15

POPULATION AND PROGRESS

CHAPTER ONE
Pessimism: Has the Bomb Exploded?   33

CHAPTER TWO
Hope: Better Placed than Ever   53

ENERGY

CHAPTER THREE
Pessimism: Slaves to Power   75

CHAPTER FOUR
Hope: Power to the People   93

FOOD

CHAPTER FIVE
Pessimism: Eating the Earth   113

CHAPTER SIX
Hope: Green Shoots   131

CLIMATE

CHAPTER SEVEN
Pessimism: Where All Roads Meet   151

CHAPTER EIGHT
Hope: Making Up for Lost Time   180

ECONOMICS

CHAPTER NINE
Pessimism: Counting the Costs   201

CHAPTER TEN
Hope: Valuing the Future   227

EPILOGUE
Pessimism: Are We Almost at the End?   247

EPILOGUE
Hope: The Grass Is Greener   261

Notes   271

Acknowledgements   351

About the Author   352

Fraudulent hope is one of the greatest malefactors,
even enervators, of the human race, concretely
genuine hope its most dedicated benefactor.

ERNST BLOCH, *The Principle of Hope*, 1954

Living successfully in a world of complex systems means...
expanding the horizons of caring. No part of the human race
is separate either from other human beings or the global
ecosystem. It will not be possible in this integrated world... for
the rich in Los Angeles to succeed if the poor in Los Angeles
fail, or for Europe to succeed if Africa fails, or for the global
economy to succeed if the global environment fails.

DONELLA MEADOWS, *Thinking in Systems*, 2008

The problem with the secular narrative is not that it assumes progress
is inevitable... It is the belief that the sort of advance that has been
achieved in science can be reproduced in ethics and politics.

JOHN GRAY, *Black Mass: Apocalyptic Religion
and the Death of Utopia*, 2007

'Do you think we're going to be okay?'

I n the past, when I was asked what my profession was, I'd say: physicist. This might have prompted a host of fun questions: Does a person running in the rain get equally wet as someone walking in the rain? Do sinks really drain anticlockwise in the southern hemisphere? If I mentioned that my master's degree was in astronomy, aliens would likely enter the conversation, or I might be asked why the solar system is flat, or if Libras are compatible with Capricorns. Communicating science has been a true joy: making complex concepts accessible to people (since the laws of physics aren't exclusive – everyone obeys them!), exploring the implications for science and society, hopefully fostering and sharing some awe for the world we live in. Now that my work is focused on environmental science, however, I tend to be asked a far more daunting question: 'Do you think we're going to be okay?'

It's a good question. But words always fail me. Should I assume the question relates to climate change, when they might mean another crisis, like species loss or microplastics? There is no straightforward answer to any of these problems. I could reframe the initial question: 'How long have we got?' I find the question so troubling that when others are arguing about it at the bar or over dinner, I sometimes try not to get involved. I overhear the all-too-familiar existential monsters: insect disappearances, plastic soups, massive wildfires, catastrophic floods, disappearing glaciers, tarmac-melting temperatures, antibiotic resistance. As we[1] live with intensifying environmental crises, these issues are inching towards the front of the newspapers. They are starting to become standard fare around the dinner table and in the media. Perhaps not every dinner table or inside every newspaper, but give it time… We gloss over how unique this is. At what other point in human history would two strangers on a blind date, within five minutes of first meeting, seriously be discussing how humanity is walking, eyes wide open, into global civilizational collapse?

At the dinner table, this First World conversation plays out quite predictably, advancing to the what-can-be-done phase, unconsciously imitating the structure and flow of reports from various international institutions like the Intergovernmental Panel on Climate Change (IPCC). Solutions are contested: we all have to learn to consume less; corporations need to be regulated, as they don't have the next generation's interests at heart; we need to stop having babies; nuclear war could render this whole discussion moot. The conversation spirals, becoming either melancholic or histrionic until someone suggests that we're living in the best time in human history: that the average global life expectancy is now seventy; that child mortality rates are at an all-time low; that food, energy and commodities are more plentiful than ever. Perhaps, a little relieved, people concede that things are changing – that, yes, it will be a difficult few decades – but we'll figure it out. Look how far we've come in only a few hundred years. Solutions *will be discovered*. We are a resourceful species. The conversation lurches back and forth from pessimism to hope, winner to loser.

When people look to me for information, I'm painfully aware that giving misleading assurances is dangerous – and potentially catastrophic. So too is sending people to their cars and bikes in a state of despair. I face two big challenges in delivering this information. The first is that the problems humanity faces are *systems problems*: complex, vast and distributed (rather than the complicated but more linear problem-solving of, say, going to the moon). This means that conversations that start with environmental issues quickly veer off into tangents on economics, politics and society. The second challenge is that the reality of the situation can be overwhelming. It's hard to grasp and communicate the speed and scale of the changes humanity needs to make. The same is true for the speed and scale of the destruction and suffering if action is delayed. Although T.S. Eliot said, 'Humankind cannot bear very much reality,'[2] it's important we do face up to this reality.

It's a big responsibility not to spoil the evening, as well as my friends' mental well-being... so I sketch out some of the grim realities. How humanity is past the point of prevention, that it's too late to avoid the suffering of millions, perhaps billions. I'm sure to note that these burdens are shouldered unequally around the world – that those in higher-income nations are beneficiaries of a system that has caused many of the problems which disproportionately fall on the poor (even though it wasn't necessarily anyone's personal decision to benefit in this way). I point out that serious alterations to the climate and ecosystems are all well under way, though the outcomes of this may take time for everyone to notice. I outline how scientists have been calling for action for decades, but that action hasn't been forthcoming. That the patterns of growth, resource exhaustion and the damage from our wastes – along with the social barriers, resistance of powerful interests and inequalities that let it happen – have led us into such a deep trap.

I do counter with many of the positive changes we're seeing, both in behaviour, like the turn to plant-based diets, and in technology, like the incredible drop in the cost of solar energy – developments that may presage a change fast enough and deep enough to confound the more pessimistic projections. I try to describe what everyday experience might look like if societies were able to make the needed changes – what a world that *thrives* might look like rather than a world that simply *survives*. I know that some listeners will interpret *thrive* as being able to eat sushi and buy a new wardrobe every year and to fly between countries regularly. What I mean by *thrive* is something very different. Thriving is a world full of nature, of clean air, water and soils, of increasing human rights, of higher levels of human fulfilment and meaning. Conversely, *simply surviving* could imply the cataclysmic collapse of society as we know it; the loss of human cultures, cities, landscapes and societies. I don't clarify *survive* at dinner parties, as I value having friends.

It's a balancing act to sustain hope while being clear about the profound problems we face. I usually sum up this little overview with something placatory: 'We have almost all the solutions we need, but we may not have the time – or will – to implement them to avoid a global catastrophe. It's a race between social change and environmental change.' But these issues shouldn't come down to a quick summary. The debate about our environmental future intersects with our jobs, food, energy, shelter, our morality... our everything. Any book about our environmental future is a book about our future.[3] Our environmental problems are *everything problems*. This debate should be taking place prominently, in the media, in the workplace, in schools and in politics. It's the single biggest challenge of our time. It involves some of the most important and difficult choices to be made and actions to be taken by our species, ever.

WHERE WE STAND

In less than 200 years, humans have completely transformed the planet's land, oceans and air through processes that will continue to impact on the environment for tens of thousands of years, *even if* we all were to disappear this very minute.[4] Some of these transformations we understand; others we don't. Already in 2005 we harnessed an astonishing 40% of global renewable fresh water, 72% of ice-free land, and more than 25% of the entire biological productivity of the planet's land (known by scientists as *human appropriation of net primary production*).[5] The things we buy influence others thousands of miles away and children who are yet to be born. Without a doubt, we have overdrawn on our natural inheritance, eliminating resources which took aeons to accrue and converting them into low-quality waste. Some of us have robbed the future in order to party in the present. How will these excesses catch up with us? Is it a question of holding out hope for dramatic technological breakthroughs (accepting the suffering that's already being experienced), or must we revolutionize the way we live? What happens if we change nothing and settle for adapting to our increasingly degraded environment and mounting deaths on the conscience of rich nations? This book will expose what these scenarios look like in reality.

To counter these arguments, the optimist will argue that this was all for a reason: that, on average, we have lives that would have been unimaginable to previous generations. The welfare of the average person alive today would have seemed absurd to the richest a century ago. No longer are people struck down by smallpox or plague. Journeys that would have taken months now take hours. We hold more information on the phones in our pockets than what would have been available to all the world's leaders two decades ago. We have made huge strides in improving the lifespan and health of billions around the world, and have begun what are likely to be massive revolutions in genomics, automation and artificial intelligence. We may debate whether happiness overall has actually improved and we can argue about what constitutes happiness, but for human beings, the last few decades have objectively been among the best to be alive in. Lifespan, education, access to healthcare and safety from conflict have all, on average, improved. This is not to excuse the serious, deep and unconscionable injustices around the world, or to gloss over the suffering and exploitation suffered by many. Rather, it is to point out that if a time machine were invented, it's safe to say that it would be more dangerous to travel to the past than to remain in the present.

As to whether or not the future would be safe to visit... pessimists would bring up data suggesting that the present has already begun to retrogress. The number of people facing hunger is now increasing, up from 607 million in 2014 to 768 million by 2020 (in part due to climate damage and Covid-19); average life expectancies in the US and UK have been dropping; modern inequality is more acute than ever; anti-scientific movements such as the anti-vaxxer campaign and climate change denial have already consigned many to suffering.[6] There is a strong sense that humans have cashed in nature (and possibly humanity's long-term future) for a short-term rush of consumption. Eco-anxiety has never been higher as temperatures break records year-on-year, wildfires break out and water sources dry up. Even if people are not keeping up with events, many *feel* deep down that this progress was always a Faustian bargain. Some have a sense of impending doom, declaring it too late, claiming that the changes wrought are happening much faster than predicted and that they are potentially irreversible; that our hope now is *deep adaptation*: a philosophy of resilience and regeneration in the face of ecological and civilizational collapse.[7]

Naturally, optimists would say that this mindset always existed, especially among scientists, about plagues and diseases, world wars, famines, the ozone layer, nuclear weapons and more. Many can remember the anxiety and brinkmanship of the Cold War, which didn't end in global destruction. Is it not the same for the environment? (And, for that matter, murderous artificial intelligence, antibiotic resistance, biological war and flu pandemics?) It isn't appropriate to project from the past to the future in the way many scientists do, they argue. Even the most optimistic researchers didn't predict the precipitous decline in solar energy costs we've witnessed these past few years. We will continue to innovate and – yes – there will be bumps along the way, but we are *resourceful*, and some combination of technologies and ideas will be the solution. What's more, surely we shouldn't be fearing artificial intelligence, but using it to monitor the environment and to implement better resource use.[8] This is, after all, an information-rich world, which, as of the start of 2020, contained roughly 59 zettabytes of stored information – forty times more than the observable stars in the universe.[9] Machine learning and automation can be used to free humans from labour, living lives full of friendship, hobbies, leisure, passions and fulfilment, all the while reducing humanity's environmental impact. Yes, this needs huge amounts of energy, but renewable technologies are already absurdly cheap. If we consider the pace of technological advancement in information and automation – which some call the second machine age[10] – perhaps this is not a race to environmental catastrophe but to a future very different and unfathomable to us as yet. Far from scarcity, the optimist might argue, we are moving to a marvellous post-scarcity society; to a world of abundance, beyond earthly constraints of energy and labour.[11]

Pessimists would counter that the problems we're facing are categorically different. We are all miniengineers of the environment. Simply eating food, switching on lights and travelling to work alters the atmosphere, oceans and the very ground beneath our feet. Given that there are billions of us, our individual miniengineering has amounted to *global geoengineering*: a planetary force in and of itself. It is not a question of technological prowess. Why are we discussing whether or not artificial intelligence will take over, or if philanthrocapitalists can get to Mars, when environmental breakdown is here, *now*? We have to focus all our energies on the deep problem of how we live our lives. We can't enjoy the luxury of failing fast and failing often, like a Silicon Valley enterprise. There is no reset button on the planet. No new investors to refuel our bank accounts. We don't get to 'clean up' after the fact, as high-income nations did with urban smog during the Industrial Revolution, acid rain, or the perforated ozone layer – which will be in the recovery room for another century at least. Unfortunately, we are incompetent, immoral engineers, aware that we have severely altered Earth's ability to host life but refusing to take responsibility for the consequences or the redesign required. While we disagree on how to fix the situation, our global experiment to drastically transform our lived reality continues unchecked.

Optimists may take this and run with it, exclaiming with pride that we are geoengineers! Humankind's ingenuity knows no bounds; the fact that we have avoided the apocalypse in the past means that we *will* overcome our future challenges. We are capable of working together to effect seismic shifts. All we need do is take responsibility for this role and commit to it – to manage nature more deeply and broadly. We might even end up elevating ourselves to planetary managers, calculatedly altering the atmosphere to counteract our impact and buy ourselves time for more technology to come to the rescue.[12]

This back and forth will continue, the rhetoric between the two camps becoming ever more extreme. The optimists are naive know-nothings, relying on unlikely and harmful technological fixes. The pessimists are dour, puritanical preachers who want to pour cold water on development and prosperity, taking us back centuries, committing those who are already poor to future penury. The pessimists are convinced human civilization as we know it is approaching its end. The optimists think that we are only a few decades away from a utopia of abundance. Unfortunately, the optimists and pessimists are often so extreme that each side ignores the other, using separate data sets to bolster their claims.

BALANCING DESPAIR AND HOPE

We mustn't shy away from the reality of the path we are on, nor down-play the *enormous* work we have to do if we want a liveable future. This book is not an attempt to settle this debate between pessimists and optimists. Instead, it will present readers with two very different and *possible* trajectories – one unflinchingly (but realistically) bleak, the other hopeful – without protecting readers from understanding the all too real, radically different futures that are ahead of us. In so doing, this book will address crucial questions: What changes must be made? Where and how must they be made? What are the consequences if we do not act?

To avoid a diluting of each scenario with numerous counterarguments, the pessimistic and hopeful perspectives are separated into paired chap-ters on key challenges: population and progress; energy; food; climate change; and the economy. Ethical, cultural, corporate and geopolitical factors are woven throughout since, as we shall see, their dysfunctions are directly related to our ecological problems and blockades. The structure of alternating pessimistic and hopeful chapters will allow us to fathom the scale of the problems faced and the urgency of the changes required.[13]

There is fervent discussion in the scientific literature about responsible ways of discussing pessimism and hope at this unique moment in human history.[14] Some vehemently reject too much doom and gloom, calling out a tendency to cherry-pick the worst scientific studies. Those people also argue that negativity can be disempowering to people: instead of taking action, citizens will resign to a 'time's up' nihilism. Others maintain that we cannot possibly prompt the massive social upheaval needed without first scaring citizens into action. The response we have to such messages is likely to depend more on the type of reader than the message itself. In my view, the dangers don't need exaggeration. Even middle-of-the-road estimates are deeply frightening. Equally, the world we can build if we are to tackle these dangers is truly hopeful. I will deliver the information with no specific ulterior motive to push you towards action, although both the pessimistic and hopeful chapters may have you shifting in your seat.

QUALIFYING PESSIMISM

The pessimistic chapters in this book relate how we got to this point, documenting the changes in society over time. They examine what will happen to ecosystems and the environment if current rates of change remain constant. Essentially, the pessimist's scenario is what is beginning to play out already and what will happen to Earth and the human species through insufficient action; this includes deep, *structural problems* such as political barriers, inequality and nationalism, but also *natural uncertainties* such as critical transitions (more commonly termed *tipping points*).

Every future scenario must include the grim provision that it takes many years for the impacts of our activity to be noticed by enough people to provoke change. For example, many of the fastest changes are seen in the Arctic, a region which for most is out of sight, out of mind. Also, it takes around a decade for a molecule of carbon released today to have its full impact on atmospheric temperatures.[15] These sorts of blind spots, combined with active disinformation campaigns, constitute a significant delay before the warning sirens sound in society. Scientists have been shocked by how much faster the impacts of climate change have been in the last few years. And things *will* get worse. This delay in the signal that things need to change can be catastrophic in complex systems like society and nature. By the time societies get a strong enough signal to react – for *enough* people to believe the signs – it may already be too late to save organized civilization as it struggles to maintain civic order under food and water shortages driven by environmental damage.

QUALIFYING HOPE

The hopeful chapters in this book are not necessarily optimistic. Optimism doesn't cut it. As professor of Environmental Studies and Politics David Orr puts it, 'Optimism has this confident look, feet up on the table. Hope is a verb with its sleeves rolled up.'[16] There is no chance that we will thrive without massive, unprecedented *effort*, technologically and socially. So rather than the common dichotomy of balancing techno-optimistic solutions against social behaviour change, pessimism will be pitted against *hope*, implying a more active set of decisions to be made, technologies to be deployed, systems to be developed and action to be taken. Importantly, the hope chapters will offer a picture of a better world for humans altogether – one with less pollution, more equality, more chances for meaningful fulfilment in life, better health and stronger ethics.

There is a danger of expressing hope by overstating one-off proto-types or gimmicks that can't possibly scale to the level needed to avert the worst.[17] A classic example is fusion power, the joke being that it's always just a few decades away. Given the scale of our problems and the fact that we need to act immediately, hope should be largely based on solutions available today, or almost certainly available in the very near future. Still other solutions need to be excluded on the basis of physi-cal or social limits (e.g. crops used as biofuels or self-imposed national rationing). Another reason to avoid too many one-off examples is the fact that the problems we are facing relate to systems – economic, nat-ural, social and more. Systems problems should largely be fought with a systems-thinking approach.

Another danger is the temptation to embrace a *qualified* hope, which conceals a deeper despair. 'If only we can find the will to ban fossil fuels,' for example. When 'finding the will' is invoked, it's clear there is no mechanism by which to make it happen. This is not hope but blind optimism, and blind optimism is dangerous. 'Finding the will' typically implies that government and business leaders have to make the 'right' decision, to conjure up the leadership that the public will follow – a public that is implied to have no agency. This citizen versus government action is a false dichotomy, as we shall see. Where businesses wait and see, politics acts. And politics only acts deeply enough in the interests of environmental hope once existing barriers to change are dismantled by public and legal contestation.

LOOKING FORWARD AS A SCIENTIST: SCENARIOS AND PREDICTIONS

At this stage, you may wonder how pessimism and hope might fit into the scenarios and forecasts produced by scientists. You may wonder how scientists think about the future in the first place: is it any more sophisticated than a guessing game?

Scientists try to avoid the word *prediction* because it implies too much certainty in a highly complex world. I can't predict what I'll want for dinner next Thursday, or how many solar panels will sit on my office rooftop next year. But I can make reasonable dinner-based *scenarios*, given the food I like and what I'm capable of cooking (there's a high probability of beans on toast). Similarly, I can say with reasonably high confidence that solar and wind will make up the majority of renewable energy supplies over the coming decades – not because I have insider knowledge about solar gadgetry, but due to the physics of the situation, the known limits. (I will expand on this in the 'Energy' chapters.) Rather than this counting as prediction or forecasting, scientists tend to extrapolate – or model – the future based on constraining assumptions. But as we have heard, much comes down to the pace of change. We can split future scenarios into pessimistic and hopeful by the rate of positive change we might see.

Since we know that the rate of change today is not enough to avoid catastrophic environmental changes, the pessimistic perspectives many scientists consider are 'no policy change' scenarios. 'No policy change' assumes a rate of change similar to today – a change that is too slow to prevent irreversible damage to the planet's life-support systems. The consequences of following these pathways are unfathomably terrifying, resulting in natural and civilizational ruin. Some scientists contend that we are currently suffering the early stages of just this. Although 'no policy change' might sound like the most *likely* scenario, it isn't. By definition, this is a picture of the future in which we continue behaving as we are today but with gradual changes built in. But we won't do that. Large changes are a certainty. In fact, changes continue to accelerate. We live in a period of dizzying transformation and, although we are uncertain about many things, one thing we can be certain about is that deep change is on the horizon.

If the pessimistic chapters steer closer towards 'no policy change', change animates the hopeful chapters. Serendipitously, many of the things we need to do to fix our environmental problems are also on our to-do list for improving equality, well-being and general welfare. For example, redesigning urban centres improves the environment, health and communities. Systemically, improving the education and well-being of the poorest around the world will alleviate suffering and reduce environmental pressures way into the future. Perhaps the hardest thing on the to-do list, the task easiest to postpone, is shifting the structure of the economy itself. There is no escaping it: we have to create a new definition of value, one not based on endless consumption or described by flawed statistics like Gross Domestic Product (GDP) – a statistic that omits many of the most important things in life like environmental and social health. Lest you think this is some tree-hugging ideology whose ulterior political motives pervade this book, even the International Monetary Fund (IMF) – hardly known for its extreme environmental views – argues that our global economy must change.[18]

In short, the hopeful perspective holds that not only must we address the issues with the planet, we must transform society for the better too. For reasons we'll explore, we simply will not manage in this unequal world where wealthy individuals and powerful institutions are insulated from the suffering and damage their actions inflict. They must have skin in the game. We shall see that the economy and equality are intrinsically intertwined with the environmental catastrophe we are experiencing. This is why hope for the future includes a vision of humanity as a whole thriving.

PREDICTION IS HARD, ESPECIALLY ABOUT THE FUTURE[19]

Scientists see many of our problems as emerging from the unprecedented interactions of two complex systems: society and nature. Complex systems are not only complicated in that they have many parts, but these parts interact in a way that is incredibly difficult to predict. The whole is greater than the sum of the parts. For instance, society and culture are *emergent* from the unpredictable ways humans interact together – in 2010 you'd have been hard-pressed to predict that 'Gangnam Style' would go viral or that Donald Trump would become president. Complex systems can flip from one state to another: just two years before the fall of the Berlin Wall, many experts thought it was as permanent a European fixture as the Alps[20]... until it wasn't. The climate system can also flip. Scientists call these flips *critical transitions*,[21] but they're more popularly known as *tipping points*. Tipping points often begin with feedbacks. A good example is Arctic ice melt. As the Arctic warms, the ice recedes and the region changes from reflective white to darker blue, thereby absorbing even more heat and melting more ice. As the general region continues to warm via this feedback, it may trigger other phenomena – further warming the frozen lands surrounding the Arctic, which releases more greenhouse gases, resulting in yet more warming. With more warming, other systems may also flip, resulting in a cascade. In a worst-case scenario dominoes may all be lined up behind one another so that the entire planet eventually ends up in a very different state.[22]

There are tipping points in society too. At the start of his term, President Obama said no to gay marriage.[23] Two years later, it was nationally protected. Equally, many people in the 1940s couldn't have imagined the successes of the civil rights movement in the 1950s. This is not to say that these sorts of beneficial tipping points come easily, or that they are the end of the story. Rights are fought for over many decades as a response to injustices or suffering. People around the world are getting the signal that the environment is changing rapidly. Are we on the cusp of a social tipping point to something far more sustainable? A more active participation in developing a philosophy for the long term? A hopeful perspective may say: look at the success of Extinction Rebellion, of school strikes and of legal actions against fossil fuel companies.

Even in the face of increasing levels of stress and damage, collapse – which is more generally thought of as a loss of complexity – doesn't have to be cataclysmic. It needn't be a downward turn towards violence and despair. No matter how bad things get, the vast majority of hopeful solutions we need to enact can give us more time to adapt, and to build resilience for the future. Even in the grip of collapse, there is a very real race between natural and social tipping points.

THE ANTHROPOCENE AND THE NEW ABNORMAL

To understand how tipping points apply on a planetary scale, it's important to grasp just how different the present day is to any other moment in our species' history. Human civilizations were born in the Holocene, a period beginning around 10,000 BCE of unprecedented hospitality for life: a stable climate, relatively minor natural disasters and abundant resources across much of the land and oceans. Before the Holocene, humans were a marginal species in the ecosystem. Now, we are among the most dominant forces on the planet. The human footprint is observable everywhere on Earth, and will be for millions of years. Our influence runs so deep and wide that some geologists suggest we have exited the Holocene entirely and are now in the Anthropocene.[24] The official use of the Anthropocene is still debated, as is the political context in which it is defined. This is because different sets of scientific evidence can be corralled, but also because the choice would tell very different stories about humanity, progress and power. If 1950 is chosen for the nuclear isotopes detectable across the planet after nuclear testing, the Anthropocene might be a story born of technological development and of nuclear war. If 1610 is chosen for the altered chemical composition of the atmosphere as a result of mass human death across the Americas, it is a story of colonialism, violence and disease.[25]

Whichever decision is made, the period we once thought of as the Holocene is over. Human emissions have altered the composition of the air, delaying the next glacial period; nuclear tests deposited traces of elements detectable in soil and sediments worldwide and can be used to age human brain cells;[26] human-manufactured plastics have found their way into every crevice, even in the deepest trenches of the ocean and the air above the Arctic.[27] In the Anthropocene, many human activities become so large they are hard to fathom. In the Anthropocene, enough plastic has been made that were it cling film it could wrap around the Earth completely.[28] In the Anthropocene, each litre of petrol burnt in a car melts over a tonne of glacial ice.[29] In the Anthropocene, humans move more material each year than all the planet's natural processes – like rivers – combined.[30] In the Anthropocene, an average European emits so much carbon dioxide each year that if you were to draw a column going from their shoulders up into the sky, right to the edge of the atmosphere, you would need to make ten such columns to contain their annual carbon emissions.[31] In the Anthropocene, humans have made enough concrete to cover the entire surface of Earth in a layer two millimetres thick.[32] In the Anthropocene, there are (probably) more mini LEGO humans than actual humans – their hard plastic will outlast any person alive today.[33] *Homo sapiens* has become a geobiological superpower, and superpowers are often self-destructive.

What's true for the scale and longevity of our impact in the Anthropocene is also true for the speed of change. We are at a point where deep geological time has accelerated to the lifespan of a single human. With the planetary-scale consequences of this in mind, perhaps we shouldn't flatter ourselves as being so significant as to augur a new geological epoch. Perhaps the Anthropocene is just a *boundary event* – a transition period not unlike the asteroid that hit Earth 66 million years ago – less of a long-lasting period indicating a new reality of human dominance than a fleeting signature in a thin band of rock indicating the expansion and collapse of a peculiarly self-destructive species.

In either eventuality, the Anthropocene requires new philosophies, both in order to reform the systems that led us to this point and to account for the degree of the response. For civilization to thrive, humanity will require philosophies not yet conceived of and governance systems as yet undeveloped. With so much changing so quickly, there are a number of discussions taking place in science and the humanities that have important ramifications for the future of civilization. It's crucial that we all become aware of these discussions and what they mean for our future. Some of the solutions scientists and policymakers are considering will change the way the world looks in the very short term. These discussions have yet to reach general awareness, even though their outcome will dictate what happens to life on the planet.

Climate change and ecosystem collapse will increasingly overshadow daily life. Having once amounted to niggling worries in the back of humanity's mind, they are intensifying like a devastating storm cloud. They reveal many of today's political struggles to be the equivalent of moving deckchairs around on the *Titanic*. It will be a long period of upheaval, during which many will look back on the Holocene with fondness.

As this book will reveal, human civilization has never faced such large-scale and enduring problems. The environmental crisis is the ultimate test of humanity's ability to work together, problem-solve and adapt. Incremental changes and moderate policies will no longer do the job. But what will? It is long past time to map out our options across the multitude of viewpoints and systems, examine what the best and worst cases really look like, and understand what it will take to keep as many people as possible above water.

POPULATION
AND PROGRESS

CHAPTER ONE

✥ PESSIMISM ✥

Has the Bomb Exploded?

Twelve hundred years had not yet passed
When the land extended and the people multiplied.
The land was bellowing like a bull,
The god got disturbed with their uproar.
Enlil heard their noise
And addressed the great gods:
The noise of mankind has become too much for me,
With their noise I am deprived of sleep.
Let there be a pestilence (upon mankind).

Atrahasis Epic, 1800 BCE

The proposition 'that the world's inhabitants will be happier, the greater their number' cannot be maintained, for as soon as the number exceeds that which our planet with all its wealth of land and water can support, they must needs starve one another out.

OTTO DIEDERICH LUTKEN, 'An enquiry into the proposition that the number of people is the happiness of the realm, or the greater the number of subjects, the more flourishing the state', 1758

We're getting handed a piece-of-shit planet, and I refuse to hand that down to my child. Until I feel like my kid would live on an earth with fish in the water, I'm not bringing in another person to deal with that.

MILEY CYRUS, *Elle* magazine, 11 July 2019

During the Second World War, a US coastguard unit set up a radio station on St Matthew Island off the coast of Alaska. The unit needed a back-up food source in case anything happened in that inhospitable sea, so they brought along twenty-nine reindeer – a live emergency pantry. The island happened to be a reindeer utopia, with abundant lichen and no predators (barring the odd polar bear).[1] After the war, the coastguard unit left the reindeer to their Eden. Thirteen years later, a biologist, curious to know what had become of the reindeer, paid a visit to the island and found that the twenty-nine had become a thriving population of 1,334! On a second visit six years later, the biologist found the population had exploded to over 6,000. But on this visit they noticed that the animals were thin and ill, and that the lichen underfoot was discoloured and patchy. On a final visit just three years later, skeletons littered the island. In all, only forty-two females and one male remained. The population finally died out in the 1980s. This may sound like a parable of boom and bust, but nature's story is not so neat and didactic.[2] In between the biologist's final two visits, the island was hit by one of the worst winter storms in decades. Already stressed by overpopulation and food depletion, the reindeer were less resilient to the weather. When adversity strikes, it leaves only a narrow window of opportunity to adapt. The (limited) resources to hand matter.

It's easy to read the reindeer example as a cautionary tale: if one consumes at an unsustainable rate, the population will struggle to adapt when disaster hits. But human populations, with their gall and creativity, are surely different? Perhaps… but if so, only very recently. For thousands of generations, from 200,000 to 12,000 years ago, humans were a marginal species, probably numbering fewer than five million globally.[3] The climate was in constant transition between frigid glacial periods and warmer interglacials – in all, there were more than fifty such interglacials during the last 2.6 million years (though the cycle has now been broken by human activities). Regionally, temperatures could swing by several degrees Celsius in just a few centuries, sometimes faster.[4] During the colder periods, the UK was covered in ice; during the warmer periods, trees grew in the northern Arctic.[5] It's no surprise that humans would have struggled in this climatic instability.[6] Already, people are finding it hard to process what just one degree of warming is occasioning right now, including Australia's wildfires and India's droughts, Indonesia's flooding and the US's hurricanes. Although modern civilization has more resources to hand, these changes are happening at least ten times faster than at any point in the last 65 million years.[7]

Had the climate continued fluctuating like this, complex societies might never have arisen and you wouldn't be here reading this book. Instead, this climatic uncertainty ended about 12,000 years ago, with the beginning of the Holocene: an extended interglacial period where temperatures rose and land emerged from beneath ice sheets. It was in this climate oasis that something unprecedented happened: groups of humans developed sustained agriculture, began irrigating land and domesticated animals. Opinions differ on whether this agricultural revolution was an effort to feed a growing population or, conversely, if it was the cause of population growth. Chicken or egg, the calories available increased dramatically. Although populations grew, there is good evidence that health actually suffered at the start of the revolution. While calories increased, giving up wild foods meant fewer vitamins and minerals. Nutrient deficiency was rife.[8] In trading quality for quantity, the human population was now so large that there was no going back, save for mass famine. The only option for humankind was to plough on.

This agricultural revolution sparked a social revolution. Humans had always been generalists, each knowing enough to get by. But with a bulging calorie surplus, people were freed up to pursue other activities like inventing pottery, building sailing boats, and developing writing. The increasing level of specialization led to new problems and new solutions, which in turn required more specialization. With surplus calories came the need for record-keeping and decisions on how to spend the calories. It is on such specialization that civilizations are built. It took a while, but humans eventually became far more organized than reindeer.

Not only did one of the earliest civilizations – Ancient Sumer, in modern-day Iraq – develop agriculture, they also developed irrigation.[9] As the best documented of the early civilizations, we also know that they developed religions and social hierarchies to organize society and decide where the food surpluses should be directed – whether to build a monument to the gods or to build more irrigation systems. With this new organization, a new type of affairs: *systemic* inequality crept into human affairs.[10] Kings and priests took the top spots, with other social castes far below. For centuries the Sumerian civilization grew, reaching a peak of perhaps a million by around 2900 BCE.[11]

Yet even while harvests thrived, an environmental tripwire was being laid down in the fields.

The irrigation of Ancient Sumer hadn't been developed with drainage in mind. As water was redirected to crops, it flowed over rocks and earth on the way, collecting salt. Where water evaporated, the salt remained and accumulated in the soils, eventually rendering them infertile. By 2000 BCE, scribes reported that the earth had turned white and crop yields had crashed by around two thirds.[12] Imagine your local supermarket's shelves just one-third full. Farmers tried to adapt by focusing on salt-tolerant crops like barley, but barley's lower nutritional value and reduced yields failed to fuel the now more populous and complex society. Outlying settlements were abandoned and communities began to build defences to protect crops from pillage. Unrest, hunger and conflict over the following centuries saw the Sumerian population shrink by three fifths.[13] The entrenched beliefs and practices of the good times proved to be a barrier to adaptation in times of stress.

The rise and fall of Sumer is one of the earliest examples of a *progress trap*. The agricultural revolution drove massive population growth and an increase in complexity; however, as soils became saline, the complex civilization could no longer be fed. Without food surpluses to pay for specialization – for the mathematicians and the toolmakers, the scribes and the builders – the civilization collapsed. The Sumerians seeded their future demise.

Similar dynamics played out in Rome around a thousand years later. The sheer scale of the Ancient Roman Empire, its culture and science, its political organization and subsequent downfall has fascinated researchers for centuries. At its peak, the city of Rome was home to over a million people – a size unmatched in Europe until London in the 1800s. Yet again, all this complexity was achieved by harnessing vast natural and human resources (often in the form of slave labour). Modern calculations demonstrate just how resource-intensive Rome was. Building the Colosseum probably required at least two thousand labourers working for 220 days a year for five years, along with 2,000 oxen to transport the materials. Feeding such a workforce would have required farmland larger than the area of the island of Manhattan.[14] This was for just one building (albeit an impressive one!) on a continent of aqueducts, roads, fora and lighthouses. Extensive deforestation made way for this farming, first in the environs of Rome and then further afield. As the forests were felled, the now-exposed soil eroded away.[15] Soil fertility also diminished. One response was to expand and lean on Rome's vast trade network. The early expansion of Rome's empire paid for itself, with the treasures and resources captured outweighing the costs of the legions. But as the expansion continued, it obeyed the laws of diminishing returns, with military adventures becoming an increasing burden. Rome also saw a similar entrenchment of inequality as Sumer: resource pressures drove civil unrest, slave riots and the transition from republic to empire. Long-term shifts in the regional climate may also have influenced agriculture, food sufficiency and the spread of plagues.[16] Many causes contributed to the collapse of Rome – it isn't simply a story of environmental degradation – but the civilization was far more fragile after centuries of resource exhaustion, overexpansion, despotic emperors and decaying institutions.

The dynamics of resource exhaustion, climatic pressures and entrenched inequalities appear to be a persistent feature across civilization collapses. The Ancient Mayans managed to survive several resource and climate crises through social adaptation before finally succumbing to an intense period of deforestation, soil erosion and extreme drought.[17] The Ancient Nazca in Peru met with demise after gradually denuding the landscape of trees to make way for crops, not realizing that the trees played a crucial role in preventing soil erosion.[18] Their fate was likely to have been sealed by several severe floods. For the Greenland Norse, after settling in Greenland during the tenth and twelfth centuries, the society failed to adapt to a cooling climate during the Little Ice Age of the fifteenth century. They had built an economy on cattle but, as the ice expanded, cattle farming became increasingly untenable. Anthropologist Thomas McGovern suggests that landed elites couldn't let go of their cattle-based power. Instead of adapting by emulating the local Inuits' food sources of seals and fish, the Greenland Norse held on to their traditional economic arrangements until they could no longer survive.[19] Long-term climatic changes on their own can spell the end of a civilization. In present-day New Mexico, the Chacoan civilization suffered deforestation and water-management problems before being overwhelmed by a fundamental change in climate and a fifty-year drought. Bear in mind that these struggles were faced in the Holocene's period of relative climatic calm. It's easy to imagine how the rapidly changing climate before the Holocene may have precluded complex civilizations such as today's.

The predominant theory of complex civilization collapse suggests that plentiful resources provide beneficial surpluses that drive up a society's size and complexity, allowing for more specialization and technological progress. With this progress, other problems arise, requiring more resources and solutions which, yet again, drive complexity. Civilization begins to resemble a gigantic pyramid scheme based on resources rather than money. As crises hit (due to resource depletion, climatic changes and other shocks), civilizations *can* adapt to the problems faced... but this adaptation is made harder by existing physical infrastructure, social arrangements and ideologies developed in the good years, in a process called *lock-in*. During crises, the powerful may attempt to maintain the status quo, protecting their privileged positions. This should sound eerily familiar to today's rampant ecological crises, which coincide with historic levels of inequality.

Although collapse sounds bad, it isn't necessarily cataclysmic. The remains of the Sumerian culture were transferred north as the society migrated to new lands. And although Rome fell, the eastern territories centred on Constantinople became the Byzantine Empire, which then took up the baton of Roman civilization. The people of the Mayan civilization continued living in more fractured, decentralized groups. Collapses result in subsequent civilizations reverting to a lower level of complexity, fragmenting into smaller social groups, reducing the sophistication of engineering projects, and losing the knowledge and culture stored by society. To give an example of this, after the fall of the Roman Empire the survivors in Britain forgot how to make nails... a technology that would continue to elude them for hundreds of years.[20] Today, we live in such a complex society that it is unlikely that a single one of us could make a modern pencil, let alone a mundane appliance like a toaster.[21] How is that for specialization and complexity? It may very well be that the higher the complexity, the longer the fall.

A GLOBAL CIVILIZATION

Globally, aside from a few major interruptions like the plague, human populations steadily increased by around 0.04% per year during the Holocene.[22] This may sound tiny, but it was quite the change from the pre-Holocene, when growth was essentially zero.

For most people, the small increase in population would have been unnoticeable; substantial new specialization and complexity weren't required or forthcoming. Things began to change in the fifteenth century, however, with what is now called the Columbian exchange (named after Christopher Columbus). Migration and trade were nothing new, but as European colonization connected Europe, the Americas and West Africa, both increased dramatically. Large volumes of species, people, diseases, technology and culture were transferred in a system that was to develop into mercantilist capitalism – the attempt to maximize a nation's exports and minimize its imports, often by force. Around 56 million people in the Americas died as a result of war, enslavement, famine and European diseases against which they had no immunity. In a depopulated continent, land that was once under cultivation reverted to wilderness, and reforested. As vegetation grew back, it drew down the levels of carbon dioxide in the atmosphere and probably helped push regional climates into the Little Ice Age the Greenland Norse faced.[23] The exchange of plant and animal species provided more staple foods in people's diets. The potato became so important that it alone contributed an estimated 25% of the population growth in Afro-Eurasia between 1700 and 1900.[24] The atmospheric signal of reduced carbon dioxide combined with the scale of species exchange lead some scientists to conclude that 1610 is a compelling argument for the onset of a new epoch – the Anthropocene.[25]

Throughout the sixteenth and seventeenth centuries, mercantilist capitalism provided a profit motive for further globalization. This was enhanced dramatically by the scientific revolution in Europe in the sixteenth century. With this new method of interpreting the world, scientists and engineers innovated at an unprecedented pace, allowing, too, for the growth of populations. On its own, the discoveries of the scientific revolution would have been sufficient to transfigure civilizations, but coupled with *some* enlightenment philosophies that prized the domination of natural resources, it was a formula that would drive society's colonization of nature. As a primary developer of the scientific method, the philosopher Francis Bacon (1561–1626) recommended: 'Let the human race recover that right over nature which belongs to it by divine bequest.'[26]

As Europe grew, fuelled by resources from conquest and technological progress, many began to fear overpopulation. The English cleric Thomas Robert Malthus (1766–1834) noticed a disparity between the rate of growth of human population and food production. He suggested that populations grow exponentially (four becomes eight, then sixteen, then thirty-two), while food production grows linearly (i.e. twenty-one, twenty-two, twenty-three, twenty-four...). To fully understand the depth of Malthus's concern, we have to understand the implications of exponential growth. The most succinct example is in the form of the following question:

A patch of lily pads is growing in a pond and it doubles in size every day. After thirty days, lily pads cover the pond entirely. On what day do they cover half the pond?

The answer is the penultimate day, since there is one more doubling left between then and the thirtieth day. This fact may feel counter-intuitive, but it is so important when we consider how quickly problems can get out of hand. It led physicist Albert Bartlett to remark that the greatest shortcoming of the human race is our inability to truly appreciate the exponential function. Exponentials are crucial to understanding how human societies grow, yet it's hard to appreciate how a modest number like a 2% annual economic growth rate results in a *doubling* of the size of an economy in just thirty-five years.

In any case, Malthus's concern wasn't that humanity would die out like the reindeer on St Matthew Island, rather that human populations would experience cycles of expansion and collapse. Populations would grow exponentially until they outgrew the environment's capacity to support them. The result would be famines and conflict, followed by population collapse (and culture loss) before a stable state allowed for expansion once again. At the close of the eighteenth century, Malthus thought human populations were already on the twenty-ninth day – that collapse was just around the corner.

While Malthus's insights about growth have been hugely influential, humanity hasn't succumbed to his predictions. What he didn't realize was that the First Industrial Revolution was already well under way, although its impacts weren't yet visible. The revolution led to an astonishing increase in the energy flowing through society in the form of steam- and water-powered machines. This led to the production of cheap iron, and the use of coal drove more production, more food, more energy and, ultimately, more mouths to feed. It became a feedback, with better steam engines providing access to more resources and more energy. In one sense, it was the first planetary progress trap, which together with the increasing trade and colonization around the world lay the foundations for an economic structure that would drive today's global climatic change. Huge tracts of land in the Americas, Africa and Asia were opened up to cultivation. From 1800 to 1900, it's estimated that 10 million square kilometres – an area the size of Europe – was opened up.[27] The huge energy surplus and agricultural expansion accelerated population growth such that, by 1830, humans broke the 1 billion barrier. By the 1870s, the Second Industrial Revolution was accelerating, bringing cheap steel, electricity, oil and chemical products. Global populations continued to grow, reaching 2 billion by 1940 and 3 billion by 1960, when a generation of parents earned the name *baby boomers*. Given the damage to the planet resulting from this single generation, some have renamed them 'baby bloopers'.[28] Population growth reached a historic peak of 2.1% per year in 1969.[29] To many, this growth was clearly untenable.

In 1968, a popular book called *The Population Bomb*, by ecologist Paul Ehrlich, reframed some of Malthus's concerns and called for a pre-emptive limit on global populations. Ehrlich predicted massive famines before the century was out.[30] With almost perfect timing, another revolution was also under way when *The Population Bomb* was published – in a way not dissimilar to the unseen First Industrial Revolution for Malthus. This time, it wasn't just the continued mechanization and energy transformation of the Industrial Revolution spreading to more of the world, but a *Green Revolution*, whereby a combination of high-yielding plant variants (of wheat and rice) and modern agricultural techniques were applied across vast tracts of land. A vital part of the Green Revolution depended on artificial fertilizer, causing a severe nutrient imbalance in natural global cycles and beginning a biodiversity disaster whose importance is often underestimated (more on this in the Food chapters). For his part, Ehrlich doubled down, suggesting that with the Green Revolution humans had delayed the onset of famine but that trouble would come eventually, and it would be even worse for the delay because populations would grow further in the interim. Apparently, the bomb has exploded. We are just waiting for the shrapnel to hit us.

SOCIETY VS NATURE

Between 1850 and 2000, the global population increased fivefold. A seventh billion was added between 1999 and 2011. We'll see the eighth billion by around 2024. Spread over the planet, there are fifty humans per square kilometre, compared to the nineteen reindeer that were on St Matthew at their peak. Although the global population was 'only' growing by 1.2% by 2018, that was on a massive base, amounting to more than the total population of Germany per year. It took thousands of years for the human population to grow to the first billion. A newborn in 2018 will see the global population grow 1 billion by her thirteenth birthday.

Population growth between 1950 and today has been so great, and its impacts on the environment so drastic, that the period has been dubbed *the Great Acceleration*. This astonishing change is embodied by the sheer quantity of biological resources shovelled into civilization's furnace. Researchers call the difference between the resources flowing through nature and those flowing through society the *metabolic rift*. The rift between the two has never been greater. As this resource extraction has grown, the total number of individuals the planet can support has grown concurrently. The maximum number of humans Earth's resources can support for the long term is called Earth's *carrying capacity*. While St Matthew Island clearly had a reindeer-carrying capacity somewhere below 6,000, the limit for humans doesn't remain static. The planet probably couldn't support much more than 100 million hunter-gatherers, but the agricultural revolution brought that limit up to an estimated 1 billion. Estimates place the limit for pre-Green Revolution societies at 3 billion humans, and the current limit is set at around 9 billion. For reference, UN population estimates suggest that the global population will be between 9 and 11 billion by 2050.

In short, Earth's carrying capacity is open to change – and this change could go in either direction. The human-carrying capacity of a world eating a plant-based diet is higher than a world of carnivores. Conversely, the deeper the climate change we provoke, the more difficult it will be to live and the lower this human-carrying capacity will go. Several well-respected climate scientists have warned that the human-carrying capacity of a four-degree-hotter world may only be 1 billion people due to effects on crops, water supplies and other climatic impacts like sea-level rise and ocean acidification.[31]

Not only has global population increased but humans have, on average, become much richer: global GDP per capita grew around five times between 1913 and 2008.[32] But since GDP has many flaws (which we'll describe later), it's better to look at the growth in the use of materials compared to population. Whenever material growth exceeds population growth, we can see the average material wealth per person rising. For example, during the twentieth century, the population roughly quadrupled, but the total amount of material used in human society – houses, roads, cars, laptops, furniture, transmission lines, etc. – grew an estimated twenty-threefold.[33] Similarly, over the same period, water use increased sixfold, fertilizer use increased almost two hundredfold, and energy use almost ninefold. Overall, carbon emissions have increased approximately eighteenfold. Across a wide range of indicators, consumption growth is outpacing population growth. To get a feel for how this consumption may increase in the future, we need to look at resource and income inequality today, and at how the gap between materially rich and poor communities might narrow.

Optimists like to cite the steep drop in *extreme poverty* around the world. In 1993, 34% of people earned less than $1.90 per day. (When corrected for the fact that a meal is cheaper in Niger than Sweden.) Twenty years later, by 2013, this extreme poverty was down to just 11% – a twenty-three percentage point decrease[34] (although the method for calculating this is controversial). This is an astonishing improvement. However, by looking more closely, we notice that 70% of this drop occurred in China alone;[35] the reduction in extreme poverty wasn't equally spread among the world's extremely poor.

So how did this one region manage to achieve something so extraordinary? Through a gargantuan and breakneck expansion of industry, urbanization and industrial agriculture.[36] China's material footprint and carbon emissions grew roughly five times between 1980 and 2014 – yet per capita emissions are still less than those of the US.[37] In other words, at an incredible cost in resources and environmental damage. To be clear, this is exactly what high-income nations did when they developed, and China has made this expansion far more efficiently than, say, Britain during the Industrial Revolution. Still, due to its size, China holds global importance. The statistics describing Chinese growth are hard to fathom; for example, China poured more concrete between 2011 and 2013 than the US did over the entire twentieth century.

It's hard to see how this development model can be followed in other countries if we are to avoid planetary destruction, especially since the historically richest countries continue to be the largest per capita polluters by far and don't seem to be making sufficient space for poorer nations by, for instance, rapidly reducing carbon emissions. A small section of global society is responsible for a vast amount of consumption and the environmental damage it causes.[38] The richest 10% of the world's population are responsible for around 50% of global carbon emissions, while the poorest half of the world's population are responsible for just 10% of total emissions. Put differently, if the top 10% of earners worldwide reduced their consumption to the level of the average European, global emissions would be cut by a third.[39] Very few people could argue that the life of an average European is any less happy or any more destitute than the very top earners.

Not only do the poorest people within and across nations contribute the least to globalized environmental damage like climate change, they are often first to experience the suffering brought on by this damage. India, Indonesia, Nigeria and several other countries with a combined population in the order of 2 billion are poised to go through a similar transition as China. So while it's common to focus on population, the bigger problem is consumption – most of all, the difference in consumption between the low and high consumers of today and the higher consumers of tomorrow. In sum, the number of people on the planet will 'only' grow by an estimated 23% between 2018 and 2050, but if the bottom half of global income earners were to live the same Western lifestyle of today (and the top half saw no change), carbon emissions would grow by *at least* 200%.[40]

INEQUALITY AND POWER

As we've heard, the collapse of ancient civilizations was often driven by resource depletion and/or climatic changes coupled with increasing social complexity, which made it hard to adapt to challenges. We saw an example of this with the Greenland Norse elites who held on to their cattle-based economy as the climate cooled, ultimately leading to their demise. This raises an important question: when the civilizational chips are down, what factors get in the way of adaptation? Perhaps even more importantly, what gets in the way of adapting fast *enough*? Unlike most animals, humans can accumulate communal wealth and knowledge in the good years, which can then be used to understand the problems in the bad years and adapt to their impacts. Yet humans also have a habit of storing wealth in an unequal way, resulting in unequal power, which can be a roadblock to the implementation of adaptive solutions. Anthropological studies and simplified theoretical models suggest that inequality in income and power can play a significant role in the process of civilizational collapse.[41] So today, with the *many* challenges humanity faces, what does our inequality say about our ability to adapt?

The picture is not good. Modern inequality is at record highs and, in some cases, even eclipses the extreme inequality of the ancient past. The Roman Empire, notorious for its unabashed plutocracy, doesn't hold a candle to modern imbalances. In 2012, the economist Branko Milanovic estimated that the businessman Carlos Slim, a Mexican billionaire, had an annual income that could buy the labour of 440,000 average-earning Mexicans every year. That made him fourteen times richer than Marcus Crassus, the richest Roman at the end of the repub-lic, whose annual income could buy the annual labour of 'only' 32,000 average-earning citizens at that time.[42] Wealth is concentrating in governments too. As of 2018, there are more than one hundred bil-lionaires in China's highest political bodies, worth an estimated $624 billion: an astonishing concentration of wealth that even beats the US, in which the top fifty lawmakers are worth over $6 billion.[43] As a quick reminder of how big a billion dollars is, imagine being handed a dollar every second by Jeff Bezos. You would become a millionaire in 11.6 days, but you would have to wait more than thirty-one years to become a billionaire (Bezos himself has a combined wealth of $109 billion. You'd need more than 3,000 years for him to hand over his wealth, a dollar a second).

Wealth is often further enlarged by avoiding communal responsi-bilities. A recent study found that the wealthiest 0.01% – those with more than $45 million in net wealth – evade 25% of all taxes they owe, compared to an average evasion across all income brackets of 2.8%.[44] This may sound like data from a notoriously unequal country like the US, but these data are in fact from Scandinavia. In the US, researchers found that the richest 400 families now pay a lower tax rate (federal, state and local) than any other income group, including the working class.[45] Estimates vary, but there is likely somewhere between $9 and $36 trillion sitting in tax havens around the world.[46] To put this in perspective, the cost to address climate change by completely decarbonizing the global economy by 2050 sits at around $19 to $130 trillion[47] (a wide estimate that could actually be considerably less than the higher bound and doesn't account for the fact that we already invest around 1.9 trillion a year on the energy system globally; we will explore this further later on in the book).

Some suggest that the rich make up for this with philanthropy, such as investing in personal charity projects. The problem is that this 'giving back' is tiny compared to simply paying taxes. Economist Gabriel Zucman estimates that the richest 400 people in the US have a total wealth of around $2.5 trillion, whereas the annual charitable giving of this group is just $10 billion, or 0.4% of their wealth.[48] (In any case, this philanthropy also goes towards elite institutions like Ivy League universities, which largely serve the rich.) These philanthropic gestures serve to release public pressure around paying fair taxes, so these donations can actually represent something quite pernicious.[49]

Evading community commitments is possible because income inequalities are closely related to power inequalities in influencing how society is structured (and how environmental costs are incurred). Donations, both corporate and individual, are great business, with every dollar spent in lobbying returning somewhere between $100 and $2500 to the interests of donors depending on the sector and the method of calculation.[50] This, in large part, explains how US lawmakers are five to six times more likely to respond to the concerns of the rich rather than the middle class. These dynamics have been shown in many studies; American political scientist Larry Bartels even goes so far as to say that 'there is no discernible evidence that the views of low-income constituents ha[ve] any effect on their senator's voting behavior.'[51] This is crucial given that environmental impacts fall unequally and heavily on low-income people (both within and between nations).[52] Those people who have most interest in addressing the problems have the least say in furthering solutions.

The environmental impacts of this complicated income–power rela-tionship is illustrated by the decades-long misinformation and political lobbying campaign of fossil fuel interests like ExxonMobil and Shell. Scientists at these companies were among the first to confirm the exist-ence of dangerous anthropogenic climate change. While these com-panies began to prepare their infrastructure for the increasing damage of extreme weather, they invested in misinformation and political lobbying campaigns. Between 2000 and 2016, as climate change became indisputable, oil companies in the US alone spent $2 billion resisting climate policy.[53] Research has shown how climate denial movements were driven by networks of elite corporate benefactors.[54] The campaign has been incredibly successful, with tactics that include *greenwashing*, by which companies extol their virtues via small initiatives with little to no environmental benefit, in order to convince the public that they are taking action. This is similar to how the rich use philanthropy as an ethical pressure valve. This confuses the public, stymies legislation, and enhances profits for carbon emitters... while floods, fires and droughts increase and the suffering of millions intensifies.

On top of this, mundane human foibles also mean we are resistant to necessary change. People generally feel better if they *don't lose* $50 than if they *find* $50 on the street. That is, we dislike loss more than we like the same amount of gain – researchers call this *loss aversion*.[55] More generally, this implies that losers are going to fight harder than potential winners. Since adaptation and transition to a more sustain-able world implies creating many losers and many winners, this is a big problem. For instance, oil companies might battle more intensely than new, upstart solar businesses. This would be true even if oil companies didn't already have the powerful institutional connections and resources to defend their interests, making change even harder. Inequality plays a role here too. Loss is much scarier in unequal societies, with fewer social safety nets than more equal ones. This makes sense if we think of the consequences of losing a coal-mining job in a country with limited healthcare or welfare support.[56] This may not totally prevent adaptation, but it can delay action long enough for the problems civilization faces to become too great to manage.

There are many other ways in which inequalities have an impact on the environment, and we'll discuss these throughout the book. While some commentators suggest that inequality can be separated from ecological action, and that mixing the two together 'distracts' from the environmental issues at hand, there is good evidence that inequality is intrinsic to our problems, and that it has been a recurring dynamic in the collapse of human civilizations.

THE OUTER LIMITS OF GLOBAL CIVILIZATION

The progress attained through the Great Acceleration means that today we face the soil infertility of the Ancient Sumerians, the deforestation of the Ancient Nazca, the entrenched power of the Greenland Norse, the drought of the Ancient Maya, the extreme wealth of the Ancient Romans and the climatic change of the Chacoans... along with two of the most intractable problems our species has ever faced: mass extinction due to biodiversity loss, and global climate change.

There are many ways in which humans informally deal with resources sustainably at a local level. Just think of the organized chaos of a children's playground. There are no laws or institutions governing the use of resources (which child gets to use which slide, and for how long), but somehow, even though the number of slides are limited, everyone gets a go. Unfortunately, as human populations grow and impacts become more global, the informal institutions that could allow for the sustainable use of resources have gone haywire. The global equivalent of a parent disciplining their child for taking another go on the slide doesn't exist.[57]

This is partly due to the scale and depth of impacts in the Anthropocene, but it is also an intrinsic feature of capitalism to seek out and develop new markets in a so-called *spatial fix*.[58] In doing so, markets offload the environmental damage of global demand on to locals (often indigenous peoples) who have less power in international trade negotiations. No one pays for the environmental and social damage of Amazonian deforestation when they buy Brazilian beef. The spatial fix generally follows gradients of power, whereby weaker groups see more damage. This dynamic of 'might is right' is prevalent at all levels. Larger, wealthier groups have more power... as the civil servants who continue to negotiate Britain's ongoing trade and political relationships with the European Union appreciate all too well.

This dynamic has developed even further in the last two decades. Richer nations, such as the US, imported massive amounts of materials and embodied emissions from poorer nations, such as China. Now that China is much richer, it is importing materials and embodied emissions from elsewhere in Asia and Africa. We might expect that, as African nations develop, they import resources from elsewhere still... but, as we've seen, there is very little 'elsewhere' left. Unlike Ancient Rome, expansion is no longer a means of delaying collapse. (And mining asteroids or moving to Mars – even if they were possible or desirable – wouldn't cut it within the timescale we've got.) After decades of globalization and increasing resource extraction, there are very few places left to import resources from, and fewer places to which environmental damage can be exported. Given the inequalities within countries and the inequalities between countries, it's very hard to see how humanity can develop the global institutions necessary fast enough to prevent collapse.

In a textbook example of trade power, good intentions gone wrong and poor global governance, the European Union – often thought of as a progressive force internationally – has driven massive deforestation in Indonesia via palm-oil plantations in order to meet biofuel targets (all in the name of sustainability).[59] Policymakers understood the damage only a few years into the biofuels scheme;[60] yet, in the interim, palm-oil biofuels had become big business. In a clear example of a social lock-in, powerful interests in the EU and Indonesia resisted concerted attempts to fully roll back the policy. While it looks as if the policy will eventually be repealed or improved, the delay has resulted in massive, largely irrecoverable damage to Indonesian biodiversity, and has released an amount of carbon equalling the yearly emissions of large nation states (as forests have been burned to clear land for palm oil).[61]

In previous civilization collapses, environmental decline often came first. By the time ancient peoples realized how bad things were, it was too late. (In fact, elites were often the last to find this out, since they were less exposed to damage and had the resources to maintain their lifestyles for longer.) As the decline begins, society is forced to spend more time and resources on coping with the environmental crises – in *defensive expenditures*. This diverts investment from the activities that previously fuelled society and fed the complexity we expect from civilization. Investments in the form of universities and innovation, hospitals and health, schools and education and other communal infrastructure – what we traditionally think of as bastions of progress. Today, defensive expenditures are increasing rapidly. Funds are pouring in to protect coastlines across the wealthier parts of the world, to carry out reconstruction post-hurricanes and to manage climate-driven diseases. In poorer regions, the only option is often migration, which can be poorly managed and brings more problems in the form of malnutrition, diseases and civic breakdown.

There is plenty of evidence that real progress is stalling. For example, climate-change-driven migration is on the rise, as is extreme hunger; diseases like malaria and Lyme disease are spreading; and water shortages are increasing. If the pattern of previous collapses unfolds today, then water and food shortages will become more widespread, perhaps when climate extremes hit the major food-producing regions at once. Where food and water scarcity hits, violence and civil disorder follow. The scale of a global collapse of this type may take time to arrive and then play out, but it will clearly be unprecedented. There are plenty of regions to be worried about, but East Asia, where the largest proportion of humanity lives, is a particular area of concern. Pakistan, India and China, all nuclear powers, each face severe water and food shortages within the next few decades. Collapse could be apocalyptic.

Much like ancient civilizations, recent global population and consumption limits were breached partly by technical know-how and partly by territorial and resource expansion. Because these changes are so extreme and far-ranging, they encourage the flattering thought (especially among Western societies) that humans show limitless capacity for innovation; that, unlike other animals, humans define the world around them; they bend nature to their will. It is perhaps the strongest form of secular faith that collective human ingenuity will continue to overcome limits. Many people today feel a level of techno-optimism that 'something will come along' to extract humanity from the progress traps it has set for itself. Yet it is clear that human systems are failing on many different levels, in some ways that are similar to previous collapses but also in different new, global ways. History may not repeat itself, but it often rhymes.

CHAPTER TWO

↝ HOPE ↜

Better Placed than Ever

Scientific progress makes moral progress a necessity;
for if man's power is increased, the checks that restrain
him from abusing it must be strengthened.

GERMAINE DE STAËL, 'De la Littérature', 1800

I magine you are sitting in a void, waiting to be born. A voice explains that you'll be born at any moment, but that first you must arrange the society into which you will enter. You must design this society without knowing what your nationality, race, sex, skills or predilections will be, or where in society or on the planet you will end up. You will be assigned a body at random, with no choice in the matter. In your conception of society, you might envision luxurious estates in the countryside... but take into account that if you build large unequal surpluses into your society, you are statistically more likely to be poor. From behind this 'veil of ignorance',[1] how might you organize the world and write its social contract?

This abstract thinking is an effective way to strip away the biases we all hold. It also gives us a better chance of evaluating civilization as it is today, how it was in the past, and what kind of society we aspire to become in the future. Most people can agree that life in general is better today than the past on the basic statistical level. In 1800, the average child born somewhere on the planet would have had a fifty-fifty chance of making it to their fifth birthday.[2] Today, that child has a 96% chance.[3] Across many of the most important values modern humans hold – education, health, personal freedom – things on average have been getting much, much better. Even being able to seriously consider thought experiments like the veil of ignorance is a sign of things getting better. The ideals of the Enlightenment – for example, those implied by the words of the US Declaration of Independence that 'all men are created equal' – fell abysmally short in their exclusion of women and enslaved peoples. Clearly, today's world is far from fair, especially in an age so well-resourced and wealthy. But we've come a long way towards improving the human condition.

Importantly, there is a broad public consensus that we need to go much further still. This public desire for more equality and enfranchisement is borne out in survey after survey. In a particularly illuminating study, the US public were asked how much of the total national wealth each 20% of the US population owned, from the poorest 20% to the wealthiest 20%. They were then asked how much each group *should* own in an ideal world. They guessed that the wealthiest 20% of Americans owned around 60% of national wealth, but that they *should* own just 30 to 35%. Many respondents didn't realize that the wealthiest 20% actually own over 84% of US national wealth. It's even starker for the poorest 20%, who own 0.1% of national wealth. (Those surveyed said this *should* be around 10%.)[4] Implementing the will of the people would result in the richest 20% handing over around 50% of the entire national wealth to the non-rich, and the poorest 20% receiving one hundred times their current share. Perhaps the most striking aspect of this finding is that it doesn't change depending on how or who you ask. The results are consistent across conservatives and liberals, rich and poor; all groups want significantly more equality.[5]

On top of the attitudes to equality revealed by recent studies, our attitudes towards the future have long been more committed than many realize. From the US to China, from Europe to Brazil, the majority of people say they want a more sustainable world, *even* at the risk of limiting economic growth.[6] As we've seen, political and institutional structures often present a challenge in translating this public sentiment into widespread action. But these issues are starting to coalesce: we can see the nexus of these concerns in civic movements like youth strikes and policy visions such as the Green New Deal. Environmental issues are being taken more seriously both in non-democratic countries like China and democratic countries like India. Before going any further, though, we have to fully address the common misconception that population is the biggest problem we face. We are already at a stage where the *amount* of consumption has a larger impact on the potential liveability of the planet than any remaining increase in population does. And, perhaps more importantly, the age of population growth may be over sooner than we think.

REACHING PEAK HUMAN

According to the United Nations, the human population is estimated to start shrinking near the end of this century. Others claim the UN is massively underestimating the speed of demographic change and that the global population may start shrinking within the next four decades.[7] Either way, peak human could be coming sooner than many think – and not because food and resources are running out, as Malthus suggested, but because women have been enabled to do something completely new: plan their families and choose to have fewer children. In 1960, the number of children per woman was slightly over five. By 1991, this had dropped to three. In 2018, it was 2.4.[8] When this drops to 2.1 – two children to replace the parents and 0.1 to account for accidents or infertility – we will reach a steady population.

Around half the nations of the world are already below replacement rate. There could still be population growth in these countries, either because of longer lives or via migration, but the long-term declines are clear. The change in Japan is particularly stark, with the native population shrinking by almost 600,000 people per year as of 2019 – the population of a mid-sized city.[9] Every nation in Europe, along with the US, Canada and Russia, is shrinking... or would be if it weren't for immigration helping to limit declines.[10]

This trend isn't limited to high-income nations. In 2017, Chinese-born women living in China had only 1.6 children on average – lower than the UK's 1.8 (it is worth noting that China's fertility trend had begun before the one-child policy was introduced in 1979 and which was enforced through fines, mandatory contraception and sterilization, with awful humanitarian consequences). More recently, in 2020, China posted a fertility rate of just 1.3 (lower than Japan at 1.34) which puts it on course for a population drop of roughly 700 million people by 2100.[11] Indian women are having more children than Chinese women, but the number is similar to Argentina, at 2.3, and is continuing to drop.[12] The only remaining region with high fertility rates is sub-Saharan Africa. But here rates are also declining rapidly. We continue to see an acceleration in the decline of fertility rates. It took around eighty years for fertility rates in the UK to reduce by a quarter, from just over six to 4.5 children per woman.[13] In Rwanda, a similar reduction took just five years.[14] The surprising speed of the decline has important implications on total population estimates. Partly as a result of these sorts of trends, in 2019 the UN revised its population estimates downwards from 10 billion to 9.7 billion by 2050.[15]

Although we won't hit peak human for another few decades, we are already very close to peak child.[16] There are 2 billion children in the world today, roughly the same number as there will be in 2100 *if* we avoid collapse. The remaining population growth over the twenty-first century will come from increased life expectancy and from the young people alive today having their own children. We could even see the numbers of children fall if trends in fertility rates continue to decline faster than expected. Covid-19 appears to have generally lowered birth rates even further. Early analysis suggests that the US has seen a decline of around 5%, with much of southern Europe seeing declines between 6% and 10%. There may be a short-term increase as the world exits Covid restrictions, but this is unlikely to make up for the decline.[17]

If the global replacement rate is not sustained, some fear an ageing population without enough young, productive humans to keep society vibrant and wealthy (for instance in maintaining pensions).[18]

But declining populations could be less of a problem and more an opportunity:[19] firstly, it will take time for these demographic pressures to play out (in contrast to today's urgent environmental crises), so we have time to adjust. Secondly, automation and machine learning might ease the problem of productivity, while retirees would be healthier and may want to work for longer; although we would have to be careful with inequality around both automation (as we'll hear more of in the Economics chapters) and retirement ages (since the poor generally don't live as long as the rich). Thirdly, regional declines in population – for example in Europe, Canada or the US – could be managed in the short term with migration (depending on political attitudes). This last option could map well onto existing ethical issues, since most shrinking nations are richer, have driven much of the climate damage (which is resulting in increasing levels of migration), and might be better placed to deal with the damage from ecological breakdown.

Growth isn't slowing because we have reached specific limits, like Malthus or Ehrlich predicted. For a more sophisticated set of reasons – including people's desires changing, along with more autonomy over family planning – it's likely that we will reach peak human within the next eighty years, if not sooner. This will happen either because populations continue to stabilize or, more pessimistically, because our environmental crises will result in a large loss of life. Perhaps now is a good time to quote biologist E.O. Wilson, who said that while populations grow, 'sustainability is but a fragile theoretical concept'.[20] With dwindling population growth, it is high time to move sustainability out of the realm of the theoretical.

GENDER EQUALITY IS A PRECONDITION

There is a widely held but mistaken belief that fertility rates are strongly correlated to income, and that poorer nations always have more children. Just one look at the data shows this to be false: In 2019, Bangladesh's fertility rate was 2.1 children per woman with an income per capita of $855 per year, while Bahrain's was also two at $23,551.[21] The most robust demographic research finds that the biggest determinant of family size is not income but women's education and autonomy.

Educated women have higher wages, upward mobility and lower maternal mortality; they can delay marriage and childbearing, have fewer babies, and raise healthier and better-educated children. In Ethiopia, the difference between no education and a high school education is four to five babies (two children, instead of six or seven).[22] Fortunately, the global literacy rate increased from 42% in 1960 to 86% in 2015, and women's education accounted for the largest share of that increase. In 1960, the global average duration of female education was just over three years, with almost 50% of all women having no education at all.[23] By 2010, it rose to eight years' worth, and the percentage of women with no education was down to 20%.[24] In 2011 it was estimated that if all nations with lower rates of female education adopted a similarly quick roll-out of women's education as seen in South Korea during the 1970s and 1980s (the historical world leader in educational development), there would be more than 500 million fewer people worldwide by 2050.[25] Those working in development note that smaller families are tightly connected to progress; that as family size contracts, more parenting and institutional effort (like healthcare) can be spread across fewer children, resulting in longer education and more development. Unfortunately, less than 1% of overseas development aid goes towards family planning and is often explicitly excluded from aid packages.[26] In a strange numerical coincidence, the annual growth of the global population is the same as the estimated number of unwanted pregnancies – around 85 million.[27]

To be abundantly clear, the overriding goal should be women's freedom, not limiting the number of children women have. But improving family planning, changing social norms with public information campaigns,[28] and improving women's education achieves both aims.

Gender equality has many impacts on sustainability beyond population growth. Some studies reveal that gender-balanced groups achieve better environmental outcomes in decision-making.[29] Studies have also found that women may focus on long-term goals more than men and that women have a higher aversion to inequality.[30] Other work finds evidence that countries with a higher proportion of women decision makers are more likely to protect land, ratify international environmental treaties, and have stricter climate-change policies[31] (these studies control for other factors, like political freedom, which may have an impact on environmental policy). In fact, several studies have found evidence that countries with higher levels of women's political and social freedom have lower carbon emissions, all other things being equal.[32] Women appear to also have greater climate-change knowledge and express slightly greater concern than men do.[33]

None of the above is necessarily to do with females being a greener sex. Researchers suggest that these findings are likely to be a result of increased perception and appreciation of risk.[34] This makes intuitive sense when we consider how women have historically been – and are in many places today – the predominant providers of food, water and fuel for families. It's for this reason that UN figures show that women are far more disproportionately affected by climate change than men (as women have to travel further to provide, are more likely to be in poverty, and have lower autonomy over spending than men).[35] The flip side is also true: research in New Zealand, Sweden, Brazil, the US and other countries has found that climate denialists are overwhelmingly conservative white men.[36] Again, this is less due to conservative white men being any more mendacious than other demographics, but because they are over-represented in leadership positions in current systems and perhaps most exposed to the risk of change (in an inversion of how women are more exposed to risk now).

If female participation in society improves environmental outcomes, then how are things progressing on this score? Women have had a tough run for the last several thousand years, but things are changing. As of 2019, more than 144 countries have now passed laws on domestic violence, while 154 countries have passed laws on sexual harassment.[37] Overall, 96% of countries now have paid maternity leave (the remaining 4% include Suriname and the US),[38] while 44% offer paid paternity leave. This may sound like thin gruel. (And yes, these figures should be at 100% by now.) That the change is so inadequate is partly because it is so recent: until the 1920s and 1930s, most women couldn't vote. Swiss women were withheld that right all the way up to 1971, and Saudi Arabian women still aren't allowed to vote today. It wasn't until the 1970s that women in the UK were allowed to open their own credit card account independently, as women were deemed 'too risky' and therefore needed the signature of a husband or father. The first UK anti-discrimination act wasn't passed until 1975. And while it appears that firms with better gender diversity at board level experience fewer environmental lawsuits,[39] women are still gallingly under-represented across leadership positions. The first female CEO of a Fortune 500 company was appointed in 1972. Today, thirty-two of these Fortune 500 CEOs are women (6.5%) – a paltry figure, but an improvement on zero.[40] We are also seeing a greater proportion of women in politics, finally, with Rwanda, Cuba, Bolivia, Mexico, Grenada and the Nordic countries all having more than 40% female representation.

However, there is some evidence of retrogression in recent years: one could point to the fact that the number of female CEOs in the UK has been declining over the last few years, evidence of a broad misogynistic backlash across many countries,[41] and the fact that, systematically, the World Economic Forum estimates that it will still take roughly around a hundred years to close the gender gap.[42] Also, while the percentages of women in public office look good in some countries, the reality on the ground can be less rosy. One study based on US and New Zealand data showed that as the proportion of female policymakers increases, those policymakers face more verbally aggressive and controlling behaviour.[43] This sort of backlash, although worrying, could provide an indication that things are changing; that, much like the build-up of resistance leading to women's suffrage in many countries, we *may* be at the cusp of significant changes in gender equality.

To quote physicist Charles Fourier, 'extension of women's rights is the basic principle of all social progress',[44] and, as we have seen, this is increasingly true for environmental progress. Project Drawdown – a project to map out the most beneficial solutions for addressing climate breakdown – estimates that improving women's education and family planning together would rank ahead of ninety-nine other climate solutions, above wind energy, plant-based diets and reductions in flying (though these are essential too).[45] Women's rights, dignity and autonomy are not only vital for a better world but are a *precondition* for a more sustainable world.[46]

A CHANGING PLANET: HEALTH AND SECURITY

If we appear to be reaching peak human, and some aspects of equality (such as gender equality) are improving, how do other indicators of human welfare look? The average human today spends twice as long on Earth than a couple of centuries ago, when life expectancy was just thirty years, due to an astonishingly high child mortality rate (42% of children died before the age of five). By 1950, life expectancy was up to forty-seven years (with 23% of children dying before the age of five). The last few decades have seen sci-fi-worthy lengthening of life expectancy, which is now 72.2 years on average, with not a single country having a lower life expectancy than fifty today. From 1980 to today, life expectancy has increased by over a decade. More than three extra months of life each year! Even countries that are sometimes considered 'failed states' have seen massive improvements. In the Democratic Republic of Congo (which was in a state of civil war from 1998 to 2003), life expectancy almost doubled from thirty-five in 1950 to sixty-eight by 2015. Perhaps more remarkably, global child mortality is now down to just 3.9%.[47] Globally, vaccination rates have gone up, and public health interventions to cut smoking and to improve maternal and child nutrition have lengthened lives. The decrease in child mortality coupled with increasing life expectancy accounts for a large part of the population growth during the late twentieth century.

These trends have improved the physiological development of humans, too. Access to nutritious food and healthcare means the average heights of modern Europeans are around ten centimetres taller than their forebears.[48] Neurological development has improved too: from 1942 to 2009 British children saw an average increase in Intelligence Quotient (IQ) scores of fourteen points – the average person today has an IQ which would have placed them in the top 98% of people in 1910.[49] This rate of IQ increase is found across the world, with some of the most rapid gains in Asia.[50] Although IQ itself is a contested indicator, many different types of intelligence test generally show an increase over time.[51] Again, researchers suggest these developments are down to improved health, nutrition, education and standards of living.

How about income and extreme poverty? It's important to know that the figures surrounding extreme poverty are far less straightforward than is usually admitted, and some of the gains in recent years have been overplayed. For instance, many reports and news headlines focus on the dramatic declines in extreme poverty as measured by an income level of $1.90 per person per day. But this is often criticized as an absurdly low level and doesn't seem to reflect trends in other indicators. For instance, while an estimated 700 million people live in extreme poverty, there are 768 million people facing hunger.[52] Further, 1.5 billion people face insufficient calories to sustain 'normal' daily activity – couldn't this also be counted as extreme poverty? In short, $1.90 is too low, and most researchers agree. Some suggest that $7.40 is a better level, while others recommend somewhere between $10 and $15, which is closer to the US poverty line of around $17.40 a day.[53] Where does that leave the narrative on extreme poverty? Looking across the data, no matter what threshold you use, the *percentage* of people living in extreme poverty has gone down – just not as much as the lower threshold implies. Using the lower threshold results in reductions of extreme poverty from 44% to 11% between 1981 and 2013. Using the $7.40 level gives a drop from 71% to 58% – a difference of twenty percentage points.[54] In all, there is definitely good news here – extreme poverty is down – but it is a bit more complicated than breathless commentaries would suggest.

But even if progress on extreme poverty is not quite as advertised, perhaps it's not all about the money? There *is* much better news for quality of life for people on even very low incomes. The availability and costs of obtaining the critical services people need in life, like education and healthcare, have dropped dramatically. Updating the metrics economist Charles Kenny presents in his book, *Getting Better*, in 2016 the US had an income per capita of $57,904, healthcare expenditure of almost $9,869 and a life expectancy of 78.5 years, yet Costa Rica, with an annual income per capita of $11,666 and health expenditure of $888, over ten times lower than the US, had a life expectancy of 79.7 years.[55] Perhaps even more impressively, Kerala, a semi-autonomous Indian state, had an income per capita below $1,500 in 2016 but a life expectancy of 74.9 years.[56] Kenny also showed that while Vietnam's current average income is the same as the UK's was in 1800, Vietnam's literacy rate in the year 2000 was 95% compared to 69% in the UK in the year 1800, life expectancy was sixty-nine years compared to forty-one in the UK in 1800, and infant mortality was 75% lower than the UK's in 1800.[57]

The costs of technologies have plummeted too. For instance, access to electricity provides services absolutely fundamental to development – storing food, lighting at night, access to information through radio and the internet, not to mention the government infrastructure that relies on electricity, such as hospitals. The percentage of the global population with access to electricity in recent years has been increasing at three times the speed of the 1990s, with 89% of all people worldwide having electricity access in 2016.[58] This has sometimes been achieved with very low carbon emissions. For example, access to electricity in India improved from 25% to between 67% and 74% of the total population (accounting for 650 million people) between 1981 and 2011, at a total carbon cost of just 50 million tonnes of CO_2 over that thirty-year period.[59] (That is six times less than the UK's emissions for just one year, with a population just 10% the size of India.) Progress like this highlights the possibility of finding a different development path altogether. This is true of developments in communications, healthcare (including the plummeting costs of drugs across many countries), and low-energy appliances such as refrigerators.[60]

In general, lives on average have also become safer over the last few decades.[61] Historically, the percentage of people killed by violence in any one group has varied widely – from 5% to 60%.[62] Between 1900 and 1960, less than 1% of the population died in armed conflicts, while factoring in both world wars. By 2007, this had dropped to just 0.05% of deaths from international violence.[63] The period from the end of the Second World War onwards has been dubbed 'the long peace'. There is a problem here, though: wars don't come along at regular intervals, especially not when there are nuclear weapons lurking around the planet. Over time, as the tools of violence become more effective, we are much more likely to see much smaller, limited wars than big ones like the world wars, since nuclear Armageddon provides a serious disincentive. This is exactly what we have seen in recent decades, with more conflicts that are smaller and less deadly. (This is not to brush away the devastating local conflicts around the world today, and it also doesn't preclude the very real risk of a global conflagration.) Researchers are cautiously optimistic but suggest we should wait a bit longer before declaring peace in our time (we will know in another century).[64]

Less ambiguous is the physical security from natural disasters. Even though storms, landslides, droughts, hurricanes, wildfires, etc. are more prevalent and are causing more economic damage,[65] their associated mortality is going down (for the time being). In the 1920s, when reliable statistics were first collected, twenty-eight people per 100,000 died annually. This continued to drop each decade, even in the 1930s when up to four million people died in a series of floods in China. By the 2000s, this rate was ten times smaller, at 1.7 people per 100,000.[66] A significant cause for this trend is the elimination of famines from drought. This is important to keep in mind, given increasing damages from climate breakdown (tropical storms alone are now three times as frequent as one hundred years ago[67]). When we get our act together, communities are capable of saving people from extreme events and famine, which is more often than not a political problem rather than a technological one (though this will not address more chronic problems with water and food availability as weather becomes more extreme). Although it's likely that future environmental damage *will* reverse some of these gains, we are in a better position to weather the consequences. Crucially, as environmental damage from storms, droughts and wildfires increases, there may be a corresponding increase in the mass movement of people across borders. Much of what we might consider future progress is dependent on how we deal with migration.

A WORLD ON THE MOVE

Humans and their antecedents have been on the move for millions of years. Migration is perhaps the oldest and most recognizable human trait. Since 2008 there has been an average of 26.5 million people migrating annually due to natural disasters, with extreme weather driving 90% of these migrations. In all, there are 258 million international migrants living worldwide, with 68.5 million forcibly displaced.[68] As signs point towards more environmentally driven migration in the future. Although migration has become politically fraught around the world, there are many reasons why welcoming people from other regions of the world is an economic, environmental and moral necessity.

Firstly, the economic case for migration appears to be unassailable. From the National Academies of Sciences in the US to the IMF, from the UK's independent migration advisory committee to reams of academic papers, study after study, country after country finds net positive economic outcomes from migration. Migrants pay more tax, make use of fewer public services and spur innovation.[69] By one estimate, every 1% increase in migrant population results in a 2% increase in national wages.[70] While there has been some concern about the (relatively small) impact of immigration on low-skilled workers in the short term, new research conducted in the UK is coming to the conclusion that, even in the short term and even for low-skilled workers, there is no negative economic impact from migration (to the surprise of even the researchers working on the topic).[71] It appears that the decline in the welfare of low-skilled workers of the past few years is almost entirely down to automation, 'outsourcing' or the offshoring of work, and increasing inequality.

Immigrants in general are more likely to be of working age, so they tend to bolster the number of productive workers. This is crucial in a time when many countries are both ageing rapidly and shrinking in size. On current trajectories, the number of people born in the EU is set to decline by at least 40 million by 2050.[72] Using a linear extrapolation, that's 800,000 people needed each year just to keep populations stable. Because the population of the EU is ageing, the number of productive workers is declining even faster, by 1.4 million people per year.[73] That's much larger than the peak in migration of refugees to the EU during the Syrian crisis.[74] The impacts of this are hard to fathom, since pension and welfare arrangements are predicated on the basis of productive workers paying for older generations. The *support ratio* – the number of workers for every person above sixty-five – has been shrinking in step with ageing and population decline. Japan, the most extreme example, saw its support ratio drop from 8.7 in 1970 to 1.9 in 2019.[75] Over the same period, the European support ratio dropped from 5.3 to 3.1, on its way to 1.9 by 2050.[76] This *pension time bomb* – the dwindling numbers of people paying into pension schemes, while the number of elderly people skyrockets – is one of the most serious long-term economic problems facing high-income economies. In the medium term, while we head towards peak human, migration could be a key opportunity.

Secondly, there is evidence to suggest that migration would help environmental sustainability. Migrants typically take on the attitudes of the country they live in over time, and, in general, trend towards the same fertility rates as the host population (which is often lower).[77] More immigration could reduce the time it takes to both hit peak human and limit the height of that peak. But it's not only migrants who benefit from migration; it's the families and communities they leave back home. Migrant remittances to lower-income nations are four times larger than the entire global foreign aid budget, at $422 billion – for example, an astonishing 4–8% of the entire Moroccan GDP is from remittances. The impacts of these remittances are vast: for every 10% increase, there is a 3.5% reduction in those living below US$1 a day.[78] Remittances are often used to pay for education, healthcare and to alleviate food shortages, especially during droughts.[79] The extra cash can allow girls to remain in education longer, reduces child labour, and improves the chance of finding a good job.[80] Clearly, this has a knock-on effect for women's equality and freedom, and could further decrease fertility rates. Finally, if we don't somehow balance out populations across the world, we'll inevitably have to build cities and infrastructure twice. Once, as we already have in developed nations (which will be shrinking and emptying), and again in emerging nations (which will be booming). Building cities requires huge amounts of energy and material investment – concrete alone comprises around 8% of the global total carbon emissions.[81] Well-managed, targeted migration could help alleviate these material and environmental pressures.

Thirdly, while the economic and environmental benefits of migration are clear, another argument is the moral imperative. The nations with the highest poverty and fertility rates are generally those that suffer most from climate change. Climate change that was in large part driven by the nations with the lowest poverty and fertility rates. For instance, taking a return flight from London to New York creates twice the carbon footprint than that of the average person in Bangladesh for an entire year, or is equivalent to the annual footprint of between sixteen and seventeen Rwandans.[82] Even a short-haul flight from London to Edinburgh emits more than the average person in Uganda or Somalia annually. The increasing environmental damage is on top of existing economic structures which have partially driven these inequalities in the first place. Many of the regions that are most sensitive to environmental changes are those struggling with the legacy of colonization and resource extraction, which continues to this day. Research estimates that the amount of wealth flowing out of poorer countries as a consequence of uneven trading and financial practices is more than twice that of the global foreign aid budget – overall, for every two dollars arriving as aid to poorer countries, five dollars flow out.[83]

In all, welcoming migrants in a well-managed way might allow economically developed nations in the midst of population crises the breathing space to avoid the negative impacts on pensions, productivity and the tax base. Migration also addresses environmental impacts in many direct and indirect ways, and could be important when we consider how much extra concrete may have to be poured in new cities around the world (not to mention steel and other construction materials). Some have concerns about the cultural aspects of migration, and it is important to note that this is often handled badly – for example, where additional help like language lessons is not forthcoming. We should also note a rise in eco-nativism, where some environmental figures mention migration as a scare tactic to provoke action on climate change. Here the argument goes that we should curb climate change lest we 'suffer' from climate-driven migration. But, as we will explore next, many of the greatest achievements of humanity have come from many different people, who may have very little in common, working together. Furthermore, truly cooperative groups need some level of diversity.[84] Cooperation will be vital, given that climate change has already been shown to play a significant role in explaining asylum-seeking and migration patterns between 2011 and 2015.[85] Humanity will have to get better at handling migration, as hundreds of millions of people will be on the move by mid-century (maybe more, if we continue to underestimate impacts such as sea-level rise).

A SAFE LANDING TOGETHER

Across many of the common measures of progress, the story is more nuanced than the purely optimistic takes imply. Without a doubt, many things on average are getting better. But it arguably wouldn't have taken much to improve on the high levels of disease, violence and subjugation of recent centuries. Given tremendous scientific breakthroughs, the rapid expansion in the use of resources and the deep changes in the way society views gender and racial equality, we might have expected a little more progress in global welfare. But perhaps even the expectation for more progress is in itself an improvement.

To address many of the problems in these pages, humanity will have to cooperate more deeply and broadly than ever in the past. Many people think this is impossible – that it amounts to Panglossian wishful thinking about human nature. Yet perhaps we are too quick to focus on the negatives. When you think about it, we cooperate with hundreds of other humans we don't know every day. People who provide food: those who grow it, who make the packaging, who operate the cargo ships, trucks and trains to get the food to other people. Then there are many others who maintain the water systems, the electricity systems and the infrastructure that makes urban communities possible. The wonder is that you will never meet a fraction of these people, yet they all work together, and for the most part peacefully – in many places in the world you don't have to be scared of strangers in the street.

These are some quotidian examples. Humans in bigger groups have made even larger monuments to cooperation. Although global cooperation only officially started with the United Nations (UN) in 1945, there is plenty of evidence of large-scale cooperation in the past. The multidisciplinary scientist Peter Turchin points out that the construction of Notre-Dame Cathedral was not based on a king's edict, but a communal effort between different communities, across generations. He suggests that it would have taken 300 people working full-time for half a century, very roughly equating to a total effort of 15,000 people-years.[86] Fast forward to today: the International Space Station represents political, financial and institutional cooperation between fifteen different nation states that required perhaps three million person-years to build. It's easy to overlook how remarkable this is. Many humans from different nations who didn't know each other, who couldn't even speak the same language, involving nations only a few years on from being on the brink of nuclear destruction, worked together to build one of the most outstanding expressions of human cooperation.

Building things is not the only activity whereby humans cooperate. Smallpox killed 300 million people over the course of the twentieth century. By 1977, it was eradicated completely through the efforts of several international health programmes and eventually the UN's World Health Organization. Another UN organization, the Food and Agriculture Organization, eradicated rinderpest – a cattle plague with an almost 100% mortality rate, which drove many large famines in history. Today, eradication is close for polio (down from around 350,000 cases in 1988 to just 33 by 2018) and Guinea worm disease (from around 3.5 million in the mid-1980s to 28 in 2018). These are just some eye-popping headline figures. The UN also provides vaccinations for around half of all children worldwide, and spearheaded the response to HIV and AIDS. Beyond medicine, the UN's various organizations are involved in providing food to famine-stricken regions, help for refugees, coordination on climate change, assistance with elections, international peacekeeping and more.[87]

Like any other institution, the UN has many severe failings, but it is remarkable to think that this level of cooperation is possible, given that it has existed for less than a century (less than 4% the length of the Mayan civilization). Perhaps the best example of international environmental cooperation was the Montreal Protocol – another UN-organized initiative. The Montreal Protocol was an international treaty for phasing out the use of chemicals called CFCs that deplete the ozone layer – the atmospheric layer that protects the planet from harmful ultraviolet radiation. Harmful is a bit of an understatement... If we were to lose the ozone layer, far more UV light would reach Earth's surface, damaging DNA, killing plants and animals alike. It's often forgotten now, but the expanding ozone hole was an existential, planetary threat. In 1989, within twenty years of researchers identifying a problem, the 197 member countries of the UN signed a protocol for rapidly phasing out ozone-depleting chemicals. In 1996, it was discovered that the phase-out wasn't happening fast enough and countries doubled down, implementing stricter enforcement and a target of zero emissions. There was also a multilateral fund to help lower-income nations meet the requirements of the protocol. The ozone layer needs around one hundred years to recover completely, but humanity avoided catastrophe relatively quickly. For this reason, the then UN Secretary-General Kofi Annan described it as 'perhaps the single most successful international agreement to date'.[88]

The ozone crisis holds many similarities with today's crises. It was scientifically complex, with different types of CFCs facing different phase-out rates. The phasing-out threatened existing interests, and companies attempted to resist the changes. Even by 1986, the Alliance for Responsible CFC Policy (founded by chemical company DuPont) was still suggesting that 'there is no imminent crisis that demands unilateral regulation', and 'at the moment, scientific evidence does not point to the need for... emission reductions'.[89] Of course, the problems we now face are both larger and different in important ways. CFCs are only used for a few sectors, whereas greenhouse-gas emissions are ubiquitous. Also, chemical companies don't have the same amounts of power and money as the big fossil fuel companies and fossil-funded governments of today. Nor did the Montreal Protocol at any point threaten the bedrock of how the economy operates in the same way climate change and ecosystem collapse does. However, if international cooperation can improve over such a short amount of time, from 1945 to today, perhaps it can continue to improve as the need for it becomes ever clearer. Hope for progress comes from knowing that it *is* possible to buy ourselves time – that we *can* work together, in small groups and international communities, by which we can move towards a fairer and more sustainable society; and, importantly, to focus on what progress actually means.

MOVING INTO THE FUTURE

Across many of the issues discussed in this book, both technological and social responses are needed. Most often the question isn't 'either/ or' but 'yes, and': reducing waste *while* promoting cleaner agricultural technologies; energy efficiency *as* we install renewables; reforming the use of economic indicators like GDP while *simultaneously* providing health and welfare services to everyone. Policies will need to be driven by societal outcomes, not by measures of resource use. This will require a dismantling of current resource-vested political interests. Suffice it to say, this cannot be a piecemeal effort. We will need to do *most* of the things discussed in these pages.

We are now in a better position than ever to make such a transformation. The majority of humankind has access to proper nutrition and at least seven years of schooling, meaning we are healthier and smarter than ever. Scientific knowledge continues to astound and arrive at solutions to the problems we face. The expansion of human rights has grown dramatically over the past centuries, with the expansion of the moral circle to people with different genders, ethnicities, sexualities; and rights are increasingly extended to other animals, too.[90] This expansion in rights is, in part, made possible by the fact that so many humans now have their physiological and safety needs met. Fulfilment of these needs means people can consider other issues and address other problems. As the playwright Bertolt Brecht said once in an interview, 'Grub first, then ethics'. It also means we can increase the understanding and cooperation between people. We can both understand and cooperate with people we barely know in remarkable ways; not only through the vast communications networks where new cultural and social adaptations can spread like wildfire (or, since this is a hope chapter, like a Mexican wave) but also through international organizations, something which would once have seemed impossible. Even the most selfish humans will now live so long that addressing environmental problems is in their own self-interest.

In short, we are better placed than ever to redefine what progress actually is, to address the challenges we currently face and to tackle the obstacles in the way. We already have many of the tools needed to build a world that looks more like what we would wish for from behind the veil of ignorance.

ENERGY

CHAPTER THREE

✒ PESSIMISM ✒

Slaves to Power

Civilisation requires slaves. The Greeks were quite right there.
Unless there are slaves to do the ugly, horrible, uninteresting work,
culture and contemplation become almost impossible. Human
slavery is wrong, insecure, and demoralising. On mechanical slavery,
on the slavery of the machine, the future of the world depends.

OSCAR WILDE, 'The Soul of Man under Socialism', 1891

Perhaps the destiny of man is to have a short, but fiery,
exciting and extravagant life rather than a long, uneventful
and vegetative existence. Let other species – the amoebas,
for example – which have no spiritual ambitions, inherit
the earth still bathed in plenty of sunshine.

NICHOLAS GEORGESCU-ROEGEN, *Energy and Economic Myths*, 1975[1]

We have seen how complex societies are fuelled by resources which create more opportunities to use yet more resources in a feedback effect. In today's more globalized societies, humans use more energy than can currently be harnessed from the tides, waves, ocean currents, rivers and geothermal reservoirs combined.[2] Every building you see, every morsel of food you eat, every train or car you travel in, every gadget you fiddle with, every field harvested, every medicine taken, every film watched... requires energy.

At night, much of the planet is so bathed in electrically powered light that it's possible to estimate economic activity from space by measuring its intensity.[3] There is so much light in cities it's often impossible to see the stars. But for most of human history, it was mostly impossible to continue to work or read past sunset, not only because lighting was expensive, but because it was of exceptionally poor quality. A six-watt LED bulb emits the same light as around thirty candles.[4]

When you fill the fuel tank of your car, the energy flowing through your hand per second is equivalent to the peak physical labour of 62,500 people. For the time it takes to fill the tank, you hold in your hand the output of an entire town's worth of work. More power than a feudal lord or than any queen or king in pre-industrial history. All at the cost of a few tens of pounds. Most people don't give this a second thought. We are so blasé about our apparent domination of nature that, while the tank fills, we might use our spare hand to turn down the heating in our home a hundred miles away with an app... an app on a phone that could be charged for more than three years on the amount of energy in a single litre of fuel.

Our energy nonchalance extends to the way we think about food production. Global agriculture uses around 13% of all net energy – using a further 1 to 2% in the production of fertilizer alone.[5] While it's the sun that makes plants grow, fossil fuels supersize yields and provide the harvesting, processing, transportation and storage (including refrigeration) that makes the food system viable. In general, high-income nations use around ten calories of fossil energy to produce one calorie of food energy.[6] Without these energy inputs the whole system would be much smaller, as would population; in a very real sense we have indirectly converted fossil fuels into people.

Nations have pulled pots of black gold out of the ground for decades, injecting cheap and plentiful energy into the veins of society. It has enabled people to do incredible, miraculous things – like flying around the world. But the fact that it's so cheap stops people from valuing it. Why should we exhaust ourselves cycling to work or feel cramped in a smaller car when there are bigger cars for the same price? Indeed, SUVs were the second largest contributor to growth in CO_2 emissions between 2010 and 2018, behind electricity generation and ahead of heavy industrial emissions. Their numbers on the road grew from 35 to 200 million.[7] Why take a longer train ride when flights cost less than dinner out and take less time? Before Covid, UK airline passenger numbers were increasing around 6% per year.[8] Why change our habits when we don't have to pay for the damage caused by air pollution, water use, energy wars and climate change? Why bother... when the bill gets sent to other people?

Given you are reading this book, it's likely you are energy-rich. But while you (hopefully) continue to read this book, many people cannot. Energy inequality is as much a fact of the modern world as income inequality, environmental inequality or inequality of opportunity. Just under a billion people still live without access to electricity. Three billion people cook on open fires, causing huge numbers of respiratory illnesses,[9] many of which are those who can least afford to be unwell: it is mostly mothers who do the cooking across the world. Energy inequality between nations is painfully obvious on a global scale. The average US fridge uses seven times more electricity than the average Ethiopian uses per year in total.[10] We know that in some countries people walk increasingly long distances to collect water while, in others, people drive SUVs to pick up fizzy drinks from the supermarket. But inequality is an important factor *within* nations too.

Governments wouldn't last long if they couldn't keep the power on and keep energy services cheap. In November 2018, Paris was set ablaze by riots. What ignited the riots? A fuel tax.[11] Masked protesters called the *gilets jaunes* (after the fluorescent yellow safety jackets they wore) fought police and prompted the worst riots France has seen in decades. Rising fuel prices and a new green fuel tax were the final straws for many rural poor who were already finding it hard to make ends meet and disgruntled with regional inequalities. Fuel taxes are vital in fighting climate change, but in this case the increased government tax was to be used to lower corporate taxes. Not a good look in the eyes of the rioters – regardless of your position on corporate tax rates. This highlights two important features of energy: 1) while the scale of energy use is often underappreciated, the cost is not; and 2) any attempt to address environmental issues through taxes *must* consider underlying social inequalities. The *gilets jaunes* may have welcomed the tax with open arms rather than violence had it been offset with progressive income tax cuts or increased investments in low-carbon rural transportation.[12] A general rule here might be to: 'tax what we burn, not what we earn'.

Inequality is as much a fact of life for those who provide energy as those who buy it. Protecting the jobs of coal miners was a talking point that helped Donald Trump secure votes in many swing states in America in 2016.[13] That coal is the most dangerous type of mining in the US, that it causes horrendous environmental and health damage, and that it is no longer economically viable, was immaterial.[14] Facts didn't stop Trump trying every trick in the book to bring coal back.[15] Trump's team received millions of dollars from fossil fuel companies directly and indirectly.[16] More recently, Joe Manchin, a Democrat senator in Virginia repeatedly held up the passing of President Biden's Build Back Better Plan, a legislative framework which contained a lot of positive climate policies and which would, among other things, help coal miners move into other employment as part of a 'clean energy transition'. Given that the largest coal workers' union vigorously supported the plan and implored Manchin to change his mind, it was hard to understand his resistance. That is, until one realizes he received an estimated $5 million from coal companies[17] over the last decade and in 2020 had holdings of between $1.4 and $5.8 million in those companies. Instead of providing good alternative options to communities in distress – communities wracked by unemployment and associated drug epidemics – legislators of different ideological backgrounds continue to watch their bank accounts rather than listen to their constituents.

Despite the tacit understanding that energy is important, most politicians are surprisingly oblivious to how the energy system functions, and how it might be transitioned to a more sustainable future. This ignorance, coupled with heavy lobbying, explains the UK government's massive subsidies for fossil fuels, including North Sea oil development and onshore shale gas exploration throughout the 2000s (totalling an estimated £12 billion per year in support), while subsidies for solar and other renewables are being slashed.[18] The UK ranks worst out of G7 countries for fossil fuel subsidies. In the four years following the Paris Agreement in 2015, the fossil fuel industry has spent over US$1 billion on lobbying, a figure that continues to rise.[19]

Green subsidies are not just lacking on a national scale, but on an individual scale too. Absurdly, a set of 2018 tax cuts for high earners in the United States introduced a new facility for the rich to offset the purchase of private jets against their tax bill.[20] The result is a win–win–win triple subsidy for the mega-rich but a lose–lose–lose for everyone else: 1) subsidized extraction of oil, 2) subsidized environmental damages, and 3) the subsidized purchase of the plane in the first place. These new subsidies are especially unconscionable at a time of extreme inequality, allowing the hyper-wealthy to further split themselves off from the rest of humanity.

It is the *scale* of energy use, the incredible liberties it provides, and its embedded political and societal interests which make energy reform so difficult. Historically, energy transitions take a long time. The faster the transition required, the harder the task, and the higher the risk of getting things wrong. The benefits of cheap fossil fuel energy are so great that it's been exceptionally hard to see the downside. Even if that downside is eventual mass extinction of most life on Earth.

ENERGY AND CIVILIZATION

To try and appreciate the modern energy system, let's look back in time to consider what energy has done for us to date – bearing in mind that in pre-industrial times, civilizations harnessed energy at a trickle compared to the gushing torrent today. Since it's so hard to fathom the enormous changes brought about by energy, the simple example of an everyday task should cut it down to size: the gathering of crops.

It's AD 500. The sun is coming up and it's harvesting time. A thousand people from the local village descend on the surrounding wheat fields and begin to painstakingly, back-breakingly reap the crop with scythes. The grain must then be separated (threshed) by repeatedly beating the plant with a flail or walking an animal over it, then separated from the detritus (winnowed) by throwing the mix in the air and letting the lighter chaff float away in the wind. Once the wheat is separated from the chaff, the grain must be collected from the ground ... speck by speck. If this sounds like the exercise regime of a particularly sadistic drill sergeant or personal trainer, it gets worse. These people will work all the daylight hours and return the following dawn to repeat the routine.

Some 1,500 years later, a solitary diesel-powered combine harvester ventures out over the very same land,[21] reaping, threshing and winnowing in a sequence of mechanical operations. By the time the harvester is finished, the wheat is packaged for transport to further processing facilities. All in one day, by a single farmer, in an air-conditioned cab with autopilot controls, listening to '(Don't Fear) the Reaper' by Blue Öyster Cult on the stereo. The mechanization doesn't just make life easier and safer for humans, it saves 999 people from having to head to the fields that morning. Instead, they go to school, focus on other jobs, look after children, or work on their tans. In pre-industrial times, as much as 90% of the English population worked in agriculture. Today, only 1.2% do[22] (although in the twenty-first century the UK also imports a lot of food). As societies develop, one of the first things people do is flee the fields for the city. It is the large-scale use of energy and information in the form of new technologies that has made this possible.

For the most part, the energy story is one of continuous liberation. The late Swedish epidemiologist Hans Rosling recounts the time his mother bought a washing machine. The machine freed her from many hours of hand-washing every week (prior to machines, at least a full day each week was spent cleaning clothes – a task almost universally undertaken by women and described by one housewife as 'the Herculean task which women all dread').[23] In these hours, freed from collecting water and physically washing clothes, Hans' mother read to him, sparking his interest in learning.[24] The machine wasn't a mere convenience. It translated directly into a remarkable academic who saved thousands of lives and educated millions of people.

Adding up hundreds of thousands of similar stories – across transportation, lighting, agriculture, construction, information technology and other areas – results in a staggering amount of freedom from hard labour, and an astounding amount of energy use each year. Measured in joules (an admittedly small unit equating to the amount of energy needed to lift an apple one metre), the number is so long it starts to play tricks on your eyes (520,000,000,000,000,000,000,000 joules).

WATTS IN A NUMBER?

To make sense of this number, let's return to our combine-harvester-driving farmer whose dream – incidentally – is to ride in the Tour de France. From the old cobbled roads of Roubaix to the winding slopes of the Alpe d'Huez and back up to Paris for the long straights of the Champs-Élysées, she rides 2,400 kilometres over thirty days for eight hours a day, sweating up mountains and pedalling through headwinds. Were you to add up the amount of energy she expends over the thirty days – as she burns off every croissant, all the glasses of wine, and the croque-monsieurs – it wouldn't come close to the amount of energy the average European uses each and every day. She would have to ride *Le Tour* four more times. That is to say that the average daily energy use of a European equates to 150 days of punishing physical exercise, or five Tour de Frances.[25] If you bump into your neighbour and they ask how you plan to spend your weekend, it's fair to tell them, 'Just the usual. Ten Tour de Frances, in one form or another.'

We rarely think about how much energy our lifestyles require. It's not just the obvious things like switching on the light or turning up the thermostat; it's the energy needed to provide food, build walls, floors, roofs, roads, machines, pavements, windows, hospitals, airports, bicycles and, well, everything else. There is another way to think of it: if each European uses the energy equivalent of 150 days of physical work every day, this is the equivalent of each person having 150 other imaginary people helping them. You could say that each European citizen on average is helped through life by 150 energy servants.

WHO IS GETTING THE BILL?

Given energy's value to society, which cannot be exaggerated, one would guess that it would be very expensive, but in fact energy is dirt cheap. Imagine paying those 150 energy servants the minimum wage of £8.72 an hour (for 24 hours a day). The bill would come in at over £30,000 every single day! Given that over 41% of Brits have less than £1,000 in savings, and seven in ten workers are 'chronically broke', most people would be bankrupt before breakfast.[26] Energy is so inexpensive partly because we aren't charged for all the accompanying costs to society... for the time being. We continually pay for these unseen costs because the energy system is so omnipresent that its effects are all around us. The burden of the fossil-fuel-powered energy system gets into our very bodies.

I have never heard anyone say: 'I love the smell of diesel fumes in the morning,' and yet terrible air quality is a price we seem to have accepted. Every time you even smell an exhaust, there are small particulates moving into your lungs. That exhaust and those particulates are taking time off *your* life. As of 2018, 91% of the global population has been breathing unsafe air; air which is killing seven million people a year – and not only in the smoggy East Asian cities that might first come to mind.[27] In Europe, levels of air pollution are so high that researchers estimate it's responsible for around 1,000 early deaths *per day*.[28] The vast majority of this pollution is pumped out by fossil fuels – whether power plants, trucks or petrol-powered chainsaws. There are over 70,000 scientific papers demonstrating how air pollution is affecting our health.[29] Particulate matter has been linked with aggravated asthma, lung cancer, Alzheimer's, miscarriage, diabetes, low birth weights, depression and suicide, and potentially much more.[30] It can even make a difference to a child's learning – schools downwind from highways experience lower academic achievement as opposed to schools upwind from highways.[31] In fact, there is growing evidence that air pollution could result in lower IQ.[32] Indeed, those most sensitive to air pollution's effects, babies and children, are closest to the pollutants and babies in prams can be exposed to 60% more pollution than adults.[33] The impacts are also economically and racially discriminating. Society's poorest people are more likely to live next to fossil-fuel-burning power plants. In the US, for just one transport-related pollutant – NO_2 – black and Hispanic people experience 37% higher exposures than white people.[34] Depressingly, particulate matter levels in the US haven't improved for years, in part due to a stagnation in environmental protections and particulates from the increasing size and number of wildfires.[35] While we may benefit from being able to travel from A to B with minimal effort, this hidden cost should never be forgotten.

We are literally bound up with the negative consequences of society's lifestyle choices. In one way or another, air pollution is likely to damage the health of almost everyone. It's astonishing that we don't care about this more, but the fact that lungs and hearts are deteriorating internally every day is much harder to notice than people dropping dead in the street. Over the past few decades, the health impacts of air pollution have been repeatedly and rapidly revised upwards, as more data becomes available. The total number of deaths globally each year from air pollution are fifteen times more than all wars and other violent deaths combined.[36] Recent research has shown that, as a global average, life expectancy has been reduced by one whole year due to the effects of air pollution on our lungs and hearts.[37] Dr Tedros Adhanom Ghebreyesus, Director-General of the World Health Organization, has dubbed air pollution 'the new tobacco'.[38] If you want to be actuarial about this, a recent World Health Organization report found that the cost of meeting climate-change ambitions is entirely outweighed by the health benefits of cleaner air.[39] That is to say our current air pollution, driven mostly by fossil-fuelled energy systems, costs us more (through healthcare and lives lost) than the cost of addressing climate change... driven in large part by those very same energy systems.

Besides air pollution, another huge invisible energy bill exists in our water system. The current energy system is very thirsty. Much like a car uses a radiator so that it doesn't overheat, fossil fuel and nuclear power stations need large quantities of water for cooling. A single gas power station can use an Olympic-sized swimming pool of water per minute, often freshwater. If sufficient water isn't available or if the water is too warm, the power station has to reduce, or even halt, generation. This problem of cool fresh water availability is set to increase as climate change alters the available quantity and distribution of water.[40] It's already taking a toll. During the 2016 Polish summer, extreme temperatures and drought resulted in power restrictions. Fortunately, Poland was able to rely on surplus generation available across the border in Germany.[41] But eventually, we will run out of luck. In 2018, the level of the Rhine in Germany was so low that boats could not pass – even into late November, when rain should have replenished the river. Ironically, the water level was so low that wind-turbine parts – blades that are too large to be transported by road – couldn't be delivered to their construction site.[42] In France, nuclear power stations are now regularly turned down during extreme summer heat.[43]

While we are thinking about unaccounted energy costs, we mustn't forget the global cost of energy security. Foreign aggression to secure energy resources has been a recurring theme of the twentieth and twenty-first centuries. Leaders and governments are well aware that the energy tap needs to be held open – forcibly, if need be. The concentration of fossil fuels has heavily shaped the way we treat one another geopolitically. If you took a modern map showing recent international conflict hotspots alongside another one showing oil and gas resources, the overlaps become very clear. Normally these interactions would be behind the scenes, as governments know that wheeling and dealing is not a good look. More recently, however, the co-dependencies have come front and centre. In 2018, *Washington Post* reporter Jamal Khashoggi was murdered in a Saudi Arabian consulate in Turkey on the order of Saudi Crown Prince Mohammad bin Salman. Choosing to stand by Saudi Arabia, President Donald Trump explained: 'Saudi Arabia, if we broke with them, I think your oil prices would go through the roof. I've kept them down. They've helped me keep them down.'[44]

Although it may seem obviously necessary, we almost never factor the cost of energy security into future models of the energy system. This cost of energy security comes on top of the other direct and indirect subsidies fossil fuel industries receive around the world to keep the power flowing. One IMF report estimates that these subsidies (excluding energy security) total around $5 trillion, or 6.5% of global GDP.[45] Not only does this crowd out alternatives by making fuels like oil artificially cheap, this sum, year-on-year, would be enough to completely finance the global renewable energy transition. These subsidies are often received so that domestic companies can compete with international ones – another form of energy geopolitics.

The world's three largest economies – the US, the European Union and China – are all heavily engaged in energy geopolitics. Take the 2003 invasion of Iraq. We can debate the role of oil security in the war, but it is clear from documents and public statements that it was an important factor.[46] Even if energy was, say, 15% of the reason for that war, we should allocate 15% of the human suffering and financial cost to the current energy system. Taking US costs alone, 15% of the estimated $3 trillion dollars spent in the war[47] would leave $450 billion at the doorstep of the energy system. That's $600 dollars per US citizen. This is approximately enough money to transition around 10% of the total US energy demand to renewables.[48]

In Europe, we see similar issues between the European Union and Russia. Russia supplies almost 40% of EU natural gas.[49] If Putin wanted to freeze the Czechs, Germans and Austrians in their homes over winter, he could simply close off the pipelines that stretch from Russia's gas fields all the way to the west coast of Europe. It's fairly well known that in his early years Putin was a Soviet intelligence agent working for the KGB. Less well known are his intricate links with Russian energy companies and his quest to use energy as a geopolitical tool. At one stage or another, Putin has been involved in the major decisions and direction of all the major Russian energy companies.[50] Official government websites even brag that Putin has a doctorate in strategic planning in the resource sector. Although further research later found that his thesis was plagiarized and that he never formally attended the university,[51] it does show the importance Putin places on energy politics. In 2014, Russia annexed the Crimean region of Ukraine. Again, the reason was ostensibly political and social: the Kremlin stated that it was simply complying with the 'principle of self-determination of peoples' after the pro-Russian president was deposed in a national revolution. Going relatively unnoticed by the media, Russia took control of some of the best oil and gas resources in the world off the coast of Crimea – estimates put the value of the resources in the trillions of dollars.[52]

East Asia is another hotspot for energy politics. Over the past twenty years, China has become a global superpower. With this has come a huge appetite for energy and material resources. Since China has relatively few oil and gas reserves within its own borders, it has had to look outwards to secure them. The most audacious move was China's 2009 submission to the UN to expand its territorial claim to encompass a large part of the South China Sea. The map was submitted with a crudely drawn set of nine dashes bulging out below China's southeast coastline, following the coastlines of Vietnam, Malaysia, Brunei and the Philippines. While much of the news coverage has focused on the impact this would have on commercial shipping and fishing in the region, a *vital*, overlooked aspect of this conflict is energy.

The area of the South China Sea enclosed by the so-called 'nine-dash line' probably holds well over half a trillion dollars of oil.[53] Despite an international tribunal dismissing the claim, China has already started drilling surveys in contested waters off Vietnam.[54] The artificial islands China has been constructing have destroyed at least 30% of the shallow reef habitat in the region,[55] and fishermen have been harvesting endangered species there.[56] Incidents between US and Chinese navies in the South China Sea are becoming more likely as the US conducts freedom-of-navigation operations to maintain access to the region.[57]

Due to the level of secrecy and complexity of decision-making, we may never know how profoundly energy is impacting geopolitics at the highest level. There are many reasons for the annexation of Crimea, the South China Sea expansion, and the Iraq War, but we can say for certain that energy played a part in each. Renewables may offer a way for many countries to produce their own energy, diffusing the global tensions around oil and gas supply. Without a substantial and deep renewable energy transition, we must accept that the stakes will continue to rise as conflicts over the world's remaining energy resources intensify (though the upheaval in energy politics as petrostates go out of business in the renewable energy transition will also have to be addressed).

PREDATORY DELAY

There is an active and highly effective resistance to the clean energy transition from vested interests who profit from the status quo. The leadership at oil and gas interests in particular have shown scant regard for the health or survival of organized human civilization. You might say that their aim is in fact to 'soak the planet in the maximum amount of fossil fuels'.[58] The overall impact of burning existing reserves alone would raise sea levels by 58m.[59] These fossil fuel entities operate in full knowledge of their intent and have done so for some time. Scientists at Big Oil companies like ExxonMobil were among the first to recognize the dire harm of climate change over forty years ago and internal documents show that the science was understood and accepted without fanfare.[60] Such was their cynicism, the leaders at companies like ExxonMobil and Shell quietly began preparing for a warmer world. They redesigned oil platforms for stormier seas and fortified coastal pipelines against increasing erosion.[61] At the same time, they were funnelling large sums of money towards opaque lobby groups to foster doubt among the public, to muddy politics and to delay action.[62]

Their deplorable tactics have been devastatingly effective. The active, well-documented disinformation campaigns run by the leaders of Big Oil squandered decades of potential action; decades in which millions of lives could have been saved from air pollution, geopolitical conflicts and climate change. Had we taken measures in earnest in 1970, when these companies themselves were convinced of the threat, our survival could have been so much likelier and easier. Reductions of less than 2% in carbon emissions each year would have done the trick, with time to spare. That we didn't, and that we have continued to equivocate is what journalist Alex Steffen terms 'predatory delay'. He explains that 'winning slowly is basically the same thing as losing outright.' The manufacturing of doubt is a prime example of predatory delay and, given the stakes, it may go down as one of the greatest immoral acts in history. The journalist and campaigner Bill McKibben argues that there should be a term for when you commit treason against an entire planet.

To many, it is old news that Big Oil bosses did some terrible things. There is an impression that they have backed down in recent years, as the impacts of their industry have become blindingly obvious. Not so. The lobbying has been unrelenting. In the two years following the Paris climate accords, Big Oil spent a further billion dollars on 'greening' their image – by plastering adverts on billboards, newspapers, TV screens, inflight magazines and on social media, trying to associate their brand with a newfound care (though several are already planning for some of the worst climate projections).[63] For instance, ExxonMobil would like you to know that they are working on algae biofuels. They don't mention that this research is window dressing, totally impractical, and amounts to a tiny percentage of their investments, compared with the tens of billions of dollars they invest in fossil fuels development.[64] Shell, advertising in the venerable New York Times, would like to tell you about their plan for net-zero emissions by 2070,[65] pretending not to know that this would be at least twenty years too late.[66] BP is excited to tell you that they are aiming to limit operational emissions to 2025 at the same level… as today. Meanwhile, lobbyists from these companies freely roam the halls of power in America, Europe and elsewhere, using their special access cards to drop in on politicians and policymakers for cosy chats about the latest techno-fixes, some of which are even worse than the fossil fuels themselves.[67] It is no exaggeration to say that these sordid, cynical tactics will reverberate for millennia in the geological record.

Aside from vested interests, there are more mundane reasons for delay. Every year, major energy institutions like the International Energy Agency (IEA) provide scenarios of the future energy system. These scenarios have become something of a running joke to energy scientists and clean-tech entrepreneurs. In 2010, they predicted that annual global solar capacity would be increasing by 30GW by 2030.[68] In 2019, 109GW were installed: over three times as much, and more than a decade ahead of schedule. Year after year, agencies like the IEA have systematically underestimated the importance and growth of clean technologies and overestimated the importance of fossil fuels. These scenarios drive many business decisions and government policies. If I am asking you for money to start a new solar business, would you lend me the cash if you read reports from eminent international institutions announcing that solar has a limited future?[69]

This question of finance is crucial. Estimations vary, but perhaps around $3 trillion per year globally is needed to make the clean-energy transition.[70] Here, financial institutions are also failing spectacularly, partly due to a systematic underestimation of the scale of the problem. In 2018, a well-researched report found that thirty-three of the most powerful private banks, including JPMorgan Chase and Barclays, had provided $1.9 trillion to fossil fuel companies over the two years since the Paris Agreement.[71] Government and international development banks have been doubling down on past mistakes too. The China Development Bank and the Export-Import Bank of China are intent on repeating the grave misdeeds of the past, investing in coal plants across lower-income nations;[72] committing them to the very mistakes high-income countries made.

These trillions of dollars of investment are a complete waste and simply won't pay off. Either the fossil energies are used and ecosystems collapse (no shareholder value then), or we make the clean energy transition and it will have been money down the drain. This is the problem of *stranded assets*, of which there are an estimated $11 to $14 trillion worth worldwide (to put this in perspective, these losses are of a similar size of the total losses of the 2008 financial crash). Just imagine if this money had been used productively.[73]

TRANSITION TIME: RESOLUTIONS WE NEVER KEEP

If we could put an end to all these delays, how fast could we roll out the clean-energy future in practice? This is where optimists will drop facts about breakneck technological transformations: Google was founded twenty years ago; the internet became publicly available twenty-five years ago; Uber has been around for less than a decade; mobile phone ownership in the UK has leaped from 16% to 79% in just ten years.[74] Optimists love to say that while resources are limited, creativity is infinite; we simply need to be smarter and to believe in technology ('believe' is a more optimistic word than 'invest'). But the energy system is a very different beast to Silicon Valley innovations, artificial intelligence, or block chains. Energy systems are vast infrastructural affairs that represent decades- or centuries-long investments. Their material foundations – the steel, aluminium, copper and concrete – literally run into the bedrock of civilization. From the massive spinning metal in power-plant generators to the thousands of kilometres of electrical transmission and gas pipes connecting continents. We may ditch a smartphone for a newer model every few years, but power plants often last for more than forty years.[75] The current energy system has huge inertia embodied within it. Like an oil tanker, it needs time to change course.

Perhaps we could build a bridge to the future by using a better fossil fuel in the short term, while we make the transition to renewables? Optimists say that natural gas is the solution we've been looking for: the metha(ne)done to wean us off our carbon addiction. This is really persuasive because gas power stations are able to fill in quickly for renewables when the sun is obscured, or the wind stops blowing. On top of that, it's much cleaner than coal. The problem is that natural gas has a nasty corollary: methane. Methane is a much stronger climate-altering gas than carbon dioxide. It is far more damaging in the short term. *If* we could keep natural gas in the pipes before we burn it (converting the methane to carbon), it might be usable; *however*, gas networks can be very leaky. Only 3.2% has to leak through the pipes and machinery before it's worse for climate change than coal over the short term.[76] Current estimates put the leakage in the US somewhere around 2.3%, but with some installations emitting as much as 4% – right in the danger zone.[77] Some would still argue that switching to gas is better because this effect diminishes over several decades – yet we know we don't have that long. In the meantime, billions of dollars of investments are being sunk into natural-gas infrastructure. Natural gas is not a bridge to a cleaner future. It is the minefield before the cliff's edge.

There is no way around it: we must restrict fossil fuels wherever possible *and* pull off a fast transition to low-carbon energy sources. It's a matter of survival. No ifs, no buts. Unfortunately, history suggests that energy transitions consistently take between fifty and seventy-five years: it took sixty years for coal to make up 50% of all energy used. It was an almost identical rate for oil and natural gas, which came later.[78] To give a sense, then, of how long we have, the 2018 IPCC special report on limiting global warming to 1.5 degrees suggests that we must transition to a 100% carbon-free economy globally within twenty years *if* we are to avoid using speculative geoengineering technologies. Even for a 2 degree limit, we must transition in just over thirty-five years. This is a Herculean task. We have never managed an energy transition anything like this before. And with nonsensical, regressive politics consuming governments, and trenchant vested interests thwarting change, the pessimist's case is grimly convincing.

Look at any plot of carbon emissions over time versus how quickly and how far they need to fall in the future, and you'll see what looks like a cliff edge. For a 1.5-degree scenario, without a *deus ex machina* materializing, we would need to decarbonize by around 7 to 10% year-on-year into the future. For reference, the drop in emissions over 2020, which included that at the height of Covid lockdowns, was 6.3%.[79] This is just flatly impossible. What's more, global growth in GDP is around one-for-one with carbon emissions – a 1% increase in GDP results in a 1% increase in emissions.[80] Pre-pandemic, in 2018, the global economy increased 3.7% (carbon emissions increased 3.4%), so the rate of decarbonization needs to be roughly 8–11% to keep pace (this casts a pall on the concept of future economic growth, something we will explore further in the Economics chapters).

If you have the stomach for more bad news, here are two more paragraphs: say we *are* able to muster the political and global will to transition to a renewable energy system, even *building* that new, green system will involve significant emissions. Think of all the concrete needed for wind turbine foundations, copper for solar panels, lithium for batteries. It's hard to calculate, but a very rough estimate suggests another 1–3% of total annual emissions.[81] Thus, the 8–11% rate of decarbonization would have to become, charitably, 12%... *per year*. These calculations are done off the basis of models which are increasingly revealed to be too conservative. If we want to avoid these temperatures in practice, it probably needs to be even faster than this.

To get an idea of how seriously corporate and global leaders are taking this issue, look no further than the UN climate talks at COP26, held in 2021. Diplomats, policymakers and scientists met following the Covid-19 lockdowns to discuss global climate policy. After a year which saw Earth's warmest month in recorded history, record-breaking European flooding (the costliest weather disaster in European history), unprecedented dust storms in China, and mounting evidence proving that climate change is progressing faster than anticipated, delegates gathered at a conference centre in Glasgow.[82] Corporate sponsors of COP26 included Reckitt Benckiser, the event's 'Hygiene Partner' (who the World Wildlife Fund's 2020 scorecard report ranked number eighteen out of twenty companies making progress towards phasing out unsustainable palm oil); Unilever, a COP26 'Partner' who the Rainforest Action Network found in 2020 to be performing inadequately in avoiding 'conflict palm oil', a type of palm oil that leads to deforestation, loss of habitats and exploitation of workers or Indigenous peoples);[83] and Jaguar Land Rover, a company producing the large, SUV-style vehicles that were the second-largest contributors to CO2 emission growth between 2010 and 2018.[84]

The conference hall and negotiation rooms were windowless and separated from the public by a huge police presence.[85] In these rooms, distanced from the child activists marching through the surrounding streets, delegates had very little contact with the environment and the communities they were ostensibly there to protect. The number of delegates at the conference who were associated with fossil fuels outnumbered the largest national delegation from Brazil.[86] Political representatives from rich countries blocked progress in developing a new framework for helping lower-income nations to address climate damage and cut emissions.[87]

The conference resulted in pledges absent from any of the short-term targets needed to make them a reality. As the number of largely symbolic announcements increased, Greta Thunberg dismissed them as 'blah blah blah', highlighting the grandiose talk and minuscule action. While the International Energy Agency estimated that the long-term pledges could result in a total warming of 1.8 degrees, Climate Action Tracker analysed the short-term targets connected to these pledges and found a warming of at least 2.4 degrees – a small improvement on previous declarations.[88] Indeed, emissions rebounded in 2021 during the post-COVID 'recovery'. Global emissions in 2021 increased by 4.9%. This is slightly below the 2019 record emissions high… and suffice it to say, that is nowhere close to the reductions we must see year-on-year to avoid catastrophe. Further national emission increases in the coming years are also expected.[89] It's enough to make one want to flee to New Zealand to build a fallout bunker. But… don't blow all your savings on canned food just yet. There is a page here to be turned. There is a whole chapter of hope up ahead that maps out an achievable transition to a clean energy system for everyone.

CHAPTER FOUR

∽ HOPE ∾

Power to the People

We are like tenant farmers chopping down the fence around our
house for fuel when we should be using Nature's inexhaustible
sources of energy – sun, wind and tide… I'd put my money on the
sun and solar energy. What a source of power! I hope we don't
have to wait until oil and coal run out before we tackle that.

THOMAS EDISON, 1931

We put together pieces of metal alloy and fossil fuel so that
we hurtle through the sky at close to the speed of sound; we
organize tiny signals from the spins of atomic nuclei to make
images of the neural circuits inside our brains; we organize
biological objects – enzymes – to snip tiny slivers of molecules
from DNA and paste them into bacterial cells. Two or three
centuries ago we could not have imagined these powers. And
I find them, and how we have come by them, a wonder.

W. BRIAN ARTHUR, *The Nature of Technology:
What it Is and How it Evolves*, 2009

At the turn of the twentieth century, there were over 100,000 horses
on the streets of New York. The stench of more than one million
kilos of manure and 220,000 litres of urine baking on hot summer streets
would have been overwhelming. Besides the dizzying smells, it would
have been physically treacherous. Nasty bugs could be caught from the
excrement, horrible injuries from slipping in the muck. The economy's
reliance on horses rendered cities particularly vulnerable. Something as
commonplace as an outbreak of horse flu – such as the one that hit in
1872 – could cause food shortages for city dwellers across the country.[1]

Despite these significant downsides, horses were an indispensable part of society, keeping shops well stocked, parties attended and trade running smoothly. By 1905, there were more than 13,000 horse-related businesses in New York alone, holding the economy up on four legs. Yet just seventeen years later, there were fewer than fifty horse-related businesses remaining, and the final horse-drawn trolley made its final trip. Horses had been replaced by electric streetcars and automobiles. The public health hazard had been traded out too – biological effluent for today's chemical effluent.

Today, we are on the cusp of a much larger, galloping energy transition. We have almost all the clean-energy solutions needed to make this transition, and many of them are cheaper than fossil-fuelled incumbents. Renewable energy has repeatedly beaten the most hopeful predictions for efficiency and affordability. For years, experts assumed that clean energy would cost us in the way that ethical products tend to cost more on the high street. We also feared that – due to the variability or intermittency of the sun and the wind – renewables would never be able to compensate for fossil fuels; that nuclear or some breakthrough technology would be needed. It turns out these challenges were addressed more easily than expected by, among other solutions, spreading out energy generation (as the wind is always blowing somewhere), installing ever-cheaper storage, and shifting the times at which energy is used. Today, an increasing number of countries average well above 40% renewable electricity, and a shift towards 80% of all energy is both possible and predicted. In fact, some studies suggest a 'realistic' plan for a 100% renewable grid across all nations could be achieved by 2050.[2] Until just a few years ago, battery experts thought they would need another decade of development before batteries would compete with oil for use in transport. Not only has the range of batteries increased in terms of distance, but the lifetimes have too. Only eight years ago, the batteries in electric cars needed replacing approximately every 150,000 km. Today's batteries push the distance to more than 600,000 km and are now accelerating towards one million km – longer than that beat-up 1998 Toyota Corolla that just keeps running.[3] Electric buses, cars and bikes are rolling out in their millions. We're even seeing the first commercial electric plane and ferry services, decades ahead of schedule.[4]

The next few years will reshape global energy systems. This is the beginning of the extinction of fossil fuels: one extinction worth celebrating. Coal, with its nightmarish health and environmental impacts, is already on its way out. In 2020, global capacity of coal power plants declined for the first time,[5] and financial trends show every sign of a plummet in coal over the near future. It will take longer for the same to happen to oil and gas, but demand for these is already declining, and it won't be long before clean technologies like electric bikes and cars substantially eat into their markets. The *vital* factor now is the *speed of change*. In my view, there is no question that humans would get to a clean energy future... eventually. But intensifying climate breakdown casts a looming shadow over the present. While previous transitions have been steam-engine slow, there are good reasons to believe that the clean-energy transformation will be much faster.

IT'S EASY BEING GREEN

The iron law of the clean-energy transition is to electrify everything we can and make that electricity low-carbon. Electricity only makes up 20–30% of all energy used today, with the rest embodied as oil in the petrol tank, gas in the boiler and coal in industrial processes. There will be some challenges, but we can *now* reach 75% electricity by electrifying all household energy use with heat pumps and induction cookers, electrifying short- and medium-distance transport options and electrifying many industrial processes.[6] The incentives for this are undeniable: electricity is an amazingly flexible and clean way to transport energy; it can power anything from a laptop in India to a water pump in the Netherlands; it can move a bus in China or a scooter in Scotland; best of all, electricity can be generated directly from the sun and wind instead of laboriously burning fossil fuels to produce steam that in turn drives turbines and generators.

Renewable electricity has become jaw-droppingly cheap. This is, to my mind, the most optimistic technical trend you'll find in the world today. New renewables have been cheaper than most new fossil generation for some years now.[7] In many countries, they are less than half the cost of gas or coal. Amazingly, renewables are becoming cheaper than the operating costs of existing fossil fuels; in some countries it's now cheaper to finance the land, solar panels, electrical infrastructure and the labour for installation than to put more coal in a completely paid-for power plant. The consulting firm McKinsey estimates that by around 2030, solar and wind will be cheaper than any operating fossil fuel power plant, anywhere in the world.[8] In some cases they are cheaper even if you add lithium battery storage.[9] It's hard to overstate how transformative this is. It overturns everything we previously thought possible in the twenty-first century. Because energy is at the core of the economy, it transforms the way we should think about the world.

These trends are demolishing received wisdom in the global energy system, especially in the largest nations. The Chinese use of coal peaked in 2014, ahead of schedule and surprising local experts (although it increased by 2021, it was still below the 2014 level).[10] China had set a solar power goal of 105 gigawatts by 2020, but managed to install 235 gigawatts by the close of the year.[11] The Indian economy is expecting 6% year-on-year growth while the country is on its way to becoming the most populous country in the world. Due to the power of exponentials, a 6% growth rate means the economy will double every eleven years. One might imagine that the energy needed to fuel this growth would come from coal, as it did for China during the 2000s. But again, renewable energy has ripped up the rule book. Some researchers now expect India's coal use to peak in the next few years and for its additional growth to be met with solar.[12]

Renewable trends also revolutionize what lower-income nations can do. Instead of following high-income nations down the path of a large, monolithic industrial energy system, these nations are able to skip ahead to renewables. In sub-Saharan Africa, there have been intense battles between centralized grid-based electricity and distributed generation. Both have benefits, but renewables are making the argument redundant on price alone. What started with millions of small-scale solar lanterns has expanded to low-power TVs, refrigerators and laptops, all powered by cheap solar power and batteries. This is important because where grid-based electricity can take many years to deliver, renewables can be installed within weeks. Schoolchildren don't have years to wait for the grid. These renewable installations can be connected into local networks – known as *microgrids* – on the level of the community or town. These small-scale transmission networks can offer more resilience and decentralization than national-level grids, since if one part fails the rest can step in. There is the potential for future electricity grids in Africa to be more stable than those in higher-income nations.

SPEED AND SCALE

The good news doesn't stop there. Renewable energy costs will continue to plummet as more renewable energy infrastructure is installed. The rate at which costs drop is called *the learning rate* and is measured per doubling of output. For instance, the Ford Model T dropped 14% in price for every doubling of Model Ts on the road. The learning rate for renewable technologies and batteries has been incredible, unexpectedly so. So far, solar-panel prices have dropped an astonishing 28.5% for every doubling, and lithium-ion batteries have dropped by 21.5%.[13] But now, with the benefit of hindsight, we can see there were clues.

The advantage of renewables and batteries is their small-scale modularity. The battery in a Tesla electric car isn't one big unit, but made up of 7,000 smaller batteries, each a little larger than an AA battery. These technologies look much more like the smartphones made in their millions rather than the massive coal and gas plants of yesteryear. Smaller, modular technologies see much faster improvements and quicker adoption than cumbersome alternatives. There hasn't been one huge breakthrough in these technologies, but hundreds of smaller ones that add up, and there are many more on their way. It is true that energy transitions have taken between fifty and sixty years in the past… but this one looks very different.

These dynamics also explain nuclear power's difficulties. Given its huge construction costs, nuclear will struggle to compete with the sheer flood and flexibility of solar panels, wind turbines and lithium batteries. Keeping existing nuclear online is important, but new plants will struggle. That's not to say that we shouldn't even consider nuclear power – although social resistance and the long-term storage of waste are significant issues – but nuclear simply struggles to compete economically. New reactors in the UK and the US have repeatedly blown their budgets, and it's public money that has kept them going. The Hinkley Point C nuclear plant in the UK has the dubious distinction of being one of the most expensive single structures in human history – you could build twenty-four of the tallest skyscrapers in the world for the same cost.[14] Some countries like France *might* see new reactors, but it's too soon to know whether they will definitely go ahead.[15] Asia has seen some growth in nuclear energy and there are several new Indian plants on the drawing board,[16] but many are experiencing mounting costs and public resistance. Newer technologies such as small, modular reactors may become competitive, but they are several years off regulatory approval and are unlikely to arrive on time to make a significant dent in wind and solar, at least in the short term.

Another benefit of solar and wind is that they are much better suited to climate breakdown. We have to accept that we have acted far too late to avoid serious stress (as we shall see) and that we are already struggling with disruptions that will only worsen. Around 90% of US nuclear plants are exposed to future flooding risks beyond their design specifications.[17] As we've heard, summer droughts and a shortage of cooling water are already forcing shutdowns of nuclear and fossil fuel plants across Europe.[18] Conversely, wind and solar use very little water.[19] With their modularity, renewable installations can be moved relatively quickly as climate changes intensify. They also enhance drought resilience and groundwater supplies, since they require very little water to operate.[20] Incredibly, wind turbines can substantially reduce the frequency and impact of tropical cyclones and hurricanes by extracting energy from local weather systems.[21]

You might wonder why the discussion so far has focused on solar and wind. Where are the other renewables, like geothermal energy and tidal? The answer lies in the total amount of renewable energy flowing through the Earth system. Even if you were to add up all the available hydropower, geothermal, wave and tidal power, you still wouldn't reach the energy contained in the fossil fuels we use. All clean technologies are important, but wind and solar are simply in a league of their own. With existing technologies, wind *alone* could produce ten times our current fossil use; solar, fifteen times.[22]

The growth in wind and solar has already been simply staggering. At current rates, total solar capacity doubles every 2.4 years. If this continues, solar energy alone will fulfil current global energy demand by 2040.[23] Of course, the sun is not available all the time, so we would need more storage and transmission infrastructure. But consider the fact that wind power, which often complements solar, has been ticking along at 15% year-on-year growth. These two sources combined, with projected growth rates, would bring the time needed for 100% renewable energy supply forward to 2035 or sooner. But can these growth rates realistically continue?

The truth is that most people are uncomfortable guessing, especially the experts who have been burned by previous predictions. Pessimists might suggest that recent growth rates were off a low base, that there has been some slowing of growth in some countries like Germany in recent years, and that previous energy transitions saw growth tailing off after a fast expansion (for example in the transition between coal and oil).[24] The counterargument is that previous investments represented a concerted effort to bring prices down. Now that prices are close to matching or beating the cost of operating fossil fuel plants, it may be sensible to expect another mind-blowing acceleration in clean energy. There is also plenty more development to come in terms of price and technologies. Researchers have shown that the cost of wind can be reduced by a further 24–30% by 2030 with improvements that are already in the pipeline.[25] Then there is the possibility of entirely new breakthrough technologies like perovskite, a dirt-cheap material that can be mined very quickly, refined and then layered on surfaces – even existing solar panels, improving efficiencies and slashing panel costs further.[26] If this wasn't good enough, there is yet another, more technical reason why we should expect renewables to grow fast for the foreseeable future.

FUELLING CIVILIZATION

What drives an economy? Is it the labour of workers, or the physical capital, like machines and robots? While both are important, neither humans nor machines can operate without energy. We've seen how the complexity and size of society massively increased with the opening up of new energy resources from the Industrial Revolution. In short, as all civilizations expand, they require more useful energy and resources from their surroundings. The key here is 'useful' energy. If it takes increasing amounts of energy to extract energy in the first place – for example having to drill deeper as shallow oil reservoirs dry up – then the useful energy that results from the whole process is reduced, and civilization could be in trouble. This is exactly what's happening today. As we search for fuel in more extreme places, we are extracting the same amount of energy, but it's costing us far more energy to do so. It's a growing, unsustainable burden on the economy and the environment.

To put this in perspective, the oil that once gushed from the fields of Pennsylvania had an energy returned on the energy invested, or EROI, of at least 100:1.[27] That is: it cost one energy servant to extract over a hundred more. There are ninety-nine units of useful energy left over from the process. Today, as the easy-to-access reserves are depleted, the average EROI for oil has dropped to less than 8:1 – only seven units of useful energy after extraction.[28] Unconventional oils like Canadian tar sands are much worse, at less than 5:1.[29] North American shale oil is probably lower than 2:1.[30] Sometimes, the ethanol produced by plants doesn't even break even.[31] Things can only get worse as remaining stores of oil are depleted. Some researchers suggest that anything less than 5:1 overall could present serious issues, which is to say, potential civilizational collapse.[32] This is a ratio we will reach and a collapse that will happen if we continue to subsidize oil and gas.

The good news is that as oil and gas die out, renewables are lighting up. Wind already has an estimated EROI of around 20:1, while solar is between 8.7 and 34.2:1.[33] By the time you read this book, both of these will be higher, perhaps significantly so, as renewables improve, with larger wind turbines and increasingly efficient solar panels. Instead of expending huge amounts of energy and effort in finding, digging up, transporting and burning fossil fuels, we can go straight to the source. Many still doubt that renewables can do a better job of powering modern society than fossil fuels, but they shouldn't.

Solar and wind do need more space, however. In fact, about a hundred times more space than gas power for the same energy output, and twenty to fifty times more than coal.[34] Fortunately, we can place solar panels on roofs and can continue to farm the land under wind turbines. There is also the potential for agricultural-electric farms where solar panels allow through the frequencies of light used by plants and harness the rest as electricity.[35] Solar could cast shade in a good way. Other solutions include offshore wind, which alone could supply eleven times the electricity needed by 2040.[36] Solar can even be placed on lakes as floatovoltaic installations.

Given the sheer scale of the energy system and the need to supply more energy for the poorest people in the world, we *must* attack the energy transition on two fronts simultaneously: through the massive adoption of clean-energy technologies as we've seen, *and* by reducing energy use in the first place.[37] Clean energy improves the EROI of society by providing more useful energy; efficiency makes sure we don't waste this useful energy.

EFFICIENCY AND ELECTRIFICATION

In a power cut, no one asks: 'Where have the electrons gone?' We ask where the torches are, and whether or not we should binge on the ice cream melting in the freezer. Normal people don't care about *energy itself*, but the things it lets us *do*. We don't want natural gas or lumps of coal: 'We want hot showers and cold beers.'[38] We can get what we want using far less energy, money and time. We can get what we want without forfeiting our health and the climate's. Do we care about such efficiencies for the mere sake of money, time, our health and the environment's? That remains a rhetorical question, but let's get to grips with what's possible right now.

Of all the energy pumped into society, between 10% and 25% is put to use, depending on the country.[39] Imagine filling a bath without the plug, so that 75% of the water flows straight down the drain: that's how much energy we waste, at least. Some of this waste is avoidable. It's certainly possible to put this 'hidden fuel' of society to better use. If we order the options for reducing carbon emissions by price, energy efficiency comes first… and at a *negative cost*. That is, energy efficiency pays. For instance, every tonne of carbon-dioxide emission you avoid by installing insulation pays you back around twenty-six to sixty-nine euros in saved energy bills over the long term.[40] Global energy efficiency improved by 1.8% per year during the 2010s[41] – three times faster than a decade before. The UN's Sustainable Development Goals have a target efficiency of at least 2.7% per year.[42] This is achievable.

The majority of things we can do to improve energy efficiency would result in massive improvements in well-being. For instance, properly insulating houses cuts energy demands by more than 80%, while new builds can be made entirely passive – that is, they require no energy to heat or cool. As heating and cooling make up over a quarter of energy use in many countries, this is important. It costs us in terms of health, too. The current lack of insulation costs the UK's National Health Service an estimated £1.4 billion per year in colds, flu and pneumonia.[43] Unfortunately, national programmes for improving buildings have been underwhelming so far. Current gas heating, used across much of the world, is a paragon of inefficiency. Natural gas has a flame temperature of 1,950 degrees Celsius, an incredible resource for industrial processes that need high temperatures, yet it's used to heat up homes by just a few degrees (often extremely inefficiently). The overall energy bargain is horrendous. Electric heat pumps, on the other hand, can move over four times as much heat energy as the electric energy they consume. Homes can be even more efficient by using ground sources of heat where available.

What's more, huge life improvements can be made by redesigning cities. In 1800, only 7% of people lived in cities. By 1900, the figure was around 16%.[44] In 2020, an estimated 56% of people were city-dwellers. With three million people moving to cities every week, this

may reach 68% by 2050.[45] That's a new city the size of Birmingham
or Chicago each week. We *must* make cities as liveable as possible,
and liveability means decarbonizing. Many jump to discussing electric
vehicles, but thinking about urban mobility before urban planning is
like attempting to lose weight without cutting down on cookies. In
the US, urban mobility transformations can save as much as $600 for
every tonne of carbon avoided.[46] Bike paths and pedestrianized areas
must be developed on a massive scale. Liveability also means rezoning
cities to accommodate a mix of activities. We need to get away from
the idea of a retail park at one end of town and the business district
at the other. One simple planning rule is that no one should have to
travel twenty minutes by car just to buy food. This would also reduce
food waste, as people could do smaller, more frequent shops.

As well as being necessary for decarbonization, electrification is
one of the most crucial aspects of energy efficiency. A petrol car is a
spectacularly wasteful way to move people around – they lose between
68% and 76% of the energy in the fuel, compared to just 16% for an
electric vehicle.[47] Not owning a car is better still, since cars need energy
and materials to be built, they spend 92% of their lives parked, and they
consume between 20% and 50% of urban land.[48] Although a lot of the
attention is on electric cars, e-bikes are the real unsung heroes of electric
transportation. They do much more heavy lifting than electric cars and
take up far less space. Their use is already growing at more than 20% per
year in many countries, including the US, China, India, and much of
Europe.[49] Even if you charged one with the dirtiest electricity you could
find, you'd still end up with 1,000 miles to the gallon compared to a
petrol car.[50] For longer trips and those unable to cycle, electric buses and
trains are having an incredible impact on energy efficiency, oil demand
and air pollution. The city of Chicago estimated that just two electric
buses are saving the city $24,000 in fuel costs and $110,000 in health
costs each year.[51]

The industry has exploded in recent years, with China going all-in
with over 400,000 electric buses running across major cities in 2018.
The impact of this one decision is so large, it's visible in the global oil
supply, resulting in an estimated 270,000 barrels a day being left in the

ground.[52] Between electric buses, the closure of coal power plants near urban centres and the electrification of homes the air quality in China is improving, but it is still a way off the air quality in similar cities across the UK and Europe.[53] China also has the world's longest network of high-speed trains. A train trip between Shanghai and Beijing of 1,310 km takes 4.5 hours – roughly the same as flying – and, crucially, the train is two to three times cheaper. Elsewhere, a flurry of activity is bringing new overnight sleeper train services to Europe, new investment in US rail corridors, and major developments to increase and improve rail networks in India and Latin America.

This electric revolution can have rapid impacts on health by reducing urban air pollution across the world, easing the impacts of some diseases within weeks.[54] But going quietly unnoticed is a whole other kind of benefit from clean energy: the noise revolution. As cities reorganize and cars retreat, the din city-dwellers experience is reduced enormously. It seems like a secondary consideration, but noise stimulates the central nervous system, releasing stress hormones and increasing the risks of cardiovascular diseases like stroke and heart attacks. Research suggests that the health impacts of noise could be significant and underappreciated.[55] The net result of all these improvements would be a massive improvement in well-being. The list of benefits goes on, there is good evidence that cycling makes humans happier and healthier and, as bikes begin to replace cars, quieter urban land will be freed up for communities.[56] Parks and forests in particular would help limit urban temperature rises experienced as the climate warms across much of the world.

DEEP DECARBONIZATION – THE COLOURS THAT DON'T MIX WITH GREEN

Though we need a huge acceleration in the deployment of renewables and the improvement of efficiency, many of the changes discussed so far are under way. But while 75–80% of energy use is 'easy' to electrify with renewables, the final 20–25% is far more challenging. The term for getting to 100% clean energy is *deep decarbonization*. The areas facing the biggest challenges include long-distance shipping, industrial processes involving very high temperatures, flying, and seasonal storage for electricity grids (mostly ensuring there is enough energy to keep warm throughout winter in colder regions). There are solutions to all of these problems, including, among others, hydrogen fuel, nuclear and synthetic fuels (more on these later). Some solutions are even on their way, including a plan for a large North Sea installation to produce hydrogen from wind, which is pumped onshore for use in industry. This would be paired with electrified industrial technologies like electric arc furnaces in metal smelting – reducing emissions by as much as 90%, but using astonishing amounts of electricity. Estimates suggest that by around 2030, renewably generated hydrogen could be cost-competitive with the coking coal used in many difficult-to-decarbonize industrial processes.[57] This sort of infrastructure takes time, though, and on the whole, solutions for deep decarbonization will be expensive or will require significant changes in our behaviour.

One such area is flying, which is likely to be a problem for a long time still. Although flying accounts for around 2.5% of total carbon emissions, you have to triple this when including the other climate impacts of flying (including contrails, which trap heat in the atmosphere, and the other greenhouse gases planes produce). Also, demand is growing such that it could account for 25% of the global carbon budget for 1.5 degrees Celsius by mid-century, when we will have hopefully decarbonized much of the electricity grid.[58] The emissions from flying are roughly equivalent to 120 kg per hour per person. Imagine carrying *six* 20 kg bags along with you on your flight, in addition to your own suitcase. That's what the impact of your short-haul flight is. No airline ad will tell you that. No employer will own up to it. Imagine arriving in New York from London and having to do something with *fifty* 20 kg suitcases full of carbon as they arrive on the luggage belt.[59] Those bags are your responsibility, your burden, if you fly… or at least they should be. It is farcical that airlines pretend that a couple of euros towards 'carbon offsetting' *optionally* donated will do anything to bury those bags.

The good news is that flights under 300 km *can be* and *are being* electrified. But we would need to triple battery energy-to-weight ratio before we could electrify intracontinental flights (a distance of 1,100 km – approximately London to Barcelona). It will likely be mid-century before this becomes a reality. Electrifying intercontinental flights (e.g. London to New York) is likely off the cards until the end of the century.[60] Some are looking to biofuels to help, but for reasons we will see, biofuels have been a disaster – we simply don't have the space or biological capacity to spare while feeding people and while drawing down carbon. Another possibility is to produce synthetic kerosene. Here, renewables would be used to draw carbon out of the atmosphere and hydrogen out of water to be combined into oil, turning back time's arrow. This can be done today, but does require large amounts of energy, and costs between 23% and 300% more than the fossil fuel from the ground.[61] It also doesn't address the net impact of plane contrails, which also warm the planet.[62] The plummeting costs of renewables or energy breakthroughs *may* help, but flying requires so many materials and so much energy that there is no escaping the reality: we need to stay grounded as much as possible.

This may sound restrictive, threatening the family summer holidays, but remember that this doesn't affect everyone equally. A 2014 UK government survey found that 15% of the population took 70% of the flights and 1% of English residents take 20% of overseas flights.[63] We don't have an income breakdown of flyers, but it's fair to say that many of these people are likely to have higher incomes or are flying for business (and a business-class seat has around three times the emissions of an economy seat[64]). The statistics are similar in the US, where two-thirds of aeroplane trips in 2017 were taken by those flying six or more times a year.[65] How many of these flights were truly necessary? How many were for meetings that could have been carried out as videoconferences? Looking towards a post-COVID future, how many meetings can we cancel rather than cancelling our future? Even in academia, which prides itself on the importance of collaboration, research has found no difference in research productivity between those who jet to conferences and those who stay at home.[66] These statistics provide a solid basis for suggesting a frequent-flyer levy, which would be both progressive and effective.

These numbers should soften the question of whether we might actually be happier staying grounded. Would it be so bad to slow down and enjoy the journey, not only the destination? Given the air turbulence that is already increasing due to climate breakdown, it may be more comfortable to take the train or bus.[67] This might mean longer trips, stopping off in more places, and having more experiences along the way. This becomes increasingly plausible if people have more leisure time and work less: something that needs to happen for a host of other environmental reasons, and something we'll discuss in the Economic chapters.

The incompatibility of flying with concern for the environment ultimately adds up to a market failure: we are not seeing or paying for the damage. Airlines don't pay a single dollar in fuel tax. You pay more tax filling up a Fiat Punto than Ryanair does a 737. This will change eventually, with nine countries petitioning the EU to introduce a kerosene tax.[68] In general, though, it's hard to get a handle on how much we should pay for flights because offsetting schemes are flakey, failing to remove the amount of carbon advertised, and their long-term benefits are uncertain. The only way to be certain about offsetting is to calculate how much it costs to physically draw down this carbon and bury it today. Right now, if you wanted to go online and pay a company to do this, it would come in at around €950 per tonne (though the cost is likely to come down dramatically over the coming years).[69] So add an extra couple of thousand euros to your bill for a return flight from London to New York. While this is likely to drop in price fast,[70] if companies were exposed to both a fuel tax and a carbon tax, we'd find a lot more pointless meetings cancelled and phonecalls booked in their place… much to the relief of many high-flyers.

THE BATTLE FOR POWER

The success of renewables has largely been driven by government incentives and research. Subsidies for renewables – especially those in Germany and China – drove the initial expansion, while universities and government labs drove innovation and technological breakthroughs, yet the role of government investment still goes underappreciated. Unfortunately, the suggestion that governments should act more decisively has been fought tooth and nail by interests that stand to lose from the energy transition (despite the fact that if we don't make the transition, the larger loss is incalculable). It is hard to predict whether the big fossil fuel industries will continue to resist what needs to be done, or if they will develop some scruples and aid the transition. It may be wishful thinking to believe they will change on their own. Some commentators suggest we should go as far as nationalizing them for the public good – anything we can do to rid ourselves of their influence.[71] More practical efforts involve blocking lobbyists from the halls of power, introducing strong sanctions on misleading advertising and a requirement for funding transparency, especially when it comes to opaque industry-funded think tanks. As we will see in the Climate chapters, the long-term future for these companies is bleak, any which way you look at it. They are operating under a Damocles' sword of geopolitical concerns, banking risks, legal exposure and skyrocketing extraction costs. Their brands have become so tarnished that few want to work for them any more.[72] Who wants to work for a company whose business it is to take away the future? The result is a starvation of talent in the industry and further ossification.

Across society, fossil fuels are becoming less acceptable. We may be close to a tipping point where attitudes across businesses, policy and the public shift completely. Within a decade, it may be socially unaccept-able to fund fossil fuels, take frequent flights or drive a petrol vehicle. Several countries including Denmark, India, Ireland, the Netherlands and Norway as well as many cities are set to ban petrol vehicle sales by 2030. Costa Rica will be banning all petrol vehicles by 2021. In response, many car manufacturers are going all out on electric vehicles. Volkswagen alone will have eight EV factories worldwide by 2022, and Mercedes is no longer investing money on combustion-engine research – only on electric.[73] In some cases, corporate interests in renewable energies will drive further political pressure. In others, we may have to pay fossil fuel interests to close up shop. Spanish and German governments have decided to fund transition programmes for regions dependent on coal, phasing them out while finding new work for people who lose their jobs.[74] These trends augur a power shift that is continuing to play out in the energy system.

INCREASING THE PACE

That emissions are still rising at an average of between 1 and 3% annually (recessions and flu outbreaks notwithstanding), *even with* the global expansion of renewables, is truly worrying. We *can* eventually get to a clean energy world... but the physics of climate change doesn't foresee 'eventually'. If we carry on as is, we will struggle to exist 'eventually'. Urgent action is needed. Building renewable energy like it's going out of fashion is a no-regrets option in so many different ways. Cleaner air, reduced noise pollution, better climate resilience... the list goes on.

A longer-term problem is that renewables and batteries aren't built out of thin air. They need a variety of materials in huge volumes. An even bigger challenge perhaps is the speed of the necessary transition, which will require a fast acceleration in material extraction. Both seem manageable at the moment, but this could become a problem in the future.[75] There are three potential solutions: 1) Developing new technologies from more abundant materials – for example, companies are already attempting to phase out critical materials like cobalt;[76] 2) Limiting the need for so many new materials by redesigning transport and energy systems – fewer people need electric cars in urban centres which have good public transport connections; and 3) We will have to get much, much better at recycling. While some materials in wind turbines and batteries are generally easier to recycle than other products, others such as solar panels and wind turbine blades are more challenging (although even these are arguably easier to recycle than mobile phones and computers, where materials are typically more mixed together). Given the scale and speed of the energy transition needed, it is increasingly important to start designing products for recycling today. These are all significant concerns, but at the moment they pale in comparison to climate disruption and the other impacts of dirty energy.

There is every reason to hope that amazing new technologies will come through for us, but we don't necessarily need them to. The truly hopeful (realizable) vision is of a world with such cheap and plentiful renewables that we can manage other ailments by treating them with this energy. This includes drawing down carbon with machines, recycling, producing energy-intensive 'printed meat' where meat cells are grown in a lab instead of rearing cows, and building large-scale desalination plants to ease water pressures. Perhaps most importantly, the accelerating pace of automation and machine intelligence will require much more energy but promises to revolutionize many problems. For example, machine learning and observation can automate the use of greenhouses, saving water, energy and soils. It is already being used to optimize materials for building construction, cutting huge amounts of concrete, metals and plastics while maintaining building strength. In essence, the progress in energy makes all other challenges we will discuss in this book easier to address.

FOOD

Eating the Earth

The world is in short supply. This field of goldenrod will never
be enough, and the ocean feels suddenly crossable. In every
apple an orchard waits, but who has twenty years to cultivate
it? Above our house, the contrails of the jets have turned into
actual clouds. The rain they promise is another lie. Meanwhile,
the taste of my blood implies that I am rusting, that a broken
machine lies half-submerged in the pond I carry with me.

CHARLES RAFFERTY, 'The Pond', 2018

Neither our individual or societal identities, nor the world's
economy would exist without the multiple resources, services and
livelihood systems provided by land ecosystems and biodiversity.

IPCC, 'Climate Change and Land', Special Report, 2019

Forests and meat animals compete for the same land. The
prodigious appetite of the affluent nations for meat means that
agribusiness can pay more than those who want to preserve
or restore the forest. We are, quite literally, gambling with
the future of our planet – for the sake of hamburgers.

PETER SINGER, Animal Liberation, 1975

Picture a near future in which every global citizen unifies around one
supreme goal: the global conversion to zero-carbon energy by 2050
(much sooner for high-income countries). In a series of momentous
local, national and global initiatives, and in full unanimity, humanity
builds vast amounts of wind, solar and batteries, reduces energy use, and
improves efficiency rapidly enough to avoid the very worst impacts of

climate breakdown. Social and technological solutions are exchanged *for free* between technologically developed countries and regions of the world still struggling with energy access and security. On average, people live longer, asthma rates decline, and health improves as urban areas convert to clean transport. Land that once accommodated cars, power plants, mines and refineries is given over to community developments, leafy parks and bike lanes. There are rough social transitions, but geopolitical tensions ease as energy becomes less a cause of international conflict. In this incredible future, could we dare to hope our existential worries are over?

The previous chapter covered *one* of several profound transitions we have to make. That other tastier type of energy – food – comes with its own, different planetary-scale impacts, any one of which could spell catastrophe. From soil erosion to deforestation; from pesticide accumulation to algal blooms; from antibiotic resistance to biodiversity loss… the sheer diversity of damage is dizzying. It would be a challenge to design a more effective system for ecological destruction than the modern industrial food system. It's a system virtually guaranteed to push ecosystems to the cliff edge, and to make the planet ultimately uninhabitable to humans.

Food dominates planetary statistics in a way that might come as a surprise. Today, humans harness more than 25% of the biological productivity of the land, potentially heading for 50% as populations consume more and grow (both numerically and physically).[1] It is an astonishing fact that, within a few decades and without significant changes, half of all earthly life on land, as measured by the synthesis of organic compounds using carbon dioxide, could be channelled into meeting the needs of one species. Our appropriation of resources represents a vast geological, hydrological and biological engineering of the planet. We use more than 72% of global ice-free land, about half of which is for food.[2] More than half of all global renewable fresh water has been diverted through dams and agricultural systems.[3] Bird's-eye photographs of forests over time tell the story plainly. At first, major roads start to appear, severing the dense vegetation. From these, secondary roads sprout and forest clearing begins. The forests are felled for animals to graze or to plant crops or to develop fast-growing non-endemic trees for wood or fuel. Now press repeat until we reach the modern day; of the endemic forests that once covered Europe, less than 1% remain.[4] This pattern is playing out for the last time in the few remaining endemic forests on the planet, from Borneo to Brazil.

As we divert increasing amounts of natural resources to food production and human society, there is less available for our fellow earthbound inhabitants – we are starving nature to death. To understand how severe the situation is, take the number and variety of mammals left on land. If we were to weigh every individual mammal on the planet, from giraffes to humans, 96% of the weight would belong to humans, or animals reared for humans to consume (mostly cattle, sheep and pigs).[5] It's hardly surprising that one in four mammal species are endangered.[6] While we are at least dimly aware of this mammalian problem, for smaller species like invertebrates the loss is incredibly hard to quantify. Current evidence suggests we have lost staggering numbers of insects – their biomass has shrunk by over 70% across high-income regions.[7] The fact that this biodiversity loss is less visible than the storms and droughts caused by climate change is deeply troubling to scientists.

The cause of most biodiversity threats is human consumption.[8] We need to eat. Yet many still go hungry in poor and rich countries alike. Inequality means that between 10% and 20% of households in which US or UK children live suffer from food insecurity (insecurity is when you can't obtain food that will keep you healthy, or you have to limit your food intake because of money).[9] An optimist might point out that it's great that more people are now overweight than are going hungry.[10] While that's true, this thinking is flawed in a couple of ways: firstly, you can still be nutrient deficient even if you are overweight; indeed, many people around the world suffer this 'double burden' (e.g. one can be vitamin A deficient and obese at the same time); secondly, obesity is related to many other health problems like diabetes and heart disease. Ultimately, this means that the majority of people are suffering from the way they eat – either too much of the wrong foods (globally, 39% of adults are overweight) or too little (26.4% of the global population experience moderate or severe food insecurity).[11] We have built systems that rely on perverse incentives, giving massive subsidies in many countries that encourage the yield over environmental and health outcomes (for example, a focus on raw calories rather than nutrients). In richer nations, subsidies generally channel towards less healthy foods. Fruit and vegetables become luxuries, while heart disease, diabetes, cancer and stroke are on the rise.

Over and above the hunger and biodiversity crises, there is perhaps a larger crisis right beneath our feet.

THE NATION THAT DESTROYS ITS SOIL DESTROYS ITSELF[12]

To paraphrase Mark Twain: 'Buy soil, they're not making any more of it.' Though most of us couldn't tell the difference between soil and dirt, there is a big distinction. Soil is a marvel of complex interactions; as Leonardo da Vinci remarked, 'We know more of the celestial bodies than the soils underfoot.'[13] There are more organisms in a teaspoon of good soil than there are people alive today. Soil is the scene of a cacophony of interactions between water, biology, geology and the atmosphere. But this party doesn't get started overnight. It generally takes hundreds of years to naturally develop fertile soil. Although soil can be produced faster with human intervention, at the current scale of use it's essentially a non-renewable resource we are squandering between ten and forty times faster than it can form.[14] By deforesting large areas of the planet and planting crops, we leave the ground exposed to water and wind, allowing this black gold to wash away or drift off the land into waterways. Pollutants from pesticides, fossil fuels, heavy metals and herbicides serve to degrade the soils further. Around a third of global soils have been squandered in the last four decades alone, and we are now losing soil at a rate of 3.4 tonnes per person per year – that's the weight of two cars for each of us.[15] Degrading our soil endowment comes at a huge cost. Farmers in Europe lose at least €1.25 billion each year.[16] In China, a fifth of all arable land is polluted with heavy metals, and some estimates to recover these contaminated soils reach $4 billion over the next few years.[17]

As if it doesn't have enough going for it, soil is also the scene of vast amounts of carbon storage, containing three times more carbon than the whole atmosphere.[18] When land is cleared or burned or when farmers aerate soil via tilling, carbon is released into the air. The biggest wake-up call to this fact was in 1997 when Indonesian forest fires lit several islands ablaze. The smoke from the fires driving over Indonesia, Malaysia, the Philippines, Thailand and Vietnam was visible from space. The carbon released from the fires appears as a spike on historic plots of carbon emissions, adding between 13% and 40% of total global emissions to that year.[19] Since then, the Indonesian rainforest has burned several more times. Globally, the annual carbon emissions from tropical deforestation between 2015 and 2017 were 63% higher than the annual amounts in the prior fourteen years. If it were a country, deforestation would rank third in emissions behind China and the US.[20]

Some of the burning land is used to produce biofuels like palm oil, so might we optimistically look at this as carbon offsetting (even if we put aside issues of extreme biodiversity loss)? The theory goes that as biofuel plants grow, they take carbon from the atmosphere, which is then returned to the atmosphere when the fuel is finally burned. This cycle is ostensibly 'carbon neutral'. But numerous alarming studies have shown that changing the land this way incurs a huge upfront carbon debt by altering soils and vegetation. If a rainforest is cleared to produce biofuels, it takes between 300 and 400 years to pay back the original carbon debt via the biofuels.[21] Even if abandoned cropland is put to work, it can take more than forty years.[22] In fact, agricultural biofuels might not even be carbon neutral in the first place.[23] For this and other reasons, biofuel development in this manner is now considered by most scientists to be an unmitigated disaster.[24] Of course, if the land clearing is for food rather than biofuels, you don't even get this piecemeal, long-delayed climate benefit. As with so many policies, we simply don't see the forest for the trees.

Despite deforestation from fires, biofuels and increased land required for crop cultivation, some studies show that forest area has been increasing since 2014.[25] Could this be a silver lining in the pessimistic cloud?

Alas, no. Global forest area expansion is coming at the cost of virgin forests that contain *huge* amounts of above- and below-ground carbon and biodiversity. Generally, as tropical rainforests are being cleared, it's the temperate regions like Europe that are reforested at a faster rate.[26] On balance, this amounts to swapping indigenous, rich and biodiverse forests for fast-growing commodity forests with much lower biodiversity, more exposure to disease and pests, and reduced potential for carbon storage. It's like taking away your child's beloved, timeworn, knitted-by-grandad stuffed bunny friend and giving her instead a plastic gizmo from a fast-food restaurant.

WHEN THE WELL IS DRY, WE KNOW THE WORTH OF WATER[27]

Fresh water, like soil, would appear to be renewable. After all, it literally falls from the sky. Well… as it happens, humanity has been taking increasing amounts of stored, underground water from aquifers to drink, irrigate crops and to green golf courses (in deserts). The water from these underground reservoirs generally takes thousands of years to replenish. On a generational timescale, much like soils, this water is non-renewable. An estimated 40% of global food production is dependent on irrigation water via groundwater – water that will run out in the not-too-distant future.[28]

This can have consequences that are hard to predict. For example, millions of people in India rely on food grown using groundwater. Yet it turns out that up to 40% of the precipitation that falls in East Africa is a result of this unsustainable Indian groundwater use. As the water evaporates, it changes the regional climate between the subcontinent and East Africa, resulting in more water falling in countries like Ethiopia.[29] As East African farmers have got used to more rain, farmers and pastoralists have adjusted their practices to this current water availability – another socioecological 'lock-in'. Even if India were to reduce its water application using more environmentally friendly techniques, the expanded crops in Africa would suffer. The risks in the Anthropocene are so linked by

economic integration and biogeophysical changes that this is an entirely
new kind of risk to food systems.

The quantity of global groundwater remaining is difficult to measure,
but we *can* measure the rate at which groundwater is being depleted.
Taking water from the ground reduces the mass of material underground,
thus subtly reducing the gravitational force in the region.[30] This minute
difference can be detected by satellites, and the results unfortunately
fit inconspicuously into the bad news.[31] Many important groundwater
reserves are on the way to depletion by mid-century. At present rates
of extraction, the vast Ogallala Aquifer that underwrites much of the
production in the breadbasket of the US Midwest will be depleted by
69% within the next fifty years.[32] California is already struggling and may
run out of accessible groundwater between 2030 and 2040, while India,
Spain and Italy may run out by 2050.[33] Some of the food grown using
this water isn't even making its way directly to our tables.[34] It's either
used to rear animals or to grow biofuels. Optimistically, we can hope
that better management techniques and choices will improve water use
in the future. However, the countervailing trend is that climate-change-
intensified droughts will continue to reduce above-ground water supply,
prompting further water withdrawal from groundwater.

All this means that above-ground water is under pressure too. Lakes,
rivers, ice packs and glaciers are all undergoing rapid changes. Many
regions of the world rely on glacial water. Almost 2 billion people
across South Asia – that is, 26% of humanity – rely to some extent
on Himalayan glacial water, including many in India, Pakistan and
China.[35] About 15% of this glacial ice has disappeared since 1970.[36]
As temperatures increase, glacier melt will accelerate. The tragedy is
likely to play out in two phases: around 2050 to 2060 (depending on
the rate of warming), river flows will surge as ice rapidly disintegrates,
dramatically increasing the risk of flooding. After the floods subside,
droughts will hit hard. The glaciers will finally disappear and, during the
summer, rivers will shrivel. Additional short-term problems are on their
way for South Asia: the monsoons which the region relies on for crops
are already changing, dumping extreme amounts of water at different
times of year, damaging crop harvests.[37] Remember that this is 26% of

humanity reckoning with significant food shortages by the middle of the century. Also remember, grimly, that this is a region with long-standing political tensions.

This does not bode well for the quantity of water available around the world. And because it never rains but it pours, the *quality* of water is also deteriorating. Algal blooms provoked by fertilizer and manure run-off have resulted in vast ocean dead zones covering thousands of square kilometres. On top of this, pesticide concentrations are building and, simultaneously, 80% of global waste water is dumped (mostly untreated across both high- and low-income nations) back into nature, with predictable results. By 2015, 1.8 billion people globally had no access to safe drinking water.[38] Nature is relentlessly signalling its ill health.

Soils, plants and water – the bedrock components of ecosystems are under severe threat. There are very few wild areas left on the planet, and the areas that still appear wild in richer countries are really just carefully managed simulacra of what wilderness used to look like. In some cases, species have thrived in human environments, for example within cities, but often this is still at a lower biodiversity than before urbanization.[39]

EXTINCTION IS THE RULE. SURVIVAL IS THE EXCEPTION[40]

The impacts from soil degradation, water loss, fertilizers and pesticides all add up, and are reflected in the degree to which life on Earth thrives or declines. In Europe, where food systems are reasonably well-monitored and developed, studies have shown that the German flying insect biomass has plummeted by an average of 76% over the past twenty-seven years, and French bird populations by 33%.[41] Bee populations are declining across the continent (except in Berlin, where thoughtful but inexperienced citizens have introduced so many colonies there's nowhere for all the bees to live. The bees have taken to swarming into alcoves in hallways, balconies and even motorbikes).[42]

We've had our share of warnings. In 1962, ecologist Rachel Carson warned that levels of pesticides were killing birds and would bring on a

Silent Spring.[43] It's a warning that has come true for many parts of Europe. While bird populations aren't plummeting everywhere, the trends suggest sizeable declines.

While large studies showing big population collapses occasionally make headlines, like the 2019 research that found that North American bird populations had declined by 29% since 1970,[44] many more stories are left untold: tragedies out of sight of the public and of scientists too. Insects, especially, provide the mortar between the bricks of nature, performing much of the decomposition and pollination that keeps ecosystems functioning. As certain populations blink out of existence, like a bridge without a keystone, further population collapse becomes likelier. The result can be a total ecosystem collapse and species extinction.[45]

Perhaps it sounds like a truism that human impact on the environment can extinguish entire species. But, aside from popular examples like the dodo, it's incredibly difficult to quantify our influence. This is partly because we don't know how many species there are. Elegant ecological theories suggest there are around 8.7 million species, and estimates suggest that humans are causing species loss at least 1,000 times faster than the natural rate – that is, the rate at which they would typically decline were there no humans on Earth.[46] From pesticides obliterating insect populations to fertilizer-driven dead-zones, from deforestation to make way for new crops to soil erosion on farms, the vast majority of biodiversity loss is as a consequence of human food systems.[47]

There have been five mass extinctions during which at least 50% of the known species on the planet have died. Though we are not clear on all the contributing factors, each of these extinctions were driven in part by a fundamental change in the Earth's climate. Most recent was the End Cretaceous extinction, 66 million years ago, when either an asteroid strike or violent volcanic activity (perhaps a combination) altered the composition of the atmosphere for as long as 100,000 years. The resulting inhospitable conditions rendered an estimated 75% of all species extinct.[48] Opinions vary, but even scientists who think the human-driven sixth extinction hasn't yet begun do agree that we are at the cusp of it. We are the asteroid. One silver lining for the optimists is that according to fossil records, biodiversity eventually returns after extinction events – in 10 to 30 million years.[49]

The pressures faced by nature are a little like trying to live on the poverty line. Week to week, you may just be able to balance the budget, but there's no money left if something goes wrong. This is a critical point: a relatively small increase in a heating bill or missed shift on a zero-hour contract breaks your budget. This may even result in eviction. Although pressures can be totally unrelated – a heating bill has nothing to do with the shift roster – and you might manage in the short term, eventually the lack of resources pushes you over the brink. An eviction is a fundamental change in circumstances, something scientists term a *critical transition*. As scientists are often unable to observe all the pressures a species might be experiencing, they are often surprised by the speed of these transitions. But here's the important thing to keep in mind: even if the pressures ease, one cannot easily recover from many critical transitions. For instance, after you're evicted, it's going to become harder to hold down a job, and it's only going to get more difficult to save the two months' deposit for renting another home. Scientists across many disciplines are now talking about these critical transitions (often more popularly called *tipping points*).[50] Social scientists are investigating when communities of people might 'tip' into more environmentally minded behaviour, for example ditching the car for a bike.[51]

Interestingly, it's not in society but in ecology where the most clear-cut examples of tipping points can be found. The best documented examples are shallow, inland lakes: when sunlight shines through a lake, it nourishes small plants on the lake bed, which oxygenate the water for fish. In the absence of any big changes, the lake stays in balance. But everything can change if many farms nearby add fertilizers or manure to the land (fertilizers contain plant nutrients in the form of active nitrogen and phosphorus). If these nutrients are not taken up by the crops, they can be washed off the fields by rain and can collect in the waterways. One might think this would be great for the plants on the lake beds, but it is the algae in the water column that grow *much* faster in these conditions. Fed fertilizers, algae can break out into blooms. As the alga grows, it throws shade on the plants on the water bed. Those plants die from a lack of light and, without them, the water becomes devoid of oxygen. Insects and fish die for lack of both food and oxygen. The lake becomes unliveable to life other than algae. (Since algae love warmer temperatures, climate change is driving more blooms as well.)[52] When a lake remains in this death-state for too long, algal blooms can produce an abundance of toxins, which can make swimmers seriously ill and kill pets.[53] Perhaps most importantly, it's an entire shift in the functioning of an ecosystem, with much poorer diversity for it. The solution may seem simple: dial back the fertilizers and manure making their way into streams. But the lake has changed, so this simple solution will no longer work. In scientific terms, this is called *hysteresis*. If you stress something slightly too much, new rules apply. In physics, the best example is a mechanical spring: exploit it a little and it'll bounce right back; exploit it too much and it will be permanently disfigured. From ecology to physics, from riots to evictions, it's often impossible to dial back our wrongdoing and hope for the best. In shallow lakes things can get so bad that ecologists may have to build the ecosystem back from scratch, cleaning the sediments at the bottom of the lake and introducing the right species at the right time in a closely monitored effort. This has been done in several European nations, but consider what this is like for countries with fewer resources.

What does this mean for the climate? How would we rebuild the world system if it undergoes a similar critical transition? Regrowing ice shelves would require a much larger shift towards ice-age conditions than under normal temperatures. Not even in the most feverish techno-utopian sci-fi reveries could we imagine planetary engineering on this scale. So... what could we do? We'll return to this particular transition in the Climate chapters. Needless to say, these potentially irrecoverable alterations to the entire Earth system keep scientists up at night.

SO LITTLE AND YET TOO MUCH

Much has been made of the reduction in hunger rates over the past few decades, but as we have seen the number of undernourished people worldwide has been increasing recently, up from 607 million in 2010 to 768 million in 2020.[54] Research suggests this has been driven by climatic changes along with political strife and violence in places like Syria.[55] In addition, the statistics touted are often problematic and ambiguous. We measure undernourishment in a similar way to absolute poverty: from an unreasonably low baseline. The UN's Food and Agriculture Organization (FAO) defines a minimum dietary energy requirement based on age and sex, but many of the world's poor are very active and are often involved in primary extractive industries or manual manufacture. Plus, average intakes tend to obscure temporary hunger. At some points in the month or year, many people living below the poverty line have less money or access to food than at others. What's more, this number focuses on calories and not on the variety of nutrients, like vitamins, that people need. If someone has 2,150 calories of cassava in a day, this person may not be classified as undernourished, even though she may be suffering from malnutrition. As such, we can surmise that 768 million represents only a small window into the true picture of food insecurity. This comes at a time when global food systems produce sufficient calories to sate the hunger of roughly 10 billion people.[56]

Given that the world has become much richer and far more technologically advanced over the past two decades, the moral question facing humanity is not necessarily the total number of people going hungry but how many people are going hungry given our capacity to ease that suffering. That capacity has increased dramatically. Global wealth increased almost two and a half times in the three decades between 1990 and 2020, but hunger has dropped by less than a quarter. As we've seen, the trend is worsening in recent data. Between 2016 and 2020, the world's wealth grew by around 10% but undernourishment *increased* by almost 20% (a trend that has been accelerated by Covid-19).[57] If there is increasing hunger in boom times, what does this augur for global society's looming downturn if environmental pressures continue? Under this ethical framework, could we conclude that modern human civilization is more immoral than at any other point in human history?

At the other end of the scale, more people than ever are suffering from obesity. In fact, there are comparable amounts of hunger and obesity in the world: 2.1 billion people are overweight (having a BMI over 25). Obesity (a BMI over 30) is one of the world's leading public health concerns, leading to non-communicable diseases like heart disease, strokes, diabetes and cancer.

MUCH MEAT, MUCH MALADY[58]

Part of the problem piling onto this burden of overnutrition is our taste for too much meat.[59] As of 2017, there were around 1.5 billion cattle, 2 billion pigs and 1 billion sheep living on Earth at any moment.[60] Around 80% of all agricultural land is used to support livestock, and consumes one-third of the world's abstracted water.[61] As such, livestock is a major driver of biodiversity loss across the world, while manure run-off partly drives marine dead zones.[62] Cattle farming alone was found to be the largest driver of bird biodiversity loss globally from 2000 to 2011.[63] 80% of Amazonian deforestation is to make way for cattle ranching. 80% of soy grown in the Amazon is for animal feed. Our taste for hamburgers and other cattle products may quite literally result in the death of the Amazon.[64]

When we think of the food system and its environmental impacts, animal production on its own constitutes a huge amount of the trouble. This is without discussing the ethical implications of industrial agriculture, or the fact that over a third of all crops goes to feeding animals while almost a billion people are hungry and billions more are eating so much meat it's damaging their health.[65]

Our taste for animals is responsible for around 14.5% of all greenhouse-gas emissions.[66] These emissions break down into two main types: bodily emissions like belches, flatulence and manure; and land-use change, as described above, both of which are hard to eliminate. Although there are innovations on the drawing board, it's currently very difficult to eliminate bodily emissions from animals, and land use is necessary unless we are willing to house animals in even denser conditions. As if today's scale of animal agriculture wasn't bad enough for the environment, the demand for animals is expected to increase further over the coming years – meat demand is estimated to grow 73% to 2050 and dairy by 53% – as more countries follow the meat-heavy, high-calorie diet typified by some Western countries.[67] Because large amounts of cattle are fed with soya, soya-bean yields would also have to increase quickly, at a time when yield growth is decreasing year-on-year from stagnating technological improvements and climate impacts.

Around the world, there is an increasing focus on decarbonizing the energy systems, but a quick calculation shows that even if we do, increasing emissions from livestock alone would be enough to push us into catastrophic climate change only a few decades later. It looks like we'll have to take most meat off the table if we're to safeguard our future. We either do this, starting with beef, the biggest culprit, or, in the worst-case scenario, nature will do the job for us, making food impossibly expensive. Perhaps meat will still be eaten, but in tiny amounts for the very rich, as increasing numbers of people go hungry from increasing food prices. In encouraging people to eat more plant-based diets – both for the benefit of health and the survival of the environment – governments have a vital role to play, for instance in government dietary recommendations or potentially, by way of taxes, making meat more of a luxury food rather than a staple.[68] In reality, many governments are focused on minutiae at the behest of heavy agribusiness lobbying, tying themselves in policy knots over whether veggie burgers should be called burgers or 'discs', and whether oat milk should be called oat 'drink'.[69]

We've seen how it is on land, so let's check in with the oceans. It's encouraging that plastics in the ocean have received lots of attention from politicians and the public – it's definitely a problem that needs fixing urgently. Less encouraging, though, is that an estimated 700 billion to 2 trillion fish are caught each year by fishing operations.[70] As marine biologist Daniel Pauly puts it, 'It is almost as though we use our military to fight the animals in the ocean. We are gradually winning this war to exterminate them.'[71] More than 90% of fish stocks are fully exploited (being used at their maximum rate beyond which further fishing would be unsustainable), while 33% are overfished.[72] Fishing for some species produces carbon emissions as high as beef.[73] Increasing acidification from carbon emissions will eventually clear out much of the remaining diversity in shellfish, coral and various types of plankton. Acidification will also increase stress on already overstressed fish populations, which have to expend extra energy cleaning their increasingly acidic blood. The related warming of the oceans from climate change will continue to wreak havoc on marine life, with one study that used the colour of the ocean to measure ocean productivity finding continual declines in biological productivity since 1999.[74]

INCENTIVIZING CATASTROPHE

Conventional logic suggests that we subsidize what we want and tax what we don't. By these metrics, what we want is more environmental crises, poorer social outcomes and a continuing global public health crisis. We are overtly and consistently subsidizing obesity, hunger, societal stress and environmental collapse around the world. An astonishing 37% of the total EU budget (€54.8 billion) is spent on agricultural subsidies, with an estimated 50% going to the worst environmental and health performers – the meat and dairy industries.[75] Similar backing is seen across the Atlantic, where US subsidies for energy-dense foods like corn, soya, wheat and rice have driven the price of bulk calories down, making nutritious fruit and vegetables relatively more expensive on the shelves.[76] Though it may not be immediately obvious, this doubly subsidizes the cost of animal products, since animal feed includes soya and corn. In some countries like the US, the types of foods that are subsidized are generally less healthy and have been associated with more health problems. Standing in the supermarket aisle is incredibly frustrating when the things you know you should eat more of are expensive, and the energy-dense choices are subsidized.[77] You may go in with the best of intentions to overcome all the desires in your way but fail at the last minute, trading the impact on your wallet against the impact on your waistline, and your arteries, and the planet.

Subsidy reform is one of the biggest win–wins out there. To begin with, subsidies can go to environmental projects on farms (which would still be producing food), rather than be paid just by production volume or land area. Indeed, this seems only logical. But efforts to alter the subsidy system have been scrappy and piecemeal, with EU subsidies for agriculture remaining at roughly the same level for over a decade. Today, the policy community generally understands that subsidies do a tremendous amount of harm – in fact, there is a UN goal to eliminate these subsidies – but attempts to reform them in the UK and EU have been roundly criticized for being too weak.[78] Around 15% of UK agricultural subsidies can now be used for environmental projects, but these have largely been a failure in implementation, punishing those farmers who are trying to do their best.[79]

Another cause for concern is increasing consolidation in the agricultural sector, resulting in near monopolies when buying seed, and monopsonies when selling grain. Although more than 80% of global food is produced by small-scale farms and 98% of farms are run by families, a whopping 60% of all seeds are sold by only four companies, companies of which you may never have heard (Bayer – which bought Monsanto, BASF, ChemChina and Corteva).[80] These companies are also among the largest pesticide and herbicide producers, which means some farmers have little option than to grow specific, environmentally destructive monocultures reliant on certain herbicides (for example, using the herbicide Roundup in combination with growing wheat genetically modified to cope with the Roundup). Similarly, four companies are responsible for around 75% of all grain traded (ADM, Bunge, Cargill and Louis Dreyfus).[81] The size and international reach of these companies allow for an oversized influence in policy – for example in pesticide regulation and international trade policy.[82] The result is that farmers are stuck in the middle, with fewer companies to buy seed from/sell grain to, and a reliance on subsidies to get by.[83]

This monopolistic, centralizing control of agricultural giants presents a crucial political challenge across the world. Regulatory capture and lobbying (much like in the fossil fuel industries) combined with farmers struggling to survive make it difficult to have any meaningful reform of subsidies. But if we don't drastically reform subsidies and increase competition very soon – *now* – we can expect that newer technologies like drones and robotic farming techniques will also be centralized, driving further concentration of wealth in the middle of the food chain.

FOOD SHORTAGES

There is deep, unprecedented trouble ahead for the food system. While it contributes to climate change, climate change will simultaneously push food systems to the edge in the most important producing regions. These trends will deepen the already prevalent inequalities in nutrition and wealth. The amount of food grown per hectare rose dramatically through the mid-twentieth century, at around 3.5% a year for US corn, wheat and rice; yields doubled in just twenty years. But this trend slowed dramatically to 1.5% by the year 2000, as technological and management improvements stagnated, and slowed further to just 1% by 2010. Even with projected developments of GMOs, yield growth is set to slow to just 0.75% per year through to 2030 (excluding impacts of climate change and soil erosion).[84]

These crises will change the way all of us think about food and the environment. Let's lay out a potential breakfast of the future: in front of you is a coffee pot, a dish of butter, a full English breakfast, a banana and a jug of milk. As you reach for the coffee pot, it turns to water. Over 60% of coffee species are under threat of extinction, and prices are expected to rise.[85] Half the butter in the dish and milk in the jug evaporate as prices for animal products have doubled.[86] You can forget about sausages and bacon on your plate: prices are so high, only Michelin-starred restaurants serve meat. Even the banana has gone; they now cost a fortune.[87] The tomatoes, mushrooms and beans are left, though, as are the triangles of toast.[88] Things have changed but, for the time being, this might actually be a healthier breakfast. Until prices rise even further.

A visit to the pub, then, to drown your sorrows? You order a beer and a portion of chips, and are outraged at the price. Then you remember that barley yields have declined dramatically across Europe, decreasing the global beer supply due to extreme drought and heat.[89] Your portion of chips is minuscule, but that's no surprise. In 2018, droughts and floods shortened the length of a UK chip by an inch.[90] The pub will never be the same again – if we have the time or money for a pint, that is. But don't worry, thinks the optimist: that's just in the UK. Those in Italy still have their olives and Chianti, right? In 2018 and 2019 olive crops across Italy crashed due to heavy rains, unseasonal cold, and disease, cutting about 50% of production.[91] Wine production will be moving further north too; a nice Italian Chianti costs a fortune as the leaf coverage of vines in Italy shrinks.[92]

We have been buffered by food-price fluctuations to date as we live in an internationally integrated world. For instance, olive oil from Italy can be topped up with Tunisian oil in the short term.[93] But as the issues in agriculture and fisheries pile up, if we don't change the ways we do things, prices will necessarily increase. This is beyond doubt. The onset of shortages will probably come on surprisingly quickly to many, because the speed follows the critical transition story given earlier in the chapter. To paraphrase Ernest Hemingway, how did food systems go bankrupt? Two ways. Gradually, then suddenly. Without significant changes in the way we produce and consume food, we will all be out on the figurative street.

We are at a point in history when the vast majority of people eat too much or not enough nutritional food. The way food is subsidized by governments enriches monopolies while causing massive public health problems and the death of many animals around the planet. Meanwhile, people across the emerging world want to adopt a high-meat, Western diet, and many rich people don't seem to consider environmental ethics when they make their selection from the menu. Our attitudes are driving environmental breakdown, which is undermining production in major food-producing regions. The planet's sixth extinction event continues apace (even if it is at the early stages), with insufficient appetite for change among the wealthy nations.

CHAPTER SIX

∽ HOPE ∾

Green Shoots

For truly in nature there are many operations that are far more than mechanical. Nature is not simply an organic body like a clock, which has no vital principle of motion in it: but is a living body which has life and perception, which are much more exalted than a mere mechanism or a mechanical motion.

LADY ANNE CONWAY, *The Principles of the Most Ancient and Modern Philosophy*, 1692

Humankind has not woven the web of life. We are but one thread within it. Whatever we do to the web, we do to ourselves. All things are bound together. All things connect.

Chief Seattle in the movie *Home*, 1975

I magine living on a spaceship. The spaceship is big – some 12,500 kilometres in diameter – and almost 8 billion people live on it. A few are in first class, but most are in economy or steerage. Around 50% of the spaceship is set aside for growing food,[1] but the systems that grow the food are antiquated, damaging the ship's structure and poisoning its air and water reserves. Although humans occupy 75% of the ship, the space they take up must be brought down to between 30 and 50% if the ship's filtration systems stand a chance of cleaning the air and water.[2] While this could be doable, there is a huge challenge: the number of people on the spaceship continues to rise, and many more want an upgrade. Like a spaceship, Earth ultimately has a space problem.[3]

We can't alleviate our space problem by adding another spaceship called 'Mars'. Not for a long time, at least. We're left with three options:

adopt diets that use less land; reduce waste so we don't use land need-lessly; or increase food production per unit of area. Only... we don't get to choose. Multiple large-scale research efforts find that we have to do all three, now.[4] The land we save has to be protected – either forested or returned to wilderness. We have to revolutionize food pro-duction generally, reducing and then reversing the damage we've caused. Researchers call this process The Great Food Transition.[5] It will have to be every bit as quick and extensive as the energy transition. Although the energy transition can be summed up as electrifying everything and generating that electricity with low-carbon sources, The Great Food Transition is harder to generalize. Given the sheer diversity of growing climates and the differences in cultural norms and tastes, the solutions will vary everywhere.

For as long as humans have been sowing crops, the response to rising populations and increasing wealth has been to try to grow more food by intensifying production and expanding arable land. Out of the three options to save our food system – diets, waste and production – two of them mean we must try something quite different in the modern age. They require an individual, social and cultural effort to change the way we buy, process and consume food, without the incentive of wartime rationing or (hopefully) prices going up. It's a revolution in attitudes, which is both encouraging and disturbing. On the one hand, attitudes are entirely under our control; they don't require any fantastical technologi-cal breakthroughs. On the other hand, we have so few social precedents for such a broad, deep and international change, especially for one that needs to be so quick. It would be foolhardy to assume automatically that we can do it, yet we must. On their own, revolutionizing diets and slashing waste would be almost but not quite enough to produce suffi-cient food for a larger future population while saving land and turning it back to wilderness.[6]

The third option – using food technologies like genetic modification, automation and drone-based sensors – could help to increase yields, spare land and reduce environmental impacts. But technologies can come with severe economic and social downsides if not pursued thoughtfully. In fact, it's disturbingly easy to paint pictures of how technologies might make hunger even worse. Imagine a robotic farm in Ethiopia, leased by a UK multinational and fenced off from the community.[7] Robots use precise amounts of fertilizer, do the weeding, and harvest the crops. Drones flying overhead make measurements of growth which are used to build optimization models and improve yields further. Perhaps there is one human monitoring the farm, but she's a contractor from Germany. From an environmental perspective, the amount of fertilizer, pesticide and herbicide have all been slashed while yields have improved. But it's a human catastrophe: some people on the outside of the fence are still going hungry, as the food is shipped overseas where richer consumers can pay more for it. Ultimately, hunger is more a problem of distribution than technology. This doesn't mean that we should ignore technologies, but it does mean they have to be implemented cautiously, *while* economic reforms are undertaken at the same time.

PLANT-POWERED HUMANS

However you cut the numbers, the types of food we arrange on our table have to change.[8] There is one simple rule overriding all others: remove as much meat as possible, and make meat a rare treat, especially red meat like beef and lamb. Lest the Western world get up in arms about this, remember that such intensive meat-eating is a new social phenomenon, only a few decades old, and one that is generally horrendous for our health and environment. Although we can rear animals in hundreds of different ways, some of the most environmentally friendly beef releases six times more greenhouse gases and uses thirty-six times more land than the *least* friendly plant proteins.[9] This doesn't necessarily mean the extinction of the dairy cow. Animal products will continue to be necessary in poorer regions where alternative plant products are lacking – although, in that case, it is more a question of poverty rather than some overriding necessity to eat meat. In middle- and high-income regions there may be a potential for a small amount of meat consumption through techniques such as *silvopasture*, where animals live within forests in a symbiosis and can actually end up sequestering carbon.[10] There are also new animal nutrition products that can ease their impacts, for example those that inhibit methane production in cows' stomachs by around 30%.[11] But even with heroic assumptions here, the potential for future meat consumption is vanishingly small compared to the large volume of meat consumed today.

What does a world without animal products look like? Land use for agriculture is cut by an incredible 75% – the area of the US, China, Europe and Australia combined.[12] Most of this is pastureland, which can be left to nature or actively rewilded. Rewilding is the protection and restoration of large areas of land planted with forests, or managed as grasslands, depending on what's best for the local environment. This is a world of wilderness, of parks, of walking and biking trails. It is a world that encourages biodiversity and draws down more carbon from the atmosphere.[13] The climate benefits of shifting diets can be roughly doubled by allowing the saved land from a reduction in animal agriculture to return to nature over the long term. It's a world in which antibiotics last longer before bacteria develop resistance to them – 73% of antibiotics globally are fed to animals.[14] A world in which millions of lives are saved from illnesses like colorectal cancer and cardiovascular disease – to put this in perspective, a meat tax of 20% could save hundreds of thousands of lives globally each year (and save 14% of total health costs).[15] This is a world in which water health is dramatically improved, fish populations recover, more rivers are swimmable, and in which many cities are no longer forced to spend millions to remove the additional nutrients and effluents of intensive animal farms from drinking water.[16]

A vast amount of animal suffering is eliminated in such a world, too. While there are still many research gaps, studies continue to demonstrate a hitherto underestimated complexity in animals' internal lives.[17] Pigs have rich lives and complex cognition. There is evidence that cows have a level of emotional complexity found in other intelligent mammals and have distinct personalities.[18] There is good evidence that fish, swimming against the myths of the goldfish bowl, can have memories at least a year old.[19] Evidence is piling up that sentience is much more like a scale than a series of steps, and that animals feel much more stress and pain than scientists thought possible just a few years ago.[20] Taken together, reducing meat consumption is probably the single best individual step we can make for a liveable world, a healthier body and a cleaner conscience.

In one way or another, the high- and middle-income areas of the world *must* make this personal dietary journey. Either it will be imposed upon us via climatic and biodiversity impacts driving up the price of food, or it will be an intentional drive to improve the state of the world. For some, this will be a personal choice: for health, for the planet, for the well-being of animals, or a combination of those three. For others, it could be a matter of taste – as more plant-based alternatives reach the supermarket shelves, it would be churlish not to try the new flavours. For others still, for taste or cultural reasons it might come down to technological replacements – clever plant-based facsimiles of minced meat or even lab-based printed meats.

Although the main nutritional concern surrounding meat reduction is ensuring we consume sufficient protein and trace metals, there is actually nothing to worry about in countries with a good variety of plant-based foods.[21] (As we heard above, the issues are very different for lower-income regions where diverse plant products are not available.) There's already a surplus of plant protein around the world and we don't need meat proteins. Indeed, cattle raised on soya beans destroy over half the protein they eat.[22] We could simply cut out the middle-cow and eat more legumes like peas, beans and soya. The same is true for iron, zinc and vitamin A – legumes and nuts offer much greater nutritional content. Even better, legumes fix nitrogen in the soils, so the shift wouldn't only eliminate much of the damage of animal agriculture, it would help reduce fertilizer use. A full vegan diet might be a challenge for some, but in most high-income nations even eating the nationally recommended diet would cut meat and cut impacts for many.[23]

In spite of these robust arguments, pessimists might insist that a meat-based diet is entrenched in society, that diets are sacrosanct and, when it actually comes down to it, who's really going to spare their ribs on the summer BBQ? But diets have always changed. Broccoli used to be white, carrots yellow;[24] margarine used to be a wonder food; and who'd heard of smashed avocado on toast a decade ago? While it's hard to assess directly, multiple lines of evidence suggest that diets are changing far more quickly than you'd think. Polling across higher-income nations like the UK and the US suggests that people identifying as vegetarian or vegan have at least tripled between 2014 and 2017.[25] That's on a relatively low base, but the rate of change is astonishing. This seems to be borne out on the high street, with the number of vegetarian and vegan options in restaurants mushrooming. This *might* drive a virtuous circle of dietary change: as more people shift and the market for these options grows, more diverse foods become available and, as these expand, even more people are encouraged to make the switch to plant-based diets. This appears to be confirmed in small-scale studies of university canteens: as more interesting veggie options become available, meat eaters eat more veggie food.[26]

We should be careful about too much extrapolation of poll data in isolation, though; it would be entirely believable if the odd vegetarian sneaked in a steak pie when they thought no one was looking. Yet even if we focus on those still eating meat, the trends appear to be positive. Supermarket research found that a third of UK consumers have now stopped eating meat or reduced it in their diet, and a quarter of people regularly buy plant-based milks – which is easy to understand given the variety of alternatives.[27] Over a longer period, the Organisation for Economic Co-operation and Development (OECD) found that red meat consumption per capita in the EU dropped by 10% since 2000 with bigger reductions in the US (although chicken consumption is up, overconsumption of red meat is most damaging to the environment and human health).[28] There has been a slow decrease in the number of cattle in the EU, and the European Commission thinks meat consumption will follow a slow decline through to 2030.[29] The US has seen a faster decline in meat consumption, but started from a higher level than Europe.[30]

Pessimists could also claim that plant-based diets are yet another dietary fad. But those placing the big financial bets don't think so. In 2018, the substitute meat company Beyond Meat, maker of 'the impossible burger', raised $5 billion in funding – an unheard-of amount for a food start-up. The company extracts the legume haemoglobin from soya beans (which is high in iron and proteins) and then uses this to give the plant burgers the bloody taste of animal meat. Given that more than half of

US beef is purchased as mince, mostly for burgers, this is a huge deal.[31] These meatless burgers won't be an expensive privilege, either. They are currently on sale in fast-food restaurants for the same price as their meat 'alternatives'. They are also being targeted at meat eaters, eroding the perception of tribal affiliations through diets. Still, this might not persuade some, but change is coming for the last hold-outs. Lab-based meats and fish cultured from animal cells are attracting attention and funding. Technical feasibility is no longer an issue; it's more a question of price. Some lab-meat entrepreneurs believe price-competitive printed meat is less than a decade away. This has more than a tinge of hyperbole; there's a well-founded scepticism that lab-based meats can deliver as quickly as the press releases suggest.[32] But even conservative observers think we'll see price-competitive lab-based meats at some point before mid-century – if there is still demand for it at that point.[33]

As we begin to turn away from meat, we will discover a whole world of plants and meat substitutes. It won't have to be a world of oat milk in oat porridge (although that is very tasty). Moving to plant-based diets is not a diet of abstinence, but of exploration. There are more than 250,000 species of plant; yet we have only explored 150 for food.[34] Imagine how many flavours and textures there are out there. An excellent example is Kernza, a wheatgrass crop that grows year-round, needs fewer fertilizers and less water, and has extremely deep roots, replenishing soil structures. These deep roots also protect the soil from erosion and don't need to be dug up each harvest, so they can continue to sequester carbon all year round. It's also a *drop in* replacement for wheat – it can be used to make bread and pizzas with no extra effort. The problem is that it yields 26% the amount of calories for the same tract of land under industrial wheat cultivation.[35] Efforts are under way to narrow this gap, and this is also just one example. If we are able to develop a diverse set of crops for different climates which act in concert with the soils and the ecosystem, we might have a good chance of improving resilience against climate disruption and plant diseases.

There are numerous reasons to think we are very close to a social tipping point for plant-based diets. Vegetarian options are exploding, and cultural mores around the world are changing fast. It's hard to imagine that 50% of meat eaters today would become vegetarian, but it's easier to think that meat eaters might halve their intake on average. The rising appreciation that we are in trouble – and the fact that this is an empowering action that individuals can take both for the planet and their health – is making a big difference. As this understanding takes root, the possibility of further policy measures opens up, from meat taxes to subsidy reform. This is a very hopeful part of The Great Food Transition, one that already appears to be under way and that would be an unalloyed success for humans, animals and plants.

WASTE NOT, WANT NOT

In one way or another, 35% of all food is wasted globally.[36] If food waste were a country, it would be the third largest emitter behind the US and China. We can see the scope for improvement by comparing different countries that have different food-waste results. Consider that while North American consumers waste around 28% of all fruits and vegetables, Europeans waste around 19% and Japanese, Chinese and South Korean consumers waste on average 15%.[37] Whenever we think of food waste, we should always link it in our minds to the tractors, fertilizer, land, water, trucks, ships, containers and all-round effort that goes into growing, preparing and delivering food. A reasonable goal of halving global food waste would go a long way to improving our prospects of cutting land use.

In richer countries, most food is wasted at the point of consumption: in the home or in restaurants. Part of the issue is how cheap food has become – similar to the way cheap energy also encourages waste. As a percentage of the UK household budget, food has plummeted from 45% in 1920 to only 10.5% today.[38] When the environmental damage from that food isn't priced in, it makes it easy to forget to cook up that casserole with the softening carrots and potatoes sporting Mohicans. But short of food prices rising or the economy reforming, how might we be proactive in cutting food waste?

Here, we can look to Denmark. In the five years between 2010 and 2015, Denmark reduced food waste by 25%. It did so by keeping things simple: information campaigns, shifting supermarket bulk discounts to single items, and ensuring wonky veg wasn't discarded. Restaurant-goers, many of whom previously felt embarrassed to take leftovers with them, are now encouraged to do so by staff, who ask if they'd like a 'goody bag' rather than a 'doggy bag'. The Danes hope to cut another 25% by 2030, partly by doubling down on existing efforts and partly by introducing new policies, such as reforming best-before dates.[39] Best-before dates and sell-by dates can be totally meaningless unregulated guides, giving dates often months or even years before the food would be unsafe to eat.

Other countries have banned supermarkets from dumping food, with fines of up to $4,500 per infringement.[40] At that price, supermarkets are heavily incentivized to make sure all food makes it into the hands of their customers or citizens. Often this increases coordination between supermarkets and food charities, but there has been an increasing wave of interest in food rescue apps like Too Good To Go, ResQ Club and NoFoodWasted. These apps connect hotels, restaurants and supermarkets to the public and, for a small charge, consumers can pick up food that would otherwise be discarded. All these initiatives and policies add up to big impacts on consumer waste.

Unfortunately, there are a few complications for food waste. One is that there is evidence that healthier diets including more vegetables and fruits can result in more waste.[41] Though this is nowhere near enough to counteract the benefit of moving to a plant-based diet over a meat-centric one (and meat/dairy is the second largest category of waste), it does highlight that better consumer habits around food shopping (more frequent, smaller trips to the supermarket) and cooking (having recipes to use up ageing food) would be helpful. Another wrinkle is that while most people think that composting is a good idea, it is a distant second choice after using up the food in the first place. Composting won't turn back the clock on all the resources that have gone into producing the food and, unfortunately, there is evidence it encourages more waste as people think 'well it's okay, we can compost'.[42]

In poorer regions, food is often lost on the fields or in processing and transportation rather than at the point of consumption. Across poorer regions in sub-Saharan Africa and Southeast Asia, somewhere between

20% and 50% of all food is wasted before it reaches the table.[43] Often, farmers don't have access to barns to keep harvests dry, and toxic mould grows in the damp grain, but fixing this is absurdly cheap. We could build four million smaller barns, 300 mid-sized storehouses and one hundred large warehouses, reducing food losses by 40% across sub-Saharan Africa; all for the cost of around $4 billion – the price of three or four football stadiums.[44] Further gains can be made through food processing, an expansion of refrigeration, and improved transport links.[45]

INCREASING YIELDS WHILE DECREASING DAMAGE

The Green Revolution dramatically improved yields but came at a dramatic environmental cost. The application of huge amounts of fertilizer – containing phosphorus and nitrogen – threw natural elemental cycles out of balance. On top of this, *monocropping*, a common feature of the revolution, encouraged pests, prompting increased pesticide and herbicide use, which in turn increased weed and pest resistance.[46] Today, we must continue to improve yields while cutting the damage. Researchers call this approach *sustainable intensification*, or *The Green, Green Revolution*, implicitly acknowledging that the first revolution wasn't all that green.

The difference between the maximum output for a region compared to its actual productivity is the *yield gap*. Yield gaps are particularly large in sub-Saharan Africa. Ghanaian corn crops have a potential yield of eight tonnes per hectare given the climate and water availability, but actual yields total just 1.5 tonnes. If this yield gap of 81% is closed for one acre, five more can be released from agriculture. Even in the US, thought of as an efficient producer, some regions could still see gaps closed by between 40% and 50%, halving the land needed.[47] A major approach to closing yield gaps while reducing labour costs has been to make farms homogeneous monocultures, and to spread more fertilizers and pesticides uniformly across fields.[48] The approach is akin to treating soils like food factories – not modern factories with high regulatory standards, but rather William Blake's 'dark Satanic mills' of the Industrial Revolution.

We'll have to reintroduce diverse cropping while reducing fertilizers and cutting pesticides, winding back the clock on monocultures but without sacrificing progress on yield growth. While farmers are among the first to understand how things need to change, they are often restricted by time and costs. They face extremely tight margins, so cheaper monocultures often prevail.[49] There are two main solutions: 1) reform the economy and price in the environmental damage to reflect the true cost of the food; 2) make the environmentally beneficial option cheaper. Ultimately, we'll have to develop a new structure for the economy, not only for food systems but for many other areas of life like energy systems: this is a huge subject which we'll address in the Economics chapters. For now, we'll focus on the second solution: how technology may help even within the current economic models.

The first step is using technology to improve the monitoring of farms and the growth of crops. Sensors, satellites and communication networks have become so cheap, it's possible to finely monitor plant growth.[50] Steep price reductions mean that farmers in low-income regions will soon benefit too. Bird's-eye information via drones or satellites provides a large-scale understanding of farms: which fields are currently dry, undernourished or need weeding. Plant and soil sensors report when water and nutrients are needed for each small section of field, slashing costs and environmental impacts. These data, coupled with information on yields and soil quality, can be combined with weather information and fed into machine-learning algorithms for high-resolution, real-time, predictive models of the farm, and smart shipping systems. These approaches are no longer experimental one-offs and are already being used in the field. For example, farmers in more than five countries are now using sensors, which, when applied to the stems of plants, reduce water requirements by over 20%.[51] The potential could be huge; there is some evidence that yields can be doubled in some places using satellite data to identify sustainable intensification options.[52] As we know, everything we do now has to scale quickly and widely, and the best thing about these technologies is that they do just that. Software and cheap devices can be rolled out at breakneck pace. An important side effect is that environmental impacts can be monitored on a large scale. This becomes important when we (hopefully) begin to price in the damage from agriculture within and between nations.

This world of sensors and software can tell us when to feed and water plants, and by how much. But who is going to tend to them in such a fine-tuned way? Going plant by plant will be time-consuming and, as we heard, farmers face tight margins.

In some cases, the solution will be automation. There are already a number of weeding, watering, fertilizing and planting robots on the market. These robots can discern which plant is weed and which is crop, squirting fertilizer or herbicide as needed. Some robots nearing production are all-round agriculturalists: they can apply pesticides, vacuum up pests and dig out weeds.[53] Eliminating herbicides with autonomous mechanical extraction is an impressive development. Given the cost of labour, the first generation of automated farmers are already cost-competitive for some jobs on farms in high-income regions.[54] Before, organic farmers had to trade herbicides for humans – either by laboriously hand-weeding or using human-driven machines – but now we are getting closer to the point where we can rid many farms of herbicides entirely.

This *precision agriculture* may allow for the large-scale readoption of one of the best agricultural inventions humans ever devised: inter- and multi-cropping. For centuries, fields were planted with different plants which worked in synergy. An excellent example is the *Three Sisters method*: maize, then beans, followed by pumpkins or squash.[55] The maize gives a scaffold for beans to grow along; the beans make nitrogen nutrients available to other crops; and the pumpkins provide ground cover to prevent weeds from growing. Indigenous Americans used this technique to provide all the nutrients and calories they needed for a healthy and sustainable plant-based diet. Across a huge variety of studies, this sort of intercropping, sometimes with as many as twenty-five different crops, has been found to improve yields dramatically while increasing biodiversity, pest resistance and soil health.[56] Everything, essentially. Intercropping has been dropped largely because it takes more labour in harvesting, but if labour pressures can be eased with more information and automation (or by reforming our economic systems), it could see a resurgence.

There is one last technological approach for increasing yields and reducing damage: improve the plants themselves. That is, *genetic modification*. For thousands of years, humans have been improving plants with the goal of selecting genes that provide beneficial traits. Modern genetic modification is becoming far more targeted, and there are already many types of genetically engineered crops. Examples include Bt corn, a crop which produces a natural pesticide with the aim to reduce pesticide application, and golden rice, a crop which has higher levels of vitamin A – a key nutrient deficiency in poorer countries that leads to the deaths of around 670,000 children under five annually.[57] Many regions are using genetically modified crops on a large scale, and even the Amish use genetically modified tobacco crops.[58] There is a huge amount of activity in developing crops that can be made to be resistant to climate-driven stressors such as funguses, saline water and higher temperatures, along with attempts to increase nutrient content.[59] Crops can also have several traits at once, resulting in higher resilience, nutrition and yields.

A big target for genetic modification is to remove the notorious inefficiency of photosynthesis itself.[60] When plants respire, they can only use light at certain frequencies, and waste the other parts of the light spectrum. In all, leaves generally use less than 5% of all the energy in sunlight.[61] If we are able to improve this, it will directly improve the productivity of land, sparing more land around the world. Call that hubristic thinking, but more so are other products like heavily processed spray-on cheese.

Previously, genetic modification took many years and millions of dollars to make progress. Part of the reason was the ability to know what genetic material might be useful (e.g. which part of a plant gives it its drought tolerance), and the trial and error nature of whether the target plant would incorporate this DNA. In 2013, a new technology called CRISPR was discovered, and it's revolutionizing the speed at which crops can be developed. CRISPR takes us one step closer to being able to code DNA directly and understand the impacts in a much deeper way. The pace of innovation is dramatic and has already been used to make underutilized crops like physalis (a sweet tomato-type plant, called ground cherries in the US) easier to grow.[62] CRISPR is also being used to address some of the climate-related problems faced by banana crops we heard about in the last chapter.[63] In general, the accuracy and speed of CRISPR might mean that many more combinations of traits can be introduced at the same time.

The amount of activity in the field is incredible and the effect has been democratizing, with many smaller labs starting up. Geneticist Rodolphe Barrangou estimates there may now be over a million geneticists working with the technology.[64] The potential for CRISPR is not limited to agriculture; it has the potential to revolutionize human health, with work on Huntington's disease, cystic fibrosis, and breast and ovarian cancers. Researchers have already shown that CRISPR can kick out HIV infections from cells.[65] It's a truly remarkable technology, and it's safe to expect many more incredible developments in the near future.

Ultimately, the question is becoming less about the technological feasibility of genetic modification, which only seems to be getting easier, and more about public acceptance and accessibility. From a health perspective, researchers have found no cause for concern about genetic modification directly. Every major national and international scientific society, including the National Academy of Sciences and the World Health Organization, has concluded that the genetically modified plants on the market pose no threat to human health.[66] However, there are social and environmental concerns around genetic modification: the widest used genetically modified plants are pesticide- and herbicide-tolerant crops. These crops have had two big downsides: 1) they have worked to connect genetic modification with big business (since they were developed with a focus on profit rather than, say, the higher-minded ideals of golden rice); and 2) they are increasing pest resistance. The result is a pesticide treadmill, where pest-resistant crops encourage pest immunity, necessitating the development of more genetic modification for more pest resistance, encouraging more pest immunity, etc.[67] This can be eased somewhat by reintroducing intercropping or using non-modified crops as buffers around modified crops, but it is a serious cause for concern. Looking to the future of GM regulation, while there are many regulatory requirements in developing a 'traditional' genetically modified crop, there are currently very few regulations for CRISPR crops. Given that, in 2019, there are an estimated thirty to forty crops in different stages of active development, this is concerning. Another problem is that while CRISPR is exciting, some studies are showing that the technique can introduce unanticipated DNA damage and even result in edited cells that can trigger cancers.[68]

Perhaps the most serious concern is how genetic modification can go hand in hand with commodification and big business. As mentioned, there are four huge, international seed companies. These companies have a reputation for pressuring farmers to use their products and have generally been hostile to small farms.[69] They are also becoming generally closed to open collaboration, with seeds increasingly falling under patents and academics struggling to get commercial permission to work with them. The ease and cheapness of using CRISPR may reduce this chokehold, but part of the reason it's so cheap is the current lack of regulations, which may have to change given the speed of developments. Either way, we'll need much higher funding of plant breeding at universities and other non-profit institutions to ensure that farmers aren't forced to purchase seeds from big seed conglomerates.[70]

A different tactic is to make seed development and agricultural technologies open-source. The open-source movement brought us Linux, the operating system now used on 80% of servers worldwide and significantly outperforming closed-source software like Windows. Linux brought programmers together on open platforms to collaboratively develop software, and we'll need such approaches for agriculture in the future. Organizations such as the Open Source Seed Initiative aim to bring open access to genetic resources for all plant breeders, while the Biological Innovation for Open Society extends the philosophy to all agricultural innovation including sensors and software.[71]

The concerns surrounding access to resources are related to broader issues within the global food model. Typically, economic incentives (such as subsidies) and commercial power tend to tip the scales towards the most environmentally damaging production that national regulations will allow. This is where the greatest margins from scale arise. All other things being equal, if the most damaging pesticide is banned, the incentive will then be to use the second-most damaging.[72] This effort to keep the short-term cost of food down and squeeze farmers ends up increasing the long-term price paid via the environment.

PEAK FARMLAND

In recent decades the amount of land used in agriculture has been levelling off.[73] With plant-based diets, reductions in food waste, and old farming techniques being updated, we may see peak farmland soon – even while populations increase. This Great Food Transition would give us the breathing space to ride out the demographic transition while preserving ecosystems and helping curb carbon emissions.

So much is down to whether we are able to change diets and reduce waste. We know that social tipping points are incredibly powerful. What was once unimaginable can become the social convention in just a few years. Research has shown that when a minority of people adopt a new practice, it can spread rapidly to the rest of society.[74] It's also likely that as the impacts of environmental disasters intensify, the social conventions surrounding meat will change profoundly. It's plausible that societies in the near future will view the current agricultural system as archaic and vicious. Even the way we use terms like *produce* and *consume* for sentient beings may come to be understood as linguistic violence. This is not standing in judgement of those of us who eat meat now, more to say that attitudes may change in the blink of an eye, especially in a connected world suffering from the impacts of the existing system.

Although The Great Food Transition could eventually reduce agricultural area by as much as half, the benefits would come long before then. As soon as humanity passes peak farmland and starts to remove land from agriculture, the subsequent rewilding and reforesting will help communities become more vibrant, happier and more relaxed. Research has repeatedly shown that green spaces improve human health and reduce mortality. For instance, one study showed that at least two hours a week in nature substantially reduces stress, improves health and increases well-being.[75] Another study, a meta-analysis using data from more than eight million people, found that 'interventions to increase and manage green spaces should be [...] a strategic public health intervention'.[76] The Great Food Transition will help us regenerate many of the things we've lost over the past centuries of land clearing, land degradation, and the conversion of nature into farming factories. The food revolution is not only necessary to curb biodiversity loss and ecosystem collapse; we have to do it for ourselves. We also have to do it for the climate; almost all the solutions described in this chapter would help significantly, both in bringing carbon out of the atmosphere and limiting emissions in the first place, from reducing livestock emissions to carbon sequestration in reforesting and rewilding. As we've seen throughout the book, solutions often act together to address many problems at once. Each step along the path of transition opens up more opportunities for beneficial feedbacks and the potential for a future in which humans thrive.

CLIMATE

CHAPTER SEVEN

∿ PESSIMISM ∿

Where All Roads Meet

That is why we feel
It is enough to listen
To the wind jostling lemons,
To dogs ticking across the terraces,
Knowing that while birds and warmer weather
Are forever moving north,
The cries of those who vanish
Might take years to get here.

CAROLYN FORCHÉ, 'San Onofre, California', 1977

In a moment the ashes are made, but the forest is a long time growing.

SENECA, *Naturales Quaestiones*, c.65 CE

[The draft text] asks Africa to sign a suicide pact, an incineration pact
in order to maintain the economic dominance of a few countries.
It is a solution based on values, the very same values in our opinion
that funnelled six million people in Europe into furnaces.

Reaction of Lumumba Stanislaus Di-Aping, head of
G77, to climate agreement draft text, Copenhagen
United Nations Climate Change Conference, 2009

In a very real sense, it's already too late. We have failed. Millions of people have already died from climate disruption, millions more are currently suffering, and somewhere between hundreds of millions and billions are in the firing line.[1] This is not like the other human problems we believe we can eventually get around to fixing, problems like global poverty, hunger, modern slavery – the ones we believe are slowly improving, even if they aren't. The ones we assume will eventually 'get there', wherever 'there' is. Climate change isn't a problem to clean up eventually, like a chemical spill. Climate breakdown is a trap which we have set and into which humans have decidedly stepped. It presents a retreating frontier: no longer a future of expansive human possibility, but one of struggle and limitation.

Weather systems are changing over our heads. The ground is shifting beneath our feet (for instance in the thawing permafrost of the Arctic).[2] The chemical compositions of the oceans and atmosphere are spoiling further every day. Yet in the cold, hard calculus of suffering, there is still so much action to be taken and so much inequity to correct. *Everything* now comes down to the scale and speed of change. If we can somehow build systems that can 'get it together' quickly enough and profoundly enough, we may prevent the millions suffering currently from becoming billions. But that is an outside chance. It is the limit of our hope, and those details are for a different chapter entirely.

We are in a terrifying amount of trouble. For a very long time, we could take solace in the fact that the worst news, represented by the numbers populating scientific tables and the traffic-light-coloured lines in the plots of technical reports, were just abstractions describing hazy futures; futures which were parameterized using varying assumptions of planetary response, global development and humanity's potential ingenuity. Even by the 1990s and 2000s, it was possible – with the help of some motivated reasoning and cognitive dissonance – to wave away intensifying storms and heatwaves as freak events ('It's devastating that so many people have suffered, but we can only bow to Mother Nature').

Now the scientific numbers are becoming manifest. Heatwaves, droughts, floods, hurricanes, wildfires: everywhere we look, things are changing beyond recognition. It would take several books to describe the unprecedented death and destruction in the last few years alone. Years when 3.5 million people in drought-stricken India ran out of water; when Mozambique was hit by two typhoons in as many months; when the largest wildfires on record scorched the earth, from east to west, north to south. These fires are nation-sized. The Alberta region of Canada saw an area the size of Cyprus burn, Australia, an area twice the size of Belgium, and Siberia lost an area the size of Austria to embers. That was just in 2019. The Australian fires of Christmas 2019 harmed or killed around a billion animals and torched 21% of the nation's forests.[3]

The numbers from the models run by increasingly panicky scientists used to be abstract, but they have always been more than numbers: they describe communities and lives, including our own.

These past few years, as horrendous as they've been, are the best we can hope for in the future. From now on, these years will feel like a brief reprieve. Even if we cease all emissions *today*, some changes, such as sea level rise, will worsen for centuries.[4] As a species, we have careered up to and off a cliff and are now having a mid-air, civilizational crisis, arguing about what to do as we plummet.

The climate will be changing rapidly for centuries (if not millennia) to come.[5] If we 'fix' the problem, we might only have to deal with the climate impacts for millennia. If we fail, the impacts on species and biodiversity loss will be with us for millions of years. What we must do is surprisingly simple to describe: move to a zero-carbon economy as quickly as possible, and then, without pausing to celebrate, push on rapidly to a negative-carbon economy. Carbon-dioxide levels are already far too high for a comfortable planet, as we shall see in this chapter. They must come down. In the meantime, we can only hope we don't set off chain reactions that would transform the planet no matter what we do. If we transition too slowly because vested interests obstruct change, or because public apathy or confusion stalls progress, then – at some point along the sliding scale from 'some success' to 'failure' – it won't be clear who 'we' are. Who are 'we' if civilizations collapse uncontrollably? Can we call that any sort of survival?

These are strange and unsettling paragraphs, and this is difficult to write. These might sound like excerpts from a science-fiction novel, or the online comments from some conspiracy crank. But they are not. They are the accepted, scientific and physical consequences of our past and current actions, and describe the decisions available to us in the coming years. Climate breakdown is such a vast, difficult concept that our minds almost completely rebel against it. It has been called a *hyperobject*: a term borrowed from mathematics to describe an object transcending many dimensions.[6] Climate breakdown is a hyperobject in the sense that it stretches over so much space (the entire planet, all ecosystems, all peoples) and time (right now, the next decade, millennia into the future) that it thwarts attempts by us mortals to grasp its enormity. The sheer expanse of the issue explains why people struggle to talk about it. How can you bring up such a problem in polite conversation? It's like dropping news of terminal cancer in an office canteen.

Some will say that we have faced such problems before; that nuclear war would have similar impacts. This is true: a large-scale nuclear conflagration would result in a nuclear winter which would change the atmosphere for thousands of years. While this thankfully hasn't happened yet, the mushroom clouds of climate change are already here: every day, we're adding to our atmosphere the heat of more than 400,000 nuclear bombs exploding.[7]

The problem is also very different. It's not like any of us are standing in a nuclear bunker ready to press the button, whereas, when it comes to climate change, we may as well be reading out the launch codes. Everyone contributes… but some, much, much more than others. Given the structure of the ecosystem and the economy, it is almost impossible for any individual not to contribute. We could be facing the existential collapse of everything we know, and yet no single person or organization is the culprit. Can we blame the fossil fuel companies? The super-rich? All of the consumers in 'the West'? Or would it be better to blame the Chinese for building coal power plants on such a massive scale in the early 2000s? The Republicans for their climate denialism? The Brazilians for felling the Amazon? The Russians for their gas? By the very nature of the problem, this list goes on. Perhaps there is enough blame to go round. In the end, we are still trapped in a world which most people have helped to create.

It's difficult for any individual to act because the entire global economic model is predicated on the emission of greenhouse gases. Looming climate catastrophe is a 'symptom of a larger disease'.[8] Any solution implies a complete transition of our behaviour, our technology, and the power structures that organize society. In the transition, we will see further segregation of winners from losers in a world already stratified by deep, unjust inequality. If we already struggle with generational issues like pensions and housing, we ain't seen nothing yet. What's worse, many of the worst perpetrators will be long gone by the time the costs arrive.[9]

Climate change is the ultimate intergenerational problem. If you asked a philosopher to devise a problem perfectly suited to confound humans and to shape them into immoral monsters, climate change would be it. Our problems don't stop at the climate crisis, though; many of our related ecological crises are equally difficult to grasp and to solve. Biodiversity loss and ecosystem collapse show the same properties: they are happening everywhere, and will take millions of years to recover unaided. Climate disruption will have massive impacts on biodiversity. These problems are symptoms of the same issue: the way we live our lives and build our societies.

Isn't this talk all a bit apocalyptic, especially when we consider that humans have adapted to many different climates – from the tropics to the Arctic? Temperatures have always changed, so surely we can adapt to whatever is coming our way? Well, perhaps to some extent. But there are two huge problems with climate disruption: the speed and depth of change. We humans can only adapt within our dexterities, and that level of adaptation often depends on wealth. It is true we've adapted to changes over the last few thousand years; after all, that's when the first civilizations were built. But the climate over these thousands of years of the Holocene has been incredibly peaceful. As we've seen, the years have been so remarkably calm that they have given human societies the breathing space to develop agriculture, cities, nations, science... and the television allegory of climate change, Game of Thrones.[10] If it weren't for climate change, we would be heading for a very slow cooling phase right now. Instead, over the past hundred years, we have pushed the lines on plots of temperature, carbon dioxide and methane up, far past the 'usual' variation of the Holocene. We have been so successful at this that we have changed the chemistry of the atmosphere to a state the planet hasn't seen for millions of years. Having done little to quench the flows of emissions, the atmosphere is now changing more than ten times faster than at any point in the last 55 million years.[11]

Right now, there is as much carbon dioxide as when trees grew in Antarctica three million years ago.[12] The only reason we don't have deckchairs at the South Pole yet is because it takes time for ice to melt and for seas to rise. But we're on to it. Try this: pour a cool glass of water from the tap, add ice, then place the iced water in the sun. It will take time but, once the ice melts, the temperature will rise quickly and you will have an unpleasantly warm glass of water. Since 1960, glaciers alone have lost more than nine trillion tonnes of ice, enough to cover an area the size of Spain with a twenty-metre layer of ice.[13] With the additional and accelerating melting of ice caps, even rich, well-resourced countries cannot adapt fast enough. More than $14 billion was spent upgrading the water defences after Katrina, the category 5 hurricane that hit the US East Coast in 2005, pummelling New Orleans, but it appears that these defences could be overwhelmed by storms as early as 2023.[14] The 2019 flooding of the Mississippi was already another near miss for New Orleans.[15]

Where this all ends is uncertain. With a massive overhaul of society in the next few years, we may be able to avoid some of the worst scenarios, but perhaps chain reactions are already unfolding – in which case, whatever we do now may not be enough over the very long run. We are exiting the climatic balance and calm of the Holocene at a frightening rate. This highly dynamic, non-linear state will have massive consequences for the climate and human civilization. Although it never seemed like it at the time, humanity had been swimming in a calm climatic ocean, most of us unaware that storms were encroaching on the horizon.

EXIT THROUGH THE HOLOCENE

Weather is an expression of the energy flowing through the climate. Think of fallen rain evaporating into the atmosphere by the sun's heat only to fall once again, or of wind speeding from areas of high pressure to low thanks to differences in solar energy absorbed across Earth's surface. Adding carbon dioxide, methane and other gases into this system increases the retention of energy in the atmosphere and oceans. These gases form a thin, diffuse blanket of molecules which reradiates heat down to the surface, raising the temperature. Since hotter air holds more water, the atmosphere becomes wetter. And since water vapour is a heat-trapping gas, this aids the warming of the planet as a whole. This is a *positive feedback loop*. We are already living in a climate hotter and wetter than the Holocene. With the increasing energy, the weather will continue to intensify. This is what climate scientist Wallace Broecker meant when he said that the climate is an angry beast we are poking with sticks.[16] If we keep pushing the climate, we may reach an entirely different stable state: an evil twin of the Holocene. It would be a world with almost no ice left, seas more than twenty metres higher, and a ravaged, rampant climate.[17]

We have known for more than a hundred years that climate crisis would hit us. Despite what climate deniers may say, the fundamental mechanics are undergraduate-level physics; climate change is Physics 101, and what has prevented politicians from accepting it is surely not an incapacity to understand it. In 1886, the Nobel Prize-winning chemist Svante Arrhenius, using a pencil and paper, calculated that for a doubling of atmospheric carbon dioxide we could expect a five degrees Celsius increase in global temperature. His calculation was remarkably accurate. Historically, computer models have generally estimated between three and five degrees Celsius, but early results for the latest generation of models destined to be included in the 2021 IPCC report are consistently finding five degrees Celsius or higher.[18] This increase is what we call the climate sensitivity. Arrhenius wasn't worried about the problem, though. At the rate of emissions in the nineteenth century, it would have taken thousands of years before carbon-emission-driven temperature rise would set in. But like so many others, Arrhenius couldn't have imagined the twentieth century's Great Acceleration, during which we increased annual emissions a thousandfold. The time window for a doubling of CO_2 compared to pre-industrial concentrations was cut from thousands of years to just over a hundred – and we are now just a few decades away.

Pre-industrial levels of carbon dioxide in the atmosphere were between 260 and 280 parts per million. That is: for every million molecules in the atmosphere, CO_2 comprised 280 of these. In 2019 we reached 415 parts per million for the first time.[19] It took two centuries to raise the concentration by forty parts per million, three decades to raise another forty, and then just two decades for another forty – I was born at 345 parts per million. We are still increasing concentrations by around two parts per million a year.[20] A few hundred parts per million sounds minuscule – it's only 0.04%. But as for other features of today's environment, small increases spread over vast areas of the Earth add up to huge impacts. To paraphrase Archimedes, 'Give me a lever long enough and a fulcrum on which to place it and I will move the world.' Greenhouse gases are the longest levers imaginable, and the fulcrum is the massive fossil-fuelled, growing economy of humanity.

The main cause of this civilizational threat seems so banal when expressed in simple chemistry. When carbon is bound with hydrogen in molecules of different configurations, these molecules can be heated, splitting them apart. This separation produces more energy than it takes to heat the molecules up. Whether you're burning plants with sugars (varieties based on $C_6H_{12}O_6$), natural gas (CH_4), petrol (a mix of compounds like C_8H_{18}) or coal (lots of C, some H and many nasty metals like mercury), it's all about those CH bonds. When the CH bonds split, the carbon heads into the atmosphere, combining with plentiful oxygen to form carbon dioxide, and the hydrogen generally heads off to form water. The splitting of this one bond is the deal on which the industrial world is built. If it sounds too good to be true, it probably is – the bargain we make for this energy is appalling. For each unit of energy we harness from this bond, more than 100,000 times more heat energy is added to the atmosphere (in the form of the heat trapped by the released carbon dioxide).[21] Put in terms of ice, each kilogram of carbon dioxide melts around 650 kilograms of glacier. That's around 1,200 tonnes of ice melted for your London–New York return flight, or 1.9 tonnes of ice for every day one person eats a meaty diet over a vegetarian one.[22] The ordinary splitting of carbon and hydrogen is leading inexorably to a world of suffering on a scale that is difficult for any book to depict, let alone for most of us to comprehend.

THE AGE OF THE UNPRECEDENTED

In this world of climate breakdown, reporters are repeatedly referring to a 'new normal', perhaps hoping that we've reached a new equilibrium, similar to when the term was coined after the post-2008 financial crash. But climate change is not a boom and bust cycle. There will be no equilibrium, never again in our lifetimes. It's increasingly difficult to grasp the speed and violence of the changes that have been wrought; as activist Ben See puts it, 'The planet you think you're living on no longer exists.'[23] Whenever this book is read, there will have been more disasters exceeding those described here. We will continue to invent new classifications and terminologies as the climate exceeds the Holocene-based scales of catastrophe. Discussion has already begun about extending the five-point hurricane scale to six.[24] Given enough time and emissions, there may be need of a seventh.[25] In this new abnormal, each year will feel unprecedented.

This new abnormal isn't only about extreme weather events. Climate breakdown overwhelms the capacity of animals, plants and fish to adapt, driving mass extinctions.[26] It raises sea levels and inundates coasts. It acidifies oceans, bringing on marine death.[27] It melts ice sheets, releasing the ground from their heft, increasing earthquake and volcano frequency,[28] and raising sea levels. It drives public health crises: disease spread, conflict and suicides.[29] As we move further into this grim future we will continue to roll the dice, but each year we will continue to trade in each set of dice for new ones that shift the odds ever onwards to more extremes.

There are no hard and fast rules as to how this will play out. But the short-term changes will follow a rewritten version of the Matthew principle (where the rich get richer and the poor get poorer). In this case no one gets richer, but those who already suffer most will suffer more. This means the poorest regions, the areas least responsible for emissions and least able to adapt will suffer the most. What humans are already coping with each year is astonishing. In this chapter, it's only possible to list a very small selection of recent events to give a sense of the diversity and depth of impacts:

Heatwaves and drought

East African drought, 2017

13.1 million people in East Africa (including 5.5 million in Ethiopia alone) are facing crisis, emergency or famine levels of food insecurity.[30] Climate-change-driven warming made the drought twice as likely.[31]

Heatwaves in Europe, 2003–19

The heatwave in summer 2003 led to more than 70,000 additional deaths.[32] Later heatwaves (in 2010, 2015, 2017, 2018 and 2019) caused fewer casualties due to better preparation, but disrupted power, transport and food systems. Although there was a possibility the droughts would have happened anyway, all were made much more likely by climate change.[33]

Severe drought and water shortage in Bolivia, 2016

The drought was considered the worst in forty years and the worst water shortage in twenty-five years. It resulted in a state of national emergency, livestock deaths, crop loss and water rationing.[34] Lake Poopó, once the country's second-largest freshwater lake and an important fishing resource, dried up.[35] Climate change contributed by reducing rainfall and melting glaciers.[36]

Extreme heat in Asia, 2016

Southeast Asia, southern India and northern Eurasia experienced extremely high temperatures, with 580 people dying in India alone. Extreme heat of this type is only possible due to climate change.[37]

Marine heatwaves, 2016

The year 2016 saw the most intense and long-lasting marine heatwaves over a quarter of the world's ocean surface since the beginning of satellite measurements in 1982.[38] In Northern Australia, the heatwave caused mass bleaching of coral reefs.[39] Climate change made the intensity of the heatwave more than eight times more likely, and the duration at least fifty-three times more likely.[40]

Cape Town and Day Zero, 2015–17

Three consecutive years of low rainfall in the Western Cape region led to the worst drought in more than a century. In 2018, with Day Zero approaching – the day when water supplies had to be turned off – the city government introduced water rationing and other measures for Cape Town's 3.8 million inhabitants. Day Zero was successfully avoided, but climate change made this event three times more likely, and increased the probability of more shortages.[41]

Eastern Mediterranean drought, 1998–2012

A recent study found that the drought was probably the worst in the past 900 years.[42] Drought in the region not only caused widespread crop failure and fuelled migration, but also contributed to political unrest in the region (although the extent to which it was a factor is disputed).[43] Climate change was a key driver of this severe drought.[44]

Pan-Caribbean drought, 2013–16

This was the most severe and widespread drought since 1950, causing food insecurity for more than two million people. The impacts were particularly harsh in Haiti, where 50% of crops were lost and 10% of the population experienced food insecurity. Warming significantly increased the intensity of the drought.[45]

Flooding and extreme precipitation

Extreme rainfall in Peru, 2017

This led to 177 deaths, affected 1.7 million people via landslides and flooding, and resulted in $3.1 billion of damage. Climate change made this event around 1.5 times more likely.[46]

Snowstorm in Nepal, 2014

A Himalayan blizzard dumped more than 1.7 metres of snowfall within twelve hours, triggering a series of avalanches which killed forty-three people. The alteration of the atmospheric conditions through climate change will make these previously rare events more likely.[47]

Flooding in Bangladesh, 2017

Around 850,000 households were affected, and 220,000 hectares of ready-to-harvest rice were damaged or destroyed, leading to a 30% increase in rice prices in Bangladesh.[48] The likelihood of such an event was doubled by climate change.[49]

Southeast China floods, 2015

More than fifty people lost their lives, while more than 250,000 people had to be evacuated.[50] This type of 'short-duration, intense rainfall event' has become between 23% and 66% more likely.[51]

Extreme rainfall and flooding in Brazil, Argentina and Uruguay, 2017

Brazil experienced a direct economic loss of $102 million and more than 3,500 people were displaced in Uruguay. Extreme rainfall such as this is between two and five times more likely as a result of climate change.[52]

Heavy rains and flooding in Nigeria, 2018

These had an impact on 1.9 million people, with 82,000 homes destroyed, 210,000 people displaced, and 200 people killed.[53] Climate change increased temperatures, rainfall and the sea level, exacerbating flooding.[54]

Hurricanes in the United States

Four of the five most costly weather events in US history, with a combined cost of $426 billion,[55] were exacerbated by climate change. Extreme rainfalls during Harvey were three times more likely, and rainfall was enhanced by between 4% and 9% for Hurricanes Katrina, Irma and Maria.[56] Harvey alone flooded more than 100,000 homes, while Katrina led to the deaths of more than 1,800 people.

———

Every one of the above events has been attributed to human emissions of greenhouse gases – and these are just the ones researchers have examined.[57] Not even an exhaustive list could cover the full spectrum of suffering from climate breakdown. Neither can large-scale accounting detail the many unique stories of tragedy.

In the richer areas of the world, such stories may be covered in the media. In the Greek wildfires of 2018, two dozen people trapped by the flames formed a huddle together on the beach as they perished.[58] During Hurricane Katrina, mercy killings were common across hospitals as patients, cut off from power and transport, couldn't be saved or evacuated.[59] In poorer regions, the individual suffering will continue to be underreported in the global press, while other stories may never be linked to climate breakdown in the first place. Reflecting the fact that obituaries will very rarely list climate change as cause of death, violent attacks, diseases, suicides and other community stressors may not be identifiable as climate-related but, like the lung cancer-afflicted smoker, its fingerprints are everywhere. It also acts as a *threat multiplier*: an additional stress that can push already stressed societies over the brink. To give just one example, while the civil war in Syria was a com-plicated, multifaceted conflict, research has found it was exacerbated by a drought between 2007 and 2010 that was made two to three times more likely by climate change.[60] As the country collapsed, refugees, especially women, were subjected to rampant sexual violence and abuse in their chaotic flight.[61]

CLIMATE FEEDBACKS: THE ROOF IS ON FIRE

The large-scale, general sequence of changes describing how climatic disruption would play out has been well known for decades, although the open question was how long the changes would take to arrive. The timing was always tough to estimate because today's heatwaves alter the speed of tomorrow's warming – the core essence of a *feedback*. We heard of such an example earlier. Arctic sea ice, which, as it melts, turns the reflective, cooling white surface to a deep blue, absorbing more heat. This increases the temperatures, driving more ice melt, revealing more deep blue, and so on. If the remaining Arctic sea ice is lost, it would be the equivalent of adding one trillion tonnes of CO_2 to the atmosphere on top of the 2.4 trillion emitted since the Industrial Age.[62]

It was once thought that some processes like significant ice-sheet loss would take centuries to unfold, but now it appears the loss is progressing on the timescales of a human lifespan. The acceleration is astonishing, confounding even our metaphors: a glacial pace is now torrential.

The clearest evidence can be seen in the Arctic, the long-term bell-wether for the climate. Changes in the north have always presaged major global transitions over long geological periods, so it's not surprising that it has seen more warming than the rest of the planet. What is surprising is how quickly the warming has taken place. The 2007 IPCC report estimated that we had some breathing space, that the Arctic probably wouldn't be sea ice-free during the summer until after 2100.[63] Later, the 2013 IPCC report had one scenario suggesting it could be 2050.[64] Research now suggests that it could happen as early as 2030.[65] With spring temperatures soaring to forty degrees Celsius above pre-industrial records in 2019, we will soon find out if any of these predictions were accurate – which is strange to say, given that these processes usually take between centuries and millennia to unfold.

Even more melt is happening out of sight. Underwater time-lapse photography has found that Arctic glaciers are losing ice between ten and a hundred times faster than previously thought.[66] The giant ice cap of Greenland is also melting quickly. Since 1972, 50% of the sea-level rise driven by Greenland has been in the last eight years.[67] In the summer of 2019, Greenland lost an estimated 12 to 24 billion tonnes of ice in a single day.[68] That's more than ten times more than the total mass of all the fish globally, or more than three times more than the weight of all food produced by humans in a year.[69] That's the ice lost in *a single day*. Where Arctic sea-ice melting is a few decades ahead of schedule, Greenland ice loss is more than seventy years early.[70]

Permafrost thawing is about seventy years ahead of schedule, too.[71] As the soils thaw, they expose older stretches of land, some of which were last uncovered 10,000 years ago. Microbes once dormant in the soil are reanimating, prompting further emissions and another feedback as they revive. It's estimated that every degree of temperature rise would thaw enough permafrost to discharge the same as four to six years of global coal, oil and natural-gas emissions (two to three times as much as previously thought).[72] Like a time machine, as the thaw goes deeper, the microbes released originate from further back in time. Recent history is already coming back to haunt communities as anthrax from dead cattle and reindeer graveyards re-enters the ecosystem. A release in 2016, during record high temperatures, resulted in the hospitalization of ninety-six people and the death of a twelve-year-old boy. There are more than 13,500 of these graveyards dotted across Russia alone.[73] Many of them are unmarked, representing ticking biological time bombs. Some are even thought to contain smallpox, although we just don't really know what the thaw might release.

Sea-ice and permafrost feedbacks are pushing temperatures in the Arctic even higher. Summer heat maps are now painted in bloody red swathes: images that appear like Munch's *The Scream of Nature* in the nightmares of climate scientists. Vast Arctic wildfires are now an annual occurrence – from Scandinavia to Siberia, Alaska to Canada – scorching hundreds of thousands of hectares. Historically, Arctic forests in Canada drew more carbon from the atmosphere as they grew than they released during fires and dieback. This trend appears to be reversing, in another alarming warming feedback.[74] In a disturbing synergy, lightning storms have moved further north as the world warms, igniting more frequent blazes.[75] Warming prompts invasive species like the mountain pine beetle to invade from the south, chewing through vast stretches of forest, killing trees and releasing yet more methane and carbon.[76] In yet another feedback effect, wildfire particulates can stay airborne and land on the frozen stretches of land and ocean, prompting more warming as surfaces darken.[77] Each of these on their own represents a worrying trend. The combination is terrifying.

This is the news so far from the Arctic, but it will certainly be worse by the time you read these sentences. A new and mounting concern is that the West Antarctic is warming much faster too.[78] In general, there are two main causes of worry: 1) as ice thins, the melting point of water increases; that is, ice melts faster the thinner it gets, and once regions near the coast thin below a certain thickness, the ice being held back can then slide into the ocean;[79] and 2) what is happening below the ice. In some regions across the West Antarctic, the ice sits on land below sea level. As water infiltrates the ice from below, it melts from both above and below, hastening the destabilization of the ice sheet.[80] In all, Antarctic ice loss has increased 500% in the last twenty-five years.[81] As many as 65,000 meltwater lakes have appeared on its surface, hastening the melt. Between 2015 and 2018, Antarctica lost more ice than the Arctic did in thirty years.[82] In 2021, researchers found new fissures in the Thwaites ice shelf, which holds back huge volumes of Antarctic land ice. Some scientists now think the ice shelf could disappear by 2030, with dire consequences for sea level rise. As glaciologist Richard Alley remarked, 'We just don't know what the upper boundary is for how fast this can happen.'[83] A rapid destabilization of ice sheets – once thought to be a very remote possibility – is now firmly a scientific concern.

As a result of all this melting, there has been a surge in sea-level rises.[84] Previous estimates in the 2013 IPCC report suggested a worst case of sea-level rise of about a metre by 2100. Today's worst-case estimate is above two metres, inundating land where around 187 million people currently live (the consensus view, which is still very bad, is closer to a metre. In short, the worst-case scenario reported in 2013 is now the expected scenario today).[85] This increasing rise will continue far beyond 2100, and research suggests that if atmospheric CO_2 remains above 400 parts per million we can eventually expect (over the course of centuries and millennia) an unthinkable twenty-three metres of sea-level rise.[86]

This is a small postcard's worth of news from the feedbacks at the poles, where many of the changes are more obvious. I could have chosen to discuss many other, equally worrying signs of new feedbacks across the planet: for example, the shifting of the Indian or West African monsoon; the change in El Niño patterns in the Southern Pacific; or the Gulf Stream, which is slowing as ice melts and the fresh water interferes with the great flows of heat in the northern Atlantic. The Gulf Stream is already at its slowest since 1600, and will carry on slowing, having the potential to raise sea levels even higher while causing chaos in western European and eastern US weather systems.[87] There is a very small but non-negligible chance that Arctic and Antarctic feedbacks have already kicked in a cascading procession of change across the entire planet.[88] It is important to bear in mind that this won't happen overnight. These are not the fast tipping points of a riot but the inexorable, unavoidable transformation of the entire Earth system over the coming centuries.

UNDERESTIMATION IN CLIMATE SCIENCE?

The scientific estimates of average global temperature change have been very accurate so far, yet there has been a broad underestimation of the speed at which regional feedbacks like those in the Arctic might kick in. There are reasons for this: the consensus view of science tends to be conservative and necessarily slow. It can take a very long time to collect data, write it up, and go through peer-review. Articles can take years to be published. Eventually, once a number of articles are published on a particular subject, a consensus can be reached. This consensus can then be reported in intergovernmental panels such as the IPCC. In general, this consensus 'errs on the side of least drama', since scientists strongly adhere to norms of restraint, dispassion and scepticism.[89] The sum of this system means that groups like the IPCC tend to be overly conservative in their wording and presentation of the impacts of climate change, and the speed of its unfolding. International bodies like the IPCC also have to deal with political tampering. While the technical chapters tell the unvarnished science, the Summary for Policymakers that accompanies every IPCC report has to be signed off by national representatives. Countries like Saudi Arabia, China, Brazil and the United States have all made previous efforts to dilute the reporting of the (already conservative) facts.[90]

We have also heard that feedbacks are generally very hard to model, and some geophysical parameters are tricky to estimate without solid observational data. Often the real world is messier than models can account for, and sometimes we lack long-term data to calibrate models. In the case of both ice and permafrost, the uneven melting/thawing of the surface has dramatically increased the speed of the overall melt/thaw.[91] For instance, ice melting on the surface of a glacier can form a pool of water which increases further melt in that spot. Eventually, the water can break through to the grounding line (the transition between where ice sits on land and then begins to float on the sea) and flow into the sea, increasing the rate of sea-level rise. This is all very tough to model. Another example is invasive species. The aforementioned beetle infestation is extremely hard to model, requiring knowledge of climate science, biology and ecology. In short, it's an extremely complicated world and research is made more difficult by traditional disciplinary boundaries and a lack of time in which to perform and then communicate the findings. However, the latest scientific reports are becoming more forthright. The 2019 IPCC report on global land use highlights the very real possibility that, with increasing climate stress, food systems could reach breaking point,[92] an impressively frank statement compared to previous reporting. Another such statement can be found in a 2019 report from the IPCC's sister organization, focusing on biodiversity: that we need a complete economic reform to safeguard future food supplies.[93]

JOINING THE DOTS

While scientists understand the broad climatic changes and have a sense of how these changes may have an effect on various different social systems, it has been difficult for them to examine how multiple stresses may impact on society. For several years, it's been left to journalists, bloggers and other commentators to join the dots of climate damage with cascading human impacts, but scientific articles are increasingly describing what this might look like. A late 2018 paper entitled 'Broad Threat to Humanity from Cumulative Climate Hazards Intensified by Greenhouse Gas Emissions' found 467 pathways by which 'human health, water, food, economy, infrastructure and security have been recently impacted by climate hazards'.[94] It's hard to know how different pressures may provide the spark for yet more pressures, but one area in which we may have underestimated the socially destabilizing effects of climate change is food production.

Already, one-third of the variability in food yields globally is being driven by climate breakdown (more than 60% in some major food-growing regions).[95] The growth in yields has been cut by more than half for some crops.[96] This was growth we were betting on to feed a growing population. In a depressing refrain, the places suffering first and deepest are the most food-insecure regions. Millions of people across fourteen African countries have been in almost continuous drought for several years now.[97]

Stresses will increasingly be felt in major food-producing regions too. In 2019, Europe, the US Midwest, Northern China and India were hammered with a series of floods and droughts.[98] Food systems are very sensitive to changes in climate and crops respond in a non-linear way to increasing temperatures. Just one day of extreme temperatures (up to forty degrees Celsius) can produce a 7% decline in maize, with similar results for rice, wheat and soya – and this isn't factoring in other climate impacts like flooding and disease.[99] In 2021, researchers found that a new generation of climate and crop models showed an earlier impact of climate change on agriculture around the world. They found markedly lower yield estimates for maize, soya beans and rice, with only wheat showing some gains.[100] There is the very real possibility that climate change, combined with the continual degradation of soils, water supplies and land, could result in regular multiple breadbasket failures.[101] One study found that the risk of these failures increases the most for maize crops, between 6% and 40% at 1.5 degrees Celsius of warming to 54% at two degrees Celsius.[102] Put a different way, today we'd expect a multiple breadbasket failure of maize crops on average every sixteen years (at around a degree above pre-industrial temperatures), which drops to every three years for 1.5 degrees Celsius and every two years for two degrees Celsius. The study found similar trends for wheat and soya beans. Similar results have been found in other studies.[103] Some crops, such as wheat, may see increases in yields, but of course the threat is not just from climate change alone but also from soil degradation and water use discussed earlier – remember, more than 70% of water use is for irrigation, and already a quarter of humanity is under water stress.[104] In the short term, the so-called *grain carry-over* (the stand-by grain held in storage around the world) can bridge any gaps, but the odds are that increasing global crop failures would heavily stress societies. Even if there is sufficient grain overall the restriction in supply would increase prices in a similar way to the 2007–8 food crisis when, for many reasons, including food subsidies, biofuel policies and climate changes, the price of several food staples (including wheat, soya beans, corn and rice) leaped up across global markets. The resulting unrest in the form of protest and riots by hungry people flared up across many countries including Mexico, Brazil, Egypt, India and others.

In globalized markets with some grain carry-over, how would we know if food shortages are getting worse? One signal might be increases in hunger, since, in an economy as unequal as ours, we'd expect those on the periphery of the global market to suffer first. This is exactly what we are seeing, with hunger increasing since 2011. This hunger will move more towards the centre of the globalized food system the longer time goes on. We have been warned.

THE POLITICAL RESPONSE:
SHIFTING DECKCHAIRS ON THE TITANIC

As we stare down the barrel of these changes, what are our political representatives doing? The institutional responses to climate breakdown are typically cast as either mitigation (to reduce emissions) or adaptation (to cope with the breakdown). But there is a third option: suffering.[105] We are doing all three, and the efforts being made towards mitigation and adaptation, along with the utterly inadequate speed of the response, give us a sense of what future suffering will be.

Unfortunately, mitigation efforts are still limping along. Outside years of recession and global flu outbreaks emissions are still rising year-on-year, growing in 2018 by the largest amount in history.[106] As of 2019, we are already 1.1 degrees Celsius hotter than the pre-industrial baseline, but because the land warms faster than the vast oceans of Earth, the temperature on land is already around 1.7 degrees Celsius warmer.[107] Despite all the noise surrounding climate targets and budgets, the world is heading for well over 2.5 degrees Celsius of warming[108] – a level that would drastically reduce the carrying capacity of the planet and probably consign billions of people to suffering and death. The voluntary targets of the lauded Paris climate agreement would result in an estimated 3.4 degrees Celsius by 2100.[109] At this level, the climate would almost certainly tip into a state which would eventually be game-over for complex civilization.

The idea of carbon budgets – how much we can still emit to meet certain temperature targets – suggests that we have emissions in credit to spend. But all we need do is remember the turmoil in the Arctic to fathom the extent of our debt. This debt is especially large for the wealthier countries. Although there is some evidence that a small number of high-income countries like the UK have managed to push down carbon emissions by a small amount, around 9 to 12% from 1990 to 2018 (around 31 percentage points lower than government declarations, which show closer to 43.5%. This is because government figures don't include emissions from aviation, shipping and imported goods).[110] But if the UK is such a leader, why has it consistently encouraged new oil platforms in the North Sea, cut solar subsidies for homeowners, and handed out tax-free concessions for shale-gas exploration?[111] Whichever way you cut it, even the timid reductions of the world's 'sustainability leaders' are small potatoes given the 10% year-on-year reductions needed to meet a 1.5-degree warming target,[112] to stand a good chance of limiting catastrophic damage to ecosystems around the world.

While some countries pay lip service to the desperate need to decarbonize, others don't even pretend to try. The US has, at some stage, committed to pull out of every major international climate agreement. Exiting the Paris Agreement was the choice of one intransigent government, but remember that George W. Bush did the same for Kyoto. This isn't just a Trump problem. As the sea-ice declines, Republican leaders in the US are eyeing the 'untapped resources' of the Arctic and its 'fisheries galore' (in the words of Mike Pompeo).[113] From the party that once espoused the view that if there is a 1% chance of a terrorist getting a nuke, they would treat it as a probability of 100%,[114] the likely outcome of exceeding two degrees Celsius and condemning billions to suffering is apparently not worth their attention. Meanwhile, Putin also appears to prefer a warming world, and wants to strengthen Russia's status as a 'hydrocarbon superpower'.[115] It's easy to see why he might find it preferable: thawing permafrost would open up vast areas of land across Siberia, and would constitute the largest land grab since the colonization of the North American West. The melting Arctic is also developing into a new hotspot for conflict. As it opens up, it represents undiscovered fossil fuel reserves that represent around 24% of global gas reserves and 6% of global oil reserves, and a trade route as valuable as the Panama Canal.[116]

As we continue to fail with mitigation, adaptation will be the only remaining response. Adaptation can take many forms, from air conditioners and sea-wall defences in rich countries to shade and rainwater collection in poorer countries. Adaptation requires an institutional capacity to look ahead and plan for what's coming. More often than not, this means national and local governments acting. But instead of local and national governments taking this chance to shine, climate breakdown is highlighting their many failures. Even in the most well-resourced countries, existing companies and institutions are struggling to adapt. In

2017 and 2018, California experienced unprecedented wildfires ignited by electrical gear sparking and other human activities in a months-long drought, despite the fact that the utility company had been warned for decades about increasing risks and had been implored to reduce at least some of the risk by cutting back vegetation hanging near power lines. The response in 2019? To put in place good policies for funding and rapidly rolling out solar-powered microgrids, which would both aid decarbonization and provide energy security? To set up comprehensive response planning to help vulnerable people? To complete an aggressive vegetation management programme?[117] No. The 'fix' was to retain the option to black out up to one-third of the state for as long as five days.[118] In October 2019, the Californian utility company cut the electricity off for millions of people across the state. In many cases people claimed they had no warning. Even in supposedly progressive California – the richest state in one of the richest countries in the world – the existing governance is wholly incapable of adapting in a way that doesn't further endanger its citizens.

If you do spend money on adaptation, you're not spending on mitigation and even if these defensive expenditures are effective, the most they can do is buy time. At a certain point, delaying the inevitable gets too expensive – no matter how rich and well-prepared you are. The Dutch, a people living mostly below sea level, have over the past 800 years developed unrivalled institutions and expertise for managing water – perfect for adapting to climate disruption. More than any other nation, the Dutch have the expertise and budget to make the best of a bad situation. Yet new sea-level-rise scenarios may outpace their planning. By 2100, in the worst-case scenario, measures to keep the sea at bay could now cost twenty times as much as previously thought.[119] Eventually, the harsh reality of cost–benefit analysis suggests that the only reasonable option would be a managed retreat. Although it is an open question how fast these changes can be expected, resilience experts are already openly discussing higher land in western Germany to which the Dutch could eventually move.[120]

After mitigation and adaptation, what's left is suffering. Indeed, many elements of adaptation lead to suffering, for example if you can't leave the shade in the middle of the day for the oppressive heat or your home electricity is shut off and you need access to medical appliances.[121] Billions of people will expect governments to step in and help. If a government repeatedly fails to protect its citizens, what legitimacy does it have? As nations with weaker institutions suffer greater damages and food prices rise, climate breakdown will be a catalyst for existing grievances, facilitating war and conflict. Connections between climate, conflict and forced migration have already been made, and experts think these connections will only deepen towards mid-century.[122] In a connected world, the pressures in one area spill over to others, with the Syrian refugee crisis testing the institutional preparedness and political attitudes across Europe.[123] As the stakes rise, decisions get harder, and mistakes graver. 'Maladaptations' may further erode trust in governments, hastening civic breakdown.

A PERFECT MORAL STORM

Was this level of abject failure always meant to be? Should we really be surprised, given our previous failures? Looking back in history, there appear to be five recognizable stages:

1. We make some awful mistake in ignorance, hubris or callousness (think smoking, leaded petrol, ozone layer depletion or, even further back, poor drainage in Ancient Sumer).
2. Researchers identify the problem and determine how serious it is.
3. A consensus is reached that something must be done.
4. An interregnum sets in, where politically influential vested interests delay action… in increasingly desperate ways.
5. Given enough damage and public outrage, there's a concerted attempt to clean up the mistake.

As a species operating on a planetary scale, our mistakes now have planetary consequences. This is part of the very meaning of living in the

Anthropocene. That is, global damage comes with a new set of wicked problems. The atmosphere has been a communal rubbish heap into which we've been chucking emissions, with effects on people across all countries. There's no single government to take responsibility and act alone. Who cuts emissions, and by how much? Shouldn't richer nations take on much more of the effort? Will they share technologies and solutions with poorer nations? Will they help poorer nations cope with the damage? Since we now need to draw carbon out of the atmosphere, we need a new set of agreements on who will undertake such an effort. Would any government campaigning to address these problems in a meaningful way get voted out of office?

Institutions will offer paltry responses until the public pressures them to commit meaningfully to emissions cuts. In general, people won't act until they feel a threat and deeply internalize it. Perhaps the reason the only people revolting now are schoolchildren is because they've been born into a world arguably free of responsible adults. That's a fast way to feel threatened. On the whole, humanity is manifestly underdeveloped to feel this type of systemic, global threat. Considering that modern humans are largely genetically indistinguishable from our 70,000-year-old ancestors, this isn't surprising. Were the wolf at the door a literal wolf, you'd bet we would do something about it. Climate breakdown doesn't engender the same terrified national audiences as terrorism, gun violence, the Notre Dame aflame or Covid-19 variants – but it should. The same goes for air pollution and noise from fossil-fuelled cars. Even if an individual begins to internalize the threat, humans across many cultures seem to jump from denial to despair without any intermediate stages. Instead of action, they may just leap to feelings of helplessness at one extreme or hedonistic abandon at another. Going out in a blaze of carbon-fuelled glory is currently in vogue, with 'last-chance tourism' packages for such disappearing locations as the Great Barrier Reef and the Norwegian glaciers.[124] Getting to the stage of sufficient global public outrage still seems a long way off. Poorer regions are feeling the threat viscerally but are least able to push for change. Richer nations are still in the denial phase, and many of the most astonishing changes are 'out of sight, out of mind' in the Arctic.

Meanwhile, climate breakdown is making a mockery of humanitarian pretensions. In 2009, industrialized nations made a target to provide a pitiful $100 billion annually from 2020 onwards to help poorer countries mitigate emissions and, later, to adapt to climate change. (To put this in perspective, a single export terminal for a new Australian natural gas facility recently cost at least $50 billion.)[125] This target was missed by a wide mark, with the fund estimated to reach $100 billion by 2023 at the earliest. What's worse, the money included in this 'fund' can consist of loans and investments as well as grants. Imagine being given a loan to repair devastated infrastructure in Bangladesh, which you then have to repay, with interest, to the countries most responsible for climate change. A very small amount of the fund is in the form of grants, with one study estimating that in 2017–18, only around $19 billion to 22.5 billion (out of a total reported assistance figure of $59.5 billion) was made up of grants.[126] The 'moral necessity' of this green climate fund, as it was once described by leaders, has become an immoral stain.

Say we do close this chasm in ethics, there is another constituency we are actively impoverishing: those who don't even get a say – future humans. As many have quipped throughout history, 'What has posterity ever done for me?'[127]

This intergenerational problem is so acute that it explicitly pervades all economic decision-making. Indeed, there will be many times when it's morally acceptable to help people today at the expense of those tomorrow. Take the example of air conditioners in Europe: energy use increases in response to higher temperatures, which can drive yet more emissions. But what is the moral alternative? Let people boil to death in their homes? More broadly, global estimates of the increased energy demand needed as a response to climate extremes range from 11% to 58% depending on the level of warming.[128] This is an operating ethic where each generation sees it as morally justified to offload their responsibilities onto the next.

TOTAL REVOLUTION OR TOTAL WAR?

We are at the point where climate breakdown is shifting our lived reality beyond recognition. As the climate crisis intensifies, many more people will directly experience its violence. As the hundreds of millions of people who have been directly affected in recent years share their stories with communities, families and friends, society as a whole will begin to realize just how horrifying the situation has become. For the first time, many will start to feel appropriate panic and sorrow – sorrow some scientists have been feeling for more than forty years. To this day, we are responding utterly inadequately. So what happens when things get worse, as they will?

On the darkest of days, I imagine how this plays out.

The spreading of fear will make a psychological weapon of the weather. (A third of the UK public are already suffering from eco-anxiety.)[129] The same human psychology that allowed us to ignore the unfolding crisis for so long is the same psychology that will fixate on the weather. A storm is no longer 'a seasonal storm', and a heatwave is no longer 'a sunny spell'. Weather becomes an expression of our past, of our guilt (well-deserved or not). It becomes a direct threat to communities, a bleak foreshadowing of an impoverished future. Some, when faced with the changes, will react with fear, hatred and loathing. Police are already starting to clamp down on climate protests across the world and prosecutors are punishing infringements that are normally overlooked.[130] Increasingly, we begin to dread the onset of summer, anticipating what the season will bring: how many wildfires, heatwaves, hurricanes, etc., and how large they will be. We will adapt by bolstering infrastructure... but the changes will outpace us. Higher temperatures will render people incapable of working; they are already increasing suicide rates.[131] Although this is hard to detect, rising temperatures will increase the prevalence and spread of diseases – there is already evidence of this happening with Lyme disease and malaria, along with probable links for tick-borne encephalitis, yellow fever, plague, dengue, African trypano-somiasis, influenza, cholera, haemorrhagic fevers and schistosomiasis, to mention just a few.[132] Extreme weather events encourage even faster spreading of many diseases and the impact of climatic changes on the availability and nutritional content of food could hasten vicious cycles whereby malnutrition leads to compromised immune systems driving further disease spread.[133] There will be increasing numbers of global health crises.[134] The fear of the future will continue to rise.

As the overdrawing of groundwater continues, exacerbated by soaring temperatures and reduced water flow from disappearing glaciers, water shortages will become the norm. In a dire scenario, we will continue to be surprised by the interdependency of regional climates. Given enough time, water shortages and extreme events like floods and heatwaves will regularly hit several food-producing regions at once. Global food prices will rocket. With our current trajectory, it's not unimaginable that supermarket shelves in some countries will start to get patchy, perhaps beginning somewhere between the middle and the end of this century. This is not scaremongering. As we've seen, even conservative IPCC reports say as much.[135]

The social response to this crisis is foreshadowed by how we are responding to the current damage today. Will the changes provide the impetus to revolutionize the way we live for a more equitable and humane world? Or will we embrace the false promises of security from fascists and authoritarians? The answer so far suggests the latter.

From Bolsonaro in Brazil to Orbán in Hungary, from Trump in America to Erdoğan in Turkey, we see authoritarians thriving the world over. To be clear, these trends are not necessarily anything to do with climate change, but we can imagine fascism thriving in a world of fearful people panicking as shortages start to pinch. This burgeoning fascism becomes critical in a world on the move. At current rates of warming, hundreds of millions will be looking for climatic refuge by 2050.[136] Migration will be used as a dog whistle by opportunistic politicians to push an agenda of increasingly fascist policies. Walls and fences will go up, such as those being built today between Bangladesh and India – an ominous sign, given that Bangladesh's location in a low-lying delta makes it likely to be the first place to see a multi-million-person flight as sea levels rise.[137] Detention centres for migrants in the US or EU (such as on the Greek island of Lesbos) will become permanent fixtures. Climate change has already been identified as a driving force for Central American migration.[138] In a world of deep inequality and power imbalances, when we most need courage and leadership from politicians, we will get despot-ism and time-honoured self-interest. As institutions decay and tensions mount, violence and war break out. Many nuclear-missile-wielding states stand to suffer from food and water shortages – both of which provoke reactions thought improbable under 'normal' times. The possibility of internal political collapse and nuclear conflagration is easy to imagine.

This would be happening in a global society that is already failing to protect the vulnerable, feed the hungry and alleviate the suffering of the poor. As 'the strong do what they can, the weak suffer what they must',[139] the mega-wealthy will survive within ten-foot electric fences and with armed security guards – already a booming business, with disused, fortified missile silos being snapped up for millions of dollars.[140] Meanwhile, an underclass would eke out ever more brutal lives, while a dangerous world gets used to increasing food and water scarcity.

That we are on *the road* to the death of current global civilization is no longer up for debate. There are even scientists who suggest that civilization cannot turn off this road; that billions may perish, and that we should already be preparing habitable areas in which the people who are left can live sustainably.[141] At present, the actions to prevent this reality are shockingly deficient. And if collapse is really coming, it can be managed or it can be chaotic. Chaos currently appears to be the name of the game. Even in the case where we dramatically increase mitigation efforts, the climate damage and subsequent migrations, food shortages, droughts and the other impacts we *know* are coming would be enough to challenge our institutions and our humanity. If we experience more surprises in the climate system, then the level of destruction is limited only by our imaginations.

CHAPTER EIGHT

∿ HOPE ∾

Making Up for Lost Time

It's not an option but rather an obligation to speak
out, all of us together, to demand changes.

BETTY VASQUEZ, Women in Resistance Movement, Honduras, 2018

Limiting global warming to 1.5°C will require rapid, far-reaching
and unprecedented changes in all aspects of society but bring
clear benefits to people, ecosystems and global goals.

IPCC, 'Global Warming of 1.5°C', Special Report, 2018

The best time to plant a tree was twenty years ago.
The second best time is now. Someone's sitting in the
shade today because someone planted a tree.

Anonymous

It's too late for gradual change. Whatever happens now, humanity will be in a period of global, environmental, economic, political and social upheaval for decades – if not centuries – to come. But while it's late in the day, every bit of warming *matters*, as every increment in temperature has more impact than the last. How soon we act *matters*, since every year of inaction tempts further climate transitions and only makes the future burden worse. How quickly mass movements coalesce to force change *matters*, because they presage the social tipping points and the emotional engagement so desperately needed. Every lifestyle choice *matters*,[1] as individual actions influence others and combine exponentially through networks, driving further systemic change. However bad it gets, there is no point at which the effort becomes so hard that it isn't worth making. As of 2020, humans have already warmed the planet by around 1 to 1.2 degrees Celsius compared to pre-industrial times and we are on track for further increases in the next few decades since cumulative emissions will increase for some time even if unprecedented action is taken.[2] We can't avoid the consequential suffering, but there is probably still a small window of time before several critical transitions are reached – this is what is meant when journalists, reporting on the 2019 special IPCC report on 1.5 degrees, said we had twelve years to make substantial changes (i.e. we need to start a massive transition now. Not, as some people thought, that we can still carry on with business as usual for twelve years).

Hope lives in the fact that some of the necessary transformations are already under way. The astonishing improvements in renewable energies have already been described, and the telltale signs are there for a mass dietary shift in high-income nations. But perhaps the most hopeful trend right now is the rise of environmental mass movements around the world. Thousands of small, grassroots organizations are gathering momentum, and larger ones such as Extinction Rebellion, the Divestment Movement, Fridays for Future and the Sunrise Movement are now receiving national coverage in the media, raising the salience of the climate crisis among the broader public. As the public has *begun* to see climate change as an election issue, many governments, both local and national, have declared a climate emergency, though in practice it's uncertain what this emergency means for them.[3] As calls for change amplify, commentators representing vested interests complain that the protesters' wishes are unreasonable; that the solutions would cause such great economic upheaval as to threaten the very foundations of society. But what these mass movements have internalized is the message scientists have been delivering for decades: climate physics doesn't care about human upheaval. Neither does Earth, which in the long run will be absolutely fine. It's now well understood that it's human civilizations that need saving, and this will be a 'long emergency'.[4]

Let's hypothesize that the pillars of civilization across the planet – governments, legal systems, communities and businesses – *align* to reduce emissions fast enough to avoid tipping points. It won't be enough.

Remember, not only must we staunch the flow of greenhouse gases, we must actively reduce the concentration of them in the atmosphere through *carbon sequestration*.[5] We can do this a number of ways. Many of the most promising approaches are 'natural', in the sense that they can be achieved through better management of land. These techniques are generally cheap and should start today on a huge scale, because they will take time to reach their maximum potential. These natural solutions can also address other environmental crises like biodiversity loss and, in many cases, can be more productive and equitable in food systems. Yet even these natural approaches are not enough. We'll most likely need machines to draw down carbon dioxide from the atmosphere.

Since the scale of climate breakdown is so terrifying, some researchers suggest that we might be able to manage the amount of energy the Earth's surface receives directly. For example, Solar Radiation Management (SRM) would involve spraying aerosols into the high atmosphere to reflect sunlight away and dim the light reaching the surface, and curtailing the incoming energy would take the edge off extreme weather. SRM is often thought of as a 'smash glass in case of emergency' technological resort, but it is also fundamentally flawed because, as we have seen, critical transitions can't easily be pushed back to their starting point – it often takes more effort to reverse changes once a system is in a very different state. Although many feel that more research is worthwhile, there is significant concern that this dramatic geoengineering would act as a moral hazard, stemming the incentive for mitigating emissions in the first place. It's also politically fraught, ethically dubious, and would condemn huge amounts of ocean life. A small number of people might consider SRM an optimistic tech backstop. But since discussion of this is in fact unremittingly negative, we'll wait for the pessimistic Epilogue to describe it further.

ACTING LIKE WE MEAN IT

Had we acted decades ago, reducing emissions by a few per cent per year we *might* have bought enough time to shift societies to a more sustainable way of living in the long term. Now, though, incrementalism would be a catastrophe. Many scientists agree that a warlike effort to retool the economy is needed to 'win' the war on climate change.[6] By the end of the Second World War, the Allies were spending around 50% of their national income on the military alone. They built hundreds of thousands of aircraft and tanks, and mobilized millions of people. The same could be done for the climate, by constructing millions of wind turbines, solar panels, geothermal heat pumps, insulating houses and much more. The warlike footing could be extended to the way we live and think: for example, rationing carbon-intensive activities like flying, and considering carbon emissions an evil comparable to fascism, not unlike the wording used by the African negotiator in the epigraph beginning the previous chapter.

The war analogy can be a useful way to open people's minds to the extent of action needed and to shift thinking to a necessary degree. During the war, no one questioned what needed to be done – it was an all-out battle against an existential threat. But since climate change is set to increase conflict in the twenty-first century,[7] perhaps war is not the best analogy, even if only referring to its outsized production capacity. Importantly, there is no winning condition in a war against climate change. Any success is contingent on feedbacks, total level of preventable warming in the short and long term, and ethical interpretations. It will be hard to say we 'won'. Similarly, a 'moon shot' approach (so named for the human achievement of landing on the moon) won't cut it either. The environmental problems humanity faces are not like the discrete problem-solving faced by Apollo engineers. Today we face complex systems problems, all-encompassing and amorphous in nature. While these analogies *can* be useful for an existential sense of scale, they don't describe the most concerning aspects of climate crisis. But if we grasp this scale, if we agree to stop tinkering around the edges and, instead, act like our lives depend on it, what do we do?

A non-exhaustive list includes: strict building standards to improve energy efficiency and renewable technologies; mandatory energy-efficiency building retrofits (with options for cheap loans to be paid back via the saved energy bills); a frequent flyer levy, penalizing each additional flight at an increasing rate; bans on road and airport expansion;

bans on the new development of fossil fuel resources; low-cost, effective (electric) public transport; nationwide bike infrastructure including bike networks and (e-)bike purchasing programmes; urban renovation including rezoning, pedestrianized areas and an overriding priority to reduce private car traffic (with the remaining electric only); increased national and international coordination of electricity grids, enabling more renewables and electricity storage; a tax on meat and dairy; increased research and development funding for energy and agricultural solutions, especially between the development of a product and its commercialization (the so-called 'valley of death' for many businesses); removing subsidies for fossil fuels and damaging agricultural practices; a price on carbon for domestically produced goods and the carbon embodied in imports.

Come hell or high water, humanity must do all of this... as soon as possible. More nuanced policies are also needed to address food production, food waste and deforestation. None of these interventions are expensive if the long-term benefits are weighed against all the costs. The solutions also build towards a better world that eases other environmental and social problems, from biodiversity loss to air pollution, from mobility to public health. Fortunately, most of these solutions also enjoy broad public support. However, these policies *will* create winners and losers. Jobs, money and people will be displaced. Putting policies in place to protect people from these impacts will be *as* important as the solutions themselves. If not, the decarbonization agenda could be derailed by public opposition and sociopolitical upheaval. Perhaps most importantly, any carbon price *must* be implemented in a revenue-neutral way (for example, it could be used to reduce income taxes) and to do so *progressively* in order to maintain political support. A different option may be to use the revenue from a carbon price to provide universal basic services, which we'll revisit later in the Economics chapters.

The government coordination and action needed to realize these crucial socioeconomic and infrastructure changes is dramatic... but not impossible. There is precedent for such a coordinated, large-scale action. After the world's greatest financial crash, the US government stepped in with the New Deal to provide jobs, investment and welfare programmes, and built infrastructure, from the national highway system to water-management programmes.[8] This massive effort wasn't limited to a short-term recovery; its legacy was the welfare state, national infrastructure, social reformation and a new sense of nationhood. The Green New Deal recasts the Depression-era New Deal in terms of environmental goals and includes many of the solutions mentioned above. It retains a focus on fairness and the addressing of inequalities, as did the New Deal.

Yet the Green New Deal might not go far enough. It's not sufficient to focus efforts inside borders. For global progress on this global crisis, richer nations will have to provide technical help and funding across the world. Economists call this a Green Marshall Plan,[9] referring to the investments the US made in Europe after the Second World War. Today, this would involve cooperation between nations and the exchange of technology, expertise and investment.

LEGAL ACTIONS AND SOCIAL SHIFTS

For large-scale change to be effective and rapid, the main social institutions have to move together: the legal system, the civil service, national and international banking institutions and local and national governments must all pull in the same direction. As recent years of insufficient policies attest, this won't happen by top-down diktat alone. Indeed, in democracies, it would be concerning if it did. The mandate for such large-scale change will most likely be won when community and civic groups take action; when we reach a social tipping point whereby governments can no longer delay or ignore problems or public will.

To this point, several recent developments are promising. For example, as of 2019, there have been more than 1,300 climate-related lawsuits lodged across twenty-eight countries.[10] Many actions have been taken directly against governments for not moving quickly enough to cut emissions. In 2015, the first successful case was brought against the Dutch government. Judges ruled that national emission reductions should be at least 25% by 2020, rather than the 17% target declared at the time.[11] While targets can be conveniently forgotten, fudged or ignored, legal requirements are more likely to spur substantive action. When it became clear the target would be overshot, the Dutch government was forced to close a coal-fired power plant.[12] Similar cases have been brought against governments in Belgium, Colombia, Ireland, Germany, France, New Zealand, Norway, the UK, Switzerland and the US.[13]

The world's largest fossil fuel companies are also increasingly exposed to legal risk. At least half of all anthropogenic carbon dioxide ever emitted has been *since* these companies have definitively known the science of climate change.[14] As we have heard, when they came to the conclusion that climate change was a threat in the 1980s they began retrofitting their operations to cope with sea-level rise and storm surges. With a flood of documents outlining the sinister deception, there is an increasing likelihood of legal remedy. It's a story eerily similar to the Big Tobacco litigation of the 1990s. Doctors first started linking smoking with lung cancer and heart disease by 1950,[15] and as the years rolled on the link became scientifically unassailable. Yet the big tobacco companies ran a lobbying and misinformation campaign to stymie action. Eventually, the four main US tobacco companies had to pay a record-breaking $206 billion in compensation, and were forced to curtail their advertising and lobbying activities – but not until 1998.[16] During the delay, millions of people who didn't understand the risks died from smoking-related diseases.[17] In fact, some of the same corporate tobacco lobbyists have been involved in delaying action on climate change, as documented in the book *Merchants of Doubt* and the film of the same name.[18] Eventually though, Big Tobacco had to pay up. We may be seeing something much larger play out with fossil fuels, and Big Oil may well be terrified.[19]

Historically, it has been tricky determining to what extent climate change contributes to any single weather event, or how much one company's emissions are responsible for fatalities and damage. This is because climate breakdown shifts the statistics governing the frequency and severity of extreme events in the way that smoking shifts the chance of cancer. The defence could always be that it's bad luck, like a smoker's lung cancer. But attributing damage to climate change is rapidly improving with attribution science – piercing the dog-eared defence that weather is an act of God – and instead connecting these events with acts of corporate deception.

It is surprising how many different ways fossil companies can be sued. One example is that of the Californian crabbers. As oceans continue to warm, toxic algal blooms are on the rise.[20] Some types of algae produce a neurotoxin called domoic acid, which makes its way through smaller sea life like mussels, clams, scallops and small fish, and bioaccumulates (concentrates) further up the food chain to organisms like crabs. Human consumption of such poisoned shellfish can cause seizures, vomiting, and even death. In the 2017, 2018 and 2019 seasons, Californian crabbers essentially lost their jobs as Dungeness crabs became unsafe to eat. Having been made redundant, the crabbers brought a lawsuit against thirty of the top US fossil fuel companies. In this case, as in many others, the outcome is uncertain, but the very fact of these cases being tried can highlight that culprits are involved in such ocean degradation, and can open up further opportunities for litigation. As cases advance to the hearing stage, more documents are being released in the discovery phase, documents that are useful for further litigation. There's a twist in the crabbers' case: the crabbers are, in turn, being sued by environmental groups because, as they fished further out to sea in an attempt to catch healthy crabs, they entangled whales in nets.[21] This may become a familiar story in the future, with cascading climate impacts reflected in cascading legal proceedings. In some cases, the web of legal actions may even start to resemble the complexity of ecosystems.

There are also legal proceedings based on investor and consumer fraud. The argument goes that investors in Big Oil were entitled to know about the exposure of their money to climate change, and consumers were entitled to make an informed decision about purchases. ExxonMobil has been sued in several US jurisdictions on both counts.[22] One case being heard in New York is based on allegations that Exxon used two different sets of accounts to assess its risk exposure to government action on climate change – one it kept private, and another (with lower costs) it gave to investors.[23]

Since climate change is an intergenerational issue, perhaps it's no surprise that litigation has also been brought by young adults.[24] It may also explain why some of the more high-profile figures of the climate movement are at each end of the generational spectrum, from Greta Thunberg to David Attenborough. The Fridays for Future movement, where schoolchildren skip school on Fridays to protest, has been particularly effective. 'Why should I be studying for a future that soon will be no more,' Thunberg argues, 'when no one is doing anything whatsoever to save that future? And what is the point of learning facts within the school system when the most important facts given by the finest science of that same school system clearly mean nothing to our politicians and our society?'[25] Not only can children effectively teach their parents about climate change,[26] but seeing children in such anguish about their future strikes a chord among many of the older generations who, for a long time, felt deep down that they weren't doing enough (or anything at all) to safeguard the future.

Although media coverage is woefully shy of the level of threat, Fridays for Future and other movements have increased the salience of climate change, and policymakers are *beginning* to respond. In May 2019, as a direct consequence of marches and protests of the Extinction Rebellion movement, the UK Parliament called a climate emergency (although it still may not be acting as if there is one).[27]

A second visible impact of increasing public action and media attention is the large gains in Green parties at the ballot box. Not only are they now the kingmakers in the EU Parliament, they also won the second largest number of votes in the German EU elections.[28] This matters, since Green parties advocate for most of the policies listed in the hopeful chapters of this book. Beyond influencing individual decisions and votes, this increased salience is influencing how culture is engaging with the climate crisis. We've seen how plant-based diets are exploding in popularity. In another example, Swedish flyers have decreased their trips by around 8% in 2019, and polling suggests this is almost certainly a cultural response to the shame of flying, termed *flygskam*.[29] Markets were worried about the impact on the airline sector even before Covid-19, which can only be a good thing.[30] As a response, trains in Europe are seeing a renaissance, with long-distance and night-train services starting again after years of decline.[31]

Some suggest that the system needs to change, that individual actions like stopping flying are pointless, as one hundred corporates emit 70% of all emissions.[32] Others demand ideological purity on climate consciousness, arguing that unless someone lives a carbon-neutral life, they can't have a say about what needs to change in the system. This dichotomy between the importance of individual action versus system change is deeply flawed. Those who say systems have to change first forget that systems can only change when enough people agitate for change, not to mention the fact that individuals are constrained by the sociotechnological arrangements in which they live. To take the example of those criticizing the crabbers mentioned above for suing fossil fuel companies ('they use diesel engines in their boats!'), this is clearly absurd if there are no electric alternatives available. The reason for the dearth of electric alternatives is the sociotechnological paradigm driven by fossil fuels and the lock-in of innovation policies based around such a system. Individuals have to live within a system, and the system responds to individual change. Both actions are vital. These arguments are perhaps best summed up by a popular online cartoon of a peasant and an online troll: the peasant, ill-looking and sad, lifts a bundle of sticks and suggests, 'We should improve society somewhat,' to which the online troll responds, 'Yet you participate in society. Curious!'[33] These arguments only serve as an excuse for more delay on individual and system change.

Research shows over and over that individual actions provoke change among others, as long as you talk about them.[34] As individual changes intensify, this can turn into more social pressure, resulting in political pressure and system change.[35] Equally, system changes open up new ways for individuals to express themselves and engage, either through markets (e.g. increased numbers of vegetarian options making it easier to move to a plant-based diet) or through direct civic action via climate movements.

RISKY BUSINESS

As governments and society respond, increasing pressures from investors and employees are changing global business attitudes, with the most exposed sector, insurance, showing perhaps the deepest concern. The largest of the world's reinsurers, providing insurance policies to thousands of other insurers, Munich Re has stated that a two degree warmer world would ultimately be uninsurable.[36] Reinsurers are already starting to walk away from policies covering smaller, local insurers that operate in areas with high wildfire or flood risk. For instance, average Californian wildfire losses used to be well below $5 billion each year but have now leaped to over $20 billion a year in recent fire seasons.[37]

Insurers are also starting to refuse to insure fossil fuel developments. Chubb, Allianz, Hannover Re and Lloyds have all divested themselves of coal power. Munich Re has been particularly strident, and in 2019 called for for carbon taxes of at least €115 per tonne, much more than even the highest carbon prices around the world, such as those in the European Emission Trading Scheme, which hovered around €30 per tonne during 2021.[38] Some development banks – large banks which offer non-commercial loans for infrastructure around the world (often in lower-income nations) – are changing fast. The European Bank for Reconstruction and Development announced a 'no coal, no caveats' policy in 2018, but went further and quit all fossil fuels in 2019.[39] China has now pulled out of new coal plant financing around the world.[40]

Broader changes to business are also under way. It may sound arcane, but because central banks set the rules for how commercial banks can account for assets, this has an outsized systemic influence on where money is invested in the economy. New analytic tools show that the financial system is heavily exposed to climatic stresses.[41] Banks like HSBC or Wells Fargo are legally allowed to lend a certain multiple of the assets they have on hand, in the form of loans. The central banks dictate how these assets can be valued and so have a big say about what sorts of assets are more secure. A group of forty-two central banks, including in the Netherlands, UK and France, are looking to reform these rules so that banks must assess the potential climate risk of their assets.[42] The higher the climate risk, the less banks can lend, and the lower the profits from lending. This obscure but critical development is hard to understate. Through this one effort, lines of credit would increasingly shift away from carbon-emitting projects with higher climate risk towards cleaner, more sustainable investments. Since markets often respond to anticipated policies,[43] these developments are probably already altering the pattern of investments. If banks and companies don't get ahead of these changes, they will be left with stranded assets and enormous losses. Central banks are also forcing banks to stress-test their balance sheets against abrupt climatic and technological changes in a scenario where 42% of fossil fuel value disappears in just three years.[44] The combination of these policy efforts, and other potential policies like carbon taxes, means that more companies are preparing ahead of time even though these policies haven't been implemented yet. The ultimate result of these mounting pressures alongside the plummeting costs of cleaner technologies could be a run on carbon in the economy. Like a run on the bank when people realize their money isn't safe, banks and investors may rush to extract themselves from investments exposed to fossil fuels – although central banks will probably try to avoid this upheaval.

These developments may partly explain why global investment decisions to 'green-light' coal plants have fallen by 75% in the last three years. This is a terminal decline in global investment in coal, and the level of disinvestment is seven years ahead of schedule, even under the optimistic projections the International Energy Agency made in 2018.[45] It may be hard to believe, but it looks like Big Oil is going the same way. In 2019, for the first time ever, ExxonMobil was the last commercial oil major to drop out of the top ten largest companies on the S&P 500's stock market index and in the same year was given a negative outlook by the ratings company Moody's.[46] Concurrently, the massive investment adviser firm Redburn downgraded Big Oil from a 'buy' recommendation surrounding concerns about peak demand.[47]

Peak oil demand may come sooner than many think. Major banks are realizing that renewable energy is cheaper, reliable, safer and more profitable than any of the carbon-based technologies, even without the shift in asset valuation mentioned above. In the Energy chapters, we heard how the energy return on investment (EROI) is dropping for oil. A similar calculation can be made for money – the energy return on capital invested. A recent report to investors from the world's eighth largest bank, BNP Paribas, found that the cost of electric vehicles plus renewables over the lifetime of the car is less than 17% of the cost of a petrol car and the fossil infrastructure needed for the petrol.[48] That is, as an investor, oil is six times more expensive than electric vehicles plus solar. They calculate that by shifting to electric vehicles and renewables over the next fifteen years, the world saves $20 trillion. These sorts of unanticipated findings bring into question some of the higher estimates for the cost of the energy transition we have seen so far.

You may be thinking that the oil industry is too big to fall so quickly, but bear in mind that the reserves-to-production ratio of oil is around ten years. This means that if a company doesn't invest billions in searching for oil and bringing it into production each year, it would run out of oil in around ten years. Not only do these costs continue to rise as oil companies go to greater lengths to secure reserves, but if demand plummets, prices will too. Ultimately, it's hard to say when a death spiral for oil might pick up speed, but for all these reasons many analysts predict that it will be much sooner than the oil majors think.[49]

NATURAL CLIMATE SOLUTIONS

Natural solutions for drawing carbon out of the atmosphere are 'least-regret' solutions. The majority of natural solutions improve other environmental problems along the way, such as reforestation (planting in deforested areas) and afforestation (planting trees in new areas that haven't seen forests before), which would help to halt species extinction. These solutions can benefit groundwater storage, soil quality, air quality, flood protection and human happiness.[50] But before we describe these developments, we desperately need to talk about deforestation – especially in the few remaining tropical rainforests.

The level of tropical deforestation is grim. Around the world, deforestation of the last tropical rainforests is either barely being kept in check or is continuing apace. In 2019, the Brazilian president, Jair Bolsonaro, made it quite clear that he welcomed further 'development' in the Amazon. As a result, deforestation leapt, the rainforest losing more than 729 square kilometres in just one month, the area of two football pitches per minute.[51] The only hopeful outcome of this is that at least we know about it using high-resolution satellite imagery, whereas before it might have gone unnoticed – thin gruel indeed. Brazilian and international researchers were able to use this imagery to classify accurately how much had been lost, and to quantify the amount burned when massive wildfires broke out. This information ultimately provoked an underwhelming response from governments around the world, but it does open up the possibility of international pressure as climate change becomes more important as an issue – there will be nowhere to hide. In terms of fighting deforestation, the best guardians of forests are indigenous peoples.[52] If natural climate solutions were to be pursued as a political priority, both nationally and internationally, indigenous peoples would be supported and their rights protected.

As well as protecting rainforests, reforestation and afforestation are the best carbon-sequestering approaches we have. Of the total capacity for drawing down carbon naturally, forestry comprises up to half.[53] Here, there has been some progress: between 1990 and 2015, EU countries reforested an area the size of Portugal.[54] In China, the Three-North Shelter Forest Program is a gargantuan effort to reforest the country and arrest the expansion of the Gobi Desert from the north. After forty years of tree planting, forest coverage has expanded from around 8% to 16% of China's land area. By some estimates, it drew down around six months' worth of global carbon emissions in the decade between 2001 and 2010,[55] prevented soil erosion and resisted the march of the Gobi. Alas, the figures are not all rosy: the amount of carbon being captured each year has been decreasing over time.[56] The Chinese programme also offers important lessons about what not to do in future forestation efforts: much of the forest area focuses on just a few tree species like eucalyptus, limiting benefits of biodiversity and exposing forests to risks from pests and diseases. This monocultural story is similar in forestry projects around the world. Rewilding is a preferable response to managed monocultures, as it encourages larger, more resilient ecosystems.

Although forests comprise the biggest overall opportunity in natural climate solutions, there are huge benefits to improving agricultural, grassland and wetland systems too. Among these approaches, the focus is on the carbon stored in soil. Techniques that retain soil carbon are vital, including no-till agriculture, the management of fertilizers, tree planting within croplands and, potentially, biochar (where carbon-dense charcoal is mixed in with soil, allowing the soil to increase carbon sequestration while making it more fertile). Biochar has good potential, but its total opportunity is as yet hard to estimate.[57] Perhaps it goes without saying that these techniques can all benefit biodiversity too.

For these approaches to be truly ingrained, though, we will need to start valuing many things that are currently undervalued. One good aim would be to develop 'carbon farming' as a natural extension of traditional farming, whereby farmers could receive subsidies for carbon sequestration instead of overproduction. Or they might benefit from a carbon price. Such solutions would need to be implemented *as* we are implementing other climate-mitigation strategies.

While we know that natural climate solutions are essential, even if we make the changes on a massive scale, they are *still* not enough. Enacting every natural carbon solution currently available to us at their maximum capacity would get us around halfway to net zero by 2030, if emissions remain at around 2019 levels (23 billion tonnes of CO_2 equivalent captured compared to total emissions of around 40 billion tonnes). In practice, we'd be lucky to see a fraction of this, especially in a warming world with increasing water constraints (limiting some types of forestation) and potentially more wildfires and forests turning from sinks into sources of greenhouse gases.[58] While we should be doing all of this today, and quickly, there are limits to what natural climate solutions, and forestry in particular, can do.

DEUS EX MACHINA

A *deus ex machina* of climate change would be to use machines to manage the concentration of carbon dioxide in the atmosphere directly.[59] But like the *deus ex machina* coming from nowhere to clumsily wrap up flagging theatre or film storylines, we should be suspicious. These technologies are untested on a large scale, they may provide excuses to mitigate emissions, and they will take a long time to roll out (during which all manner of catastrophes may unfold). With these caveats out in the open, we have to admit that some of these approaches are probably necessary, given today's level of inaction.

The general idea is to filter excess carbon dioxide from the atmosphere, store it in a liquid or solid form, and bury it deep underground. This would shift the stock of carbon dioxide from the atmosphere to the geosphere (as opposed to soils and organic matter with natural climate solutions). In effect, these negative-emission technologies attempt to run the global fossil fuel system backwards. This would be a vast effort, both because this system is so very large and because running any system backwards is an affront to the second law of thermodynamics. Compare the effort involved in trying to unscramble an egg versus whisking it in the first place. Under some scenarios, we'd have to build several times the current infrastructure of the global fossil fuel industry to scrub out the necessary quantity of carbon.[60] This is accident clean-up on a global scale.

There are two main methods proposed for technological carbon sequestration. The first involves plants, which absorb CO_2 as they grow and can then be burned in power stations; the CO_2 in the power station exhaust is then captured and stored underground. This is cumbersomely called Bioenergy Carbon Capture and Sequestration, or BECCS. The second, more modular approach is Direct Air Capture (DAC), which uses large fans to force air through a carbon separator and then directs the captured carbon underground. On the face of it, BECCS might seem more sensible, as the carbon dioxide billowing through smokestacks would be at much higher concentrations than the ambient air which DAC must filter. BECCS would also *produce* energy rather than consuming it – you could make money doing it. This is why BECCS appears in the vast majority of high-profile climate economy models as reported by the IPCC. When you look more closely, however, the BECCS concept has massive and insurmountable problems.

We may need to commit to removing around 20 billion tonnes of CO_2 per year from the atmosphere by 2030 (around half the current emissions).[61] The reason we don't know quite how much is needed is because we are uncertain as to what will happen in society in the meantime. For example, if we have a global recession, less will be needed. The only time carbon emissions fell in recent years was during financial crashes and pandemics (with a decrease of 1.4 to 6.3%,[62] but emissions rebounded quickly to the general trend afterwards).[63] To do this removal using BECCS, we'd need to grow plants over a cultivated area twice the size of Europe![64] This is completely unthinkable with the other challenges we face to conserve land and biodiversity, and it would preclude a significant number of the natural climate solutions already discussed. Plants simply require way too much space,[65] and Earth already has a space problem. If this wasn't enough, BECCS needs huge volumes of water to grow the plants and to operate the power plants – water that's becoming increasingly scarce.[66] BECCS also suffers from the same air pollution and ash disposal problems of other combustion technologies. Why these technologies appear in the models reported by the IPCC will be covered in the next chapter, but, put plainly, BECCS on any significant scale is a non-starter.

That doesn't mean we should give up on negative emission technologies altogether. Although Direct Air Capture consumes energy rather than produces it, cheap renewable energy paired with modular DAC units could do the job without sacrificing biodiversity or food security in their operation. As with solar cells and wind turbines, the modularity of DAC equipment is a huge advantage. Estimated technical costs for the latest carbon sequestration technology using commercial DAC technologies have fallen from around $600 per tonne of CO_2 to between $94 and $232 per tonne.[67] Research suggests that costs for DAC could drop further, to just over $50 per tonne by 2040.[68] The land use, too, is far lower with DAC rather than BECCS – instead of twice the size of Europe, it would take up to two Irelands' worth of land.[69] To put this in perspective, Saudi Arabia covers an area twenty-five times the size of Ireland, with vast storage capacity in the reservoirs from which so much oil has already been pumped (and, of course, plentiful solar energy).[70] As the fortunes in the Middle East decline with the sinking demand for oil, the region could become a leader in carbon sequestration – if there is any way to make it pay politically and monetarily, that is.

However, it's hard to see how this nascent technology might scale to the size of two Irelands if there is no market for it. But while carbon prices aren't yet high enough to provide a market for DAC directly, the technology is already being used to produce synthetic fuels in air-to-fuel processes.[71] This is another process reminiscent of rewinding the tape: air-to-fuel captures carbon from the air and then combines it with hydrogen (preferably created from water hydrolysis using renewable technology), making the CH bonds the carbon-fuelled economy craves. There are markets around the world which penalize fossil oil and gas so heavily that synthetic oil would be cheaper.[72]

This is all encouraging, since the large majority of pathways reported by the IPCC for limiting warming to 1.5 or two degrees Celsius assume these technologies are necessary. But we should keep in mind that with negative-emission technologies, you can essentially create any carbon budget you like. You can assume you'll pull all the extra carbon out of the atmosphere in the second half of this century (and hope that you don't transgress tipping points along the way). This represents a moral hazard in thinking there is a technological saviour. Ultimately, because progress on mitigation has been far too slow, we will certainly need some sort of negative-emission technology – even more so when we consider that some areas of the economy are very difficult to decarbonize. This is why governments talk in terms of *net-zero* emissions targets and not *zero* emissions. While we should use every natural solution available, it would be prudent to have DAC technologies ready to go, as long it's in addition to, rather than instead of, mitigation in the first place.

A SCARY RACE BETWEEN NATURAL AND SOCIAL TIPPING POINTS

In 1914, Winston Churchill described the lack of preparation in the run-up to the First World War as if leaders were 'all drift[ing] on in a kind of dull cataleptic trance'.[73] But what happens if hundreds of millions of people snap out of their cataleptic climate trances? Firstly, even a minority view can become universal very quickly. The threshold for this change depends on the issue, but studies suggest a threshold of between 10% and 40%.[74] The threshold for a tipping point is potentially much lower if we look at mass civic action. Researchers have found a rule of thumb that any cause has never failed which sees more than 3.5% of the population out on the streets in sustained action and that many have succeeded with much less. Promisingly, researchers found that non-violent action was more likely to succeed than violent action (although both findings are not without controversy).[75] If we were to imagine this 3.5% threshold globally, it would see around 270 million people out on the streets. For comparison, an estimated six million people joined the September 2019 climate strikes – a long way to go for sure, but this was the very first round of recent major action.[76]

It certainly feels like we are on the cusp of a tipping point for climate mobilization, and many researchers (including myself) are more hopeful than they've been for years. Some are worried about burnout, that such a movement will struggle to continue applying pressure. But this overlooks the fact that the climate won't let society relax. The longer serious climate action is delayed, the more environmental disruption we can expect, and the more civic disobedience will result. There is the possibility that natural tipping points have already been effected, but even if they have, the speed at which policies can be implemented will slow down the changes and give more time to adapt. There is still a very real race to be run between climatic and social tipping points.

The critical period of the next few years will decide if there are to be additional degrees of warming at the end of this century. It will decide whether the millions suffering today will become billions. It will decide whether malaria-carrying mosquitoes will recolonize areas of the world from which they were eradicated. It will decide how many people will fall ill from other diseases. It will decide how deep crop losses and food shortages will be. It is all in the balance.

From natural climate solutions to liveable cities, the solutions we so desperately need to implement would result in a better world for many people. But it's beyond doubt that many solutions simply aren't possible within the current economic structure. Natural climate solutions and negative-emission technologies have no clear economic pathways by which to scale to the size needed. If humanity is not able to reform economic arrangements, then even our most valiant climate response will be a game of whack-a-mole,[77] where we will simply end up facing different environmental crises in the future.

ECONOMICS

CHAPTER NINE

↜ PESSIMISM ↝

Counting the Costs

The machine turns, turns and must keep on
turning – for ever. It is death if it stands still.

ALDOUS HUXLEY, *Brave New World*, 1932

Don't be seduced into thinking that which does
not make a profit is without value.

ARTHUR MILLER, *All My Sons*, 1947

We could have saved it, but we were too doggone cheap.

KURT VONNEGUT, *Hocus Pocus*, 1990

Add just one extra ounce of stress to someone already under professional, financial or personal pressure, and it can trigger a breakdown. Society, like individuals, has stress limits, which can be expressed as protest voting, strikes, boycotts, marching, civil unrest and violence, and even war. In 2009, researchers set out to calculate the environment's limits to stress – limits beyond which Earth's systems could flip into a new state.[1] The 2015 update included nine classifications in what they called the 'planetary boundaries' framework: biosphere integrity (i.e. biodiversity), climate change, land system change, freshwater use, biogeochemical flows (like phosphorus and nitrogen), novel entities (potential unknown problems like plastic nanoparticles), acidification, atmospheric aerosol loading, and ozone depletion.[2] Already, we have exceeded *four* safe boundaries of this distressing list: climate change, biogeochemical flows, biosphere integrity and land system change.[3]

In other words, climate change is not humanity's only problem. If climate change is addressed and overcome, other problems are waiting in the wings, worsening. Humanity will be confronted with continuing, escalating threats, each of which have the potential to destabilize the current, habitable Earth system. Instead of trying to put out each fire as they start, it's usually better to address failures in a *systematic* way, using *systems thinking*. At the heart of all these problems is the culprit system that connects the complex systems of nature and society: the economy.

All the things we consume or accumulate in life – from food to shelter, washing machines to experiences – require resources. Resources either come from nature, in the form of materials, food or energy, or from other people, in the time needed to build products or provide services. As we all understand, such resources are limited or, to use the economists' term, *scarce*. Since materials and people's time are finite, we have to make decisions about how to divide and deploy these as resources. Measuring and formalizing the connection of our needs and wants to these finite resources is at the heart of the economy; its structure and the politics surrounding it set limits on who gets what, when they get it, and how much of it they get.

This makes the economy a partial reflection of what we value: both as individuals, in how we live, and as a society, in how we treat the needs of others and the community. We share the feeling that the economy should facilitate the human species to thrive, not just survive. This explains why we reach for health metaphors when describing a nation's economy: economies doing badly can be 'sick', suffering from 'chronic shortfalls' or 'anaemic growth'. Economies doing well have 'healthy indicators' or are 'nimble' or 'in good shape'. Rightly or wrongly, the modern economy has become the lens through which much of the world views progress. This is reflected by its importance in national and international politics, along with the increasing influence of economists within society's decision-making apparatus. This overriding focus was perhaps best summarized during Bill Clinton's successful 1992 presidential run: 'The economy, stupid.'[4]

Economic textbooks and models often see the economy and society as being external to nature, but this is a grave, grave error. It's only just entering mainstream thought that 'the economy is a wholly owned subsidiary of the environment, not the reverse.'[5]

We've already seen how some economic incentives drive damage to food, energy and climate systems. In this chapter, we'll see that the very structure and connections comprising the economy itself are driving our existential problems.

MEASURING VALUE

The economy's health has become entangled with an unfortunate, dominating measure of 'progress': Gross Domestic Product (GDP). Although GDP didn't enter the public consciousness until 1950,[6] it has become the key indicator to which journalists, businesses and governments hold fast. GDP's ascribed importance is clear when we compare the reaction to the 2008 financial crash with the stagnation and then decline in US and UK life expectancy from 2011 onwards.[7] The former occupied headlines for years, while the latter barely garnered mention in mainstream media. These retrogressive trends are unprecedented in the modern era, and have been linked to deaths of despair: by drugs, alcohol and suicide.[8] Here is a siren, signalling that something is terribly wrong in society. But while life expectancy declines, GDP slowly climbs... and GDP an overriding economic focus of many politicians, economists and journalists.

Take the UK recession of 1976, when preliminary GDP figures showed two consecutive quarters of falling GDP – formally defined as a recession. In response, the Labour government of the time was forced to take out large international loans and to slash public spending.[9] Labour suffered defeat by the Thatcher government three years later. When the final, corrected GDP figures came through, it turned out there had never been a recession in the first place. No matter one's politics, it's worrisome that life-changing government policies (in this case austerity) could be to some extent dependent on one number, especially when that number can be miscalculated. We'll see that the formula behind GDP is so flawed that it has always been a miscalculation – one as dangerous as an overdose.

GDP is calculated by summing all the money spent on final goods and services in an economy: all the televisions and gel pedicures, all foodstuffs and doorknobs, each grand piano and toilet brush; absolutely everything that can be bought on the formal market by consumers.[10] The flaws in GDP are apparent in its definition: *everything bought on the formal market*. If we confuse GDP with value, as many do, and we optimize policies to maximize GDP, as many economic models do, then we are blind to much of what makes life worth living. We don't buy friendship on a market, or a walk in the woods with that friend, or the existence of the forest through which we walk, or thriving communities that care for the forest. We can't purchase a safe and sustainable future on the open market – it has no price. As Robert Kennedy famously remarked, GDP is 'a measure of everything except that which is worthwhile'.[11]

GDP also omits many of the things we think of as being good to do: caring for the elderly, raising children, housework, volunteering and other unpaid work. An effort by the OECD to calculate the benefit of unpaid work found that, were it to be included, national GDP would be between 15% and 70% higher.[12] A quick way to make GDP grow in the short term would be to force all childcare into the market. In fact, you could send children into the job market too. Governments have been known to do truly dire things in a desperate attempt to keep their GDP growth 'healthy' and appear creditworthy – even if that harms long-term outcomes like education.

Unfortunately, GDP doesn't value having a planet to live on. Deforestation in the Amazon *adds* to GDP when the wood is purchased in a furniture shop. GDP grows again when someone buys beef from cows reared on soya beans produced where the forest once stood. Nowhere in the calculation is there a subtraction from GDP for nature and bio-diversity loss, decimated indigenous communities or carbon emissions. Imagine a tree felled in your city to supply wood for a sun shelter: one might argue that the tree itself was a nice shelter from the sun (which also provided biodiversity, flood control, soil benefits and much more), but the man-made structure adds to GDP, whereas the tree didn't. At its core, GDP incentivizes economic growth through displacing nature with more society.

Since it's just one number, GDP can't include the distribution of wealth or costs of inequality. As the joke goes, Bill Gates walks into a bar and everyone is, on average, a billionaire. For instance, while the UK's GDP grew three times between 1980 and 2016, cases of childhood rickets (caused by malnutrition and food insecurity) rose around seven times and the number of food banks, once unheard of in the UK, soared to more than 2,000.[13] Economies can grow, in the short term at least, as a result of impoverishment: draconian working conditions and shrinking employee benefits. GDP can grow again if this social stress is triaged with privatized mental and physical health services. This helps explain the fact that while US GDP per capita is around 50% higher than in much of Europe, it's hard to find any social measure by which Europe doesn't outperform the US – incarceration rates, maternal deaths and social mobility to mention just three.[14] GDP can go onwards and upwards as individuals, the environment and communities suffer.

How has such a poor indicator become so prominent a proxy for welfare? Even Simon Kuznets, the economist who designed the measure in 1934, remarked that 'the welfare of a nation can scarcely be inferred from a measurement of national income.'[15] There have been serious complaints about GDP for decades,[16] but the reason it has prevailed is largely because of a social lock-in and because, for a time, it was incredibly useful. Before GDP, there were very few ways to even begin quantitatively understanding how economies operated. The measure proved its worth during the Second World War, when it was used to optimize the production of tanks, planes and weapons in the US war economy. As its influence spread to other countries and was used for comparison between economies, it became the de facto measurement of economic activity, despite its clear flaws. Today, there are thousands of statisticians and policymakers worldwide working on gathering data and calculating GDP. Environmental economist Rutger Hoekstra suggests that if GDP were thought of as a company, it would be a monopolistic multinational.[17]

The employees of this multinational might not speak the same language, but they could converse in the lingua franca of GDP. Similarly, we may not be versed in macroeconomic terms, but it's easy to say whether this number is going up or down. Its embeddedness is what makes GDP so difficult to reform, and doing so would require an international effort.

At this point, you may be doubting that policymakers really place any serious emphasis on such a general measure. Unfortunately, while many appreciate that GDP is a terrible indicator, it's still used to develop and conduct national and international policy. The overarching economic goals of the United Nations' 2030 Sustainable Development Goals are given in GDP.[18] GDP is core to economists' models, from cost–benefit analyses to macroeconomic models. Troublingly, GDP represents another case whereby the stories humans tell themselves become self-fulfilling. Even if countries see a failing society and degrading environment, as long as they have 'robust' GDP growth they will see improved credit scores from ratings agencies,[19] and can attract further investment in the form of low-interest loans or foreign direct investment, incentivizing more GDP growth and damage to the society and environment.

In 1991, economist William Nordhaus provided an interesting example of how GDP messes with our ideas of value. He remarked that climate change won't be too awful, since much of the damage will be in the agricultural sector, which only makes up 3% of US national GDP.[20] This rather shocking statement confuses GDP-measured market activity with *value*. Given that everyone needs to eat, you'd think it would be obvious that the 3% share of GDP spent producing food makes the remaining 97% of GDP possible in the first place – or is the message, as economic historian Dirk Philipsen puts it, 'let them eat GDP'!?[21] In short, some measurements can warp our understanding, both of the world and of social priorities.

RECALCULATING VALUE

These failures of GDP are inherited by the market as a whole. Say oil is priced at $60 per barrel. This figure comprises the costs of extracting, refining and transporting the oil, along with profits. What it doesn't include are any costs to the climate, to biodiversity loss, to lungs damaged by the car in which it's burned, not to mention the oil-driven geopolitical tensions it embodies. Economists call these costs *externalities*. Even the language is instructive, these costs are 'outside' what we count. The size of externalities can be larger than a product's price. For instance, a carbon price of $300 per tonne, recommended by some mainstream economists, would result in a cost more than twice the current market value of oil ($130 per barrel).[22] This equates to a total market price of nearly $200 per barrel – or, very roughly, £4 per litre at the petrol pump – for 'just' climate change, not including other externalities like air quality.

Externalized damages often amount to our shared environs and resources: the atmosphere, national security, water, the health of a community. These are all termed *common pool resources* and make up many of the most important things in life. Indeed, they make life possible in the first place. In the building of economic policies and models, many common pool resources are disregarded or marginalized. What happens if we try to include some of these resources into our accounting?

Hubristic as it may be, a framework exists to assess how much nature is 'worth' to humans in terms of the services it provides, called *ecosystem services*. To calculate the worth of, say, a mountain, researchers might assign a value to the mountain's aesthetic appeal (with a view to tourism and surrounding land value increasing), to its freshwater capture and storage, to its natural water filtration that renders water safe for humans to drink (cutting out the cost of water purification technology), and to its social value, whereby humans like to (and might be willing to pay to) hike the mountain. After assessing the different services the mountain provides, researchers could arrive at an estimated value for the mountain.

One study in 1996 did exactly this, but for Earth as a whole, including all ecosystems. It found that the total value was very roughly $60 trillion – around the same as the global GDP at the time.[23] It's a valiant calculation, but it undoubtedly befuddles our sense of reality, as its authors highlight. Say we did 'cash in' that $60 trillion, then we'd have no ecosystem left. No life, no clean water, nor soil, nor air! An ecosystem itself has no value on a market. Yes, it's possible to cash in parts of it, such as the topsoil of the Alaskan tar sands, or a stand of Indonesian forest, but the functioning of the system as a whole has no value in current accounting. 'It is by definition worthless.'[24] Put a different way, many economists assume we can *substitute* natural capital (water filtration, erosion protection) with human capital (machines, infrastructure), and that we can exchange the two. But we can't exchange the global ecosystem because, once exhausted, there is no amount of money or human ingenuity that could recover it. Whether natural values should be included in economic systems via this sort of valuation is fraught with problems,[25] and we'll return to this in Chapter Nine.

In 2017, researchers for medical journal *The Lancet* took a different approach to including the environment in economic calculations, by summing some of the costs of pollution emitted by the global economy. They found a total of $4.6 trillion of damage annually – 6.2% of the total global economy as measured by GDP in 2017 (accounting for nine million deaths).[26] The annual global GDP growth at the time was 2.2%, giving a net decline of 4% per year. These numbers are broadly similar across other studies using different methods. Another study found total damages amounting to 7% of GDP *per year*.[27] Both of these studies represent minimum estimates of the total damage because it's still very hard to include the costs of some damages like biodiversity or overall ecosystem functioning – the decline could be much larger. So, including only a subset of environmental damage, we are unequivocally in a global economic–ecological recession, a recession that will worsen with further climate change and biodiversity losses.[28]

This is all very bleak, but the economy's problems don't stop with the environment. (This is the penultimate pessimism chapter in the book; stay strong!) They extend into the social domain too. Hundreds of studies have shown that acute inequality is a causal factor in increased national levels of violence, reduced child welfare, drug addiction and a myriad of other personal and social ills.[29] It's also tied to higher rates of infant and adult mortality. Society is literally sick to death of inequality. Troublingly, our inner lives suffer too, from diminishing self-worth, status anxiety (whereby we feel sad, inadequate or anxious when we compare our lives to others), depression and other mental-health problems. It's difficult to calculate the cost of all this damage, but one extremely rough estimate found that the UK would save £39 billion per year in incarceration and health costs if inequality was reduced to 'just' the average of the higher-income OECD countries.[30] Again, though, it's difficult to represent all damages in such an analysis, and this is likely to be a minimum estimate of the true harm.

There is also evidence that inequality has an impact on the environment, the consumption of unhealthy and environmentally damaging goods and reduced compliance with international agreements. Controlling for other factors, more unequal countries have been found to have higher biodiversity loss, carbon emissions and air pollution.[31] A large-scale review found that inequality drives materialistic behaviour in an effort to 'keep up with the Joneses', often called *status consumption*.[32] If you own something that makes you feel better relative to other people, then other people will need to own more again to make themselves feel better. It's a self-defeating game where the goalposts – as defined by the most obvious indicators of success in money or consumption – are constantly shifting. It is a game in which all players, including the environment, lose. There is evidence that this zero-sum dynamic makes everyone unhappy – especially the rich, who come to envy the lifestyles and consumption of the 'mega-rich',[33] or come to fear that their wealth won't be enough should there be tax reform, a recession, or... let's face it, climate catastrophe. A panic room doesn't offer *quite* so much peace of mind as a fallout shelter. Surely nothing is more peaceful or forward-thinking to own than a bunker... and a private army. Sickening as it may be, status anxiety is a problem wherever you are on the income ladder.

Some researchers have attempted to recalculate GDP using both the environmental and social costs mentioned above – they call this measure the Genuine Progress Indicator (GPI).[34] A review of GPI research in the scientific literature found that, across seventeen high-income countries, economic well-being has ceased to increase with economic growth.[35] In fact, the data show that well-being as measured by GPI peaked in the latter 1970s, while global GDP grew ten times during the same period.[36]

WHAT ARE WE GROWING? AND WHY?

Despite these problems, most policymakers, businesses, and (implicitly) society as a whole are scrambling for as much economic growth as possible. Growth in TVs, airline flights, cars, insurance offerings, everything on the market. If other aspects of life can be drawn into the market, all the better! If it were possible to commodify clean air, it would be a new market. Growth! This is not as far-fetched as it sounds: due to the high levels of air pollution from China's breakneck industrialization, household air purification units became all the rage in the richer echelons of Chinese society, helping to grow GDP.[37]

Perhaps economic growth and the damage it can inflict is the price we pay for human happiness. While this is undoubtedly true for countries with low national income – for instance, Mali or Rwanda, where more growth would be welcome – beyond a certain income level, happiness appears to stagnate. This goes to the heart of what researchers call the Easterlin Paradox (named after economist Richard Easterlin), by which short-term GDP growth appears to correlate with a short-term growth in happiness (think of the relief of a new job after being unemployed during a recession), while long-term economic growth has no correlation to long-term happiness. One study used surveys from seventeen Latin American countries, seventeen developed countries, eleven countries transitioning from socialism to capitalism, and nine developing countries, and found that economic growth didn't result in greater happiness over the long term. Three countries (China, South Korea and Chile) grew so quickly that, as the authors of the paper suggest, 'you'd expect there to be dancing in the streets' – they doubled GDP in under ten, thirteen and eighteen years respectively – yet there was *no* statistically significant increase in happiness.[38]

A partial explanation may be the different modes of consumption society engages in at different levels of income. At low incomes, it makes a lot of sense to grow: consumption is necessary for shelter, food, water, electricity, education, bikes, library cards, children's crayons, etc. Beyond a certain level of income, research suggests that it is our *relative* place in society that becomes more important than our absolute income (even at the top), and that, beyond a certain level of income, GDP growth is not a telling factor for civic or environmental well-being.[39] Yet even if economic growth did make us happy, what good would growing the economy be if we (or our fellow humans, or our children) don't have a decent planet to live on?

IS 'GREEN GROWTH' NAIVE?

Reforming concepts of economic growth could make solving our environmental and social problems easier. *Green growth* is the idea that environmental damage can be *decoupled* from economic growth (as measured by GDP): we produce more, but it has lower impacts. We already see a lot of *relative decoupling* around the world – for example, when GDP grows by 5% but carbon emissions by 'only' 2%. Ultimately, the environment doesn't notice this pyrrhic victory – *absolute decoupling*, where GDP grows while carbon emissions shrink, is the only way this could work. While plenty of countries have seen absolute decoupling between happiness and growth, only a few have seen absolute decoupling between growth and environmental damage. Taking the example of climate change, a number of high-income nations like the UK have managed to grow modestly (1–2% annually) while seeing small decreases in emissions (around 2–3% annually), *but* the speed of this decoupling is nowhere near fast enough to avoid environmental collapse, and can be partially explained by low growth in these countries.[40] Remember that we need reductions of around 5% to 10% *annually* from now on, depending on the level of negative emissions using natural climate solutions and carbon capture.

At least there is evidence that GDP *can* decouple from carbon emissions. In other economic measures such as resource use, we see no evidence that absolute decoupling is happening or likely to happen. As economies have become better at reducing the material impact per unit of consumption – for instance, the industrialization of China in the 1980s to 2010s was far more efficient than the Industrial Revolution in England – this has been completely counteracted by the absolute growth of economies. One study on decoupling economic growth from material footprint (the mass of materials flowing through nations) showed that while higher-income countries saw small amounts of material decoupling in the late twentieth century, this effect disappeared once imports from poorer nations were taken into account.[41] Another study found that this effect happens within countries too: for instance, richer Chinese provinces import materials from poorer regions (due to the size, population and fast growth of China, some Chinese provinces, like Jiangsu, have similar material footprints to large, rich countries like Germany).[42]

As high-quality resources are extracted, the quality of the remaining resources reduces. While mining is more efficient than ever, miners move three times more rock and material for the same quantity of metal extraction a century ago.[43] The net result is that we use the same amount of energy today per tonne of some metals as we did in the 1980s. There is a broader phenomenon here, which was identified more than a hundred years ago, during the peak of the Industrial Revolution: economist William Jevons noticed that as industries used coal more efficiently, more coal was used overall. This went against common sense. Surely improved efficiency would reduce coal usage? Jevons surmised that as efficiency improved, the price of coal fell, so more people bought coal and – importantly – developed new uses for it.[44] Today, we call this *the rebound effect*.

When you switch to LED bulbs (if you haven't already), you save around 90% of the money you have been spending on electricity to light your home. So why not light up your house like a stadium? This would be called a *direct rebound*. Many studies estimate this effect to be quite small – who does that?! The *indirect rebound* is more concerning: you save money on electricity, so you fly to Paris for a holiday. This might be worse for the environment than if you'd continued paying for inefficient lighting – a nasty *backfire effect*. Much like the question of growth, there is a noisy academic debate on the magnitude of this effect. Some academics find rebounds of between 5% and 30%.[45] That is, for every kilogram of carbon avoided by energy-efficiency measures like LED bulbs, we might expect fifty to 300 grams emitted from the new purchases people make with the money they save. Others argue that if you analyse the entire economy, the rebound effect is essentially 100%.[46] That is: the reductions in environmental impact of improving efficiency will always be cancelled out when the money saved from the efficiency solution is spent elsewhere. The truth is probably somewhere towards the lower estimates, but still puts a dampener on efforts to improve efficencies.

Economic growth has been questioned for almost as long as it's been coveted,[47] but there has been almost no movement in addressing these devastating design failures. While the concept of growth is coming under increasing attack as experts both inside and outside the environmental sciences are realizing the depth of our predicament, the political and social challenges involved in taking our feet off the growth accelerator are significant.[48] Society and economics have been arranged around assumptions of growth and the primacy of the market, from pensions to taxes, amounting to a social lock-in.

MARKETS AND INFORMATION BREAKDOWN

It seems we live in a confused economy where *price* has little relation to *value* and *growth* doesn't necessarily translate into increased *welfare* or *happiness*. Even if they did, they wouldn't count for much if the whole system leads us towards environmental ruin. To describe this as an information breakdown is to put it mildly. As we've heard throughout this book, part of the explanation for this lies in the market's failure to transmit important social and environmental information via its prices. The big question is: why have markets increasingly been given the reins?

Humans have been taking part in markets for thousands of years, formally and informally. But recently, many policymakers and economists have embraced a philosophy of market fundamentalism, premised on the belief that free markets are the most efficient way of organizing *all* resources. Some suggest that this level of primacy is intrinsic to capitalism as an organizational structure, that capitalism would always trend towards markets freed from any regulation in order to transmit the purest information on value, worth and welfare through prices, wages and what economists call *utility*.[49] Stylistically, free markets are an extreme opposite of centralized government planning like that practised by the ex-Soviet Union – a system that resulted in horrendous levels of suffering. The Soviet system produced such mangled information on value that, by 1990, 40% of Russia's territory showed severe symptoms of ecological stress, including deforestation, air pollution, soil pollution, water pollution and nuclear waste.[50] Despite the fact that central planning can also be beneficial (for example in the US in the 1930s and 1940s and in post-war Europe), the cautionary tale of the Soviet Union was used to make the argument, either naively or self-servingly, that moving to market fundamentalism, and assuming that *any* public ownership was inefficient, we'd live in the best of all possible worlds. The result was waves of privatization from 1970 onwards through much of the world's largest economies. Expansion of the markets extended into the welfare state and public services like water and electricity. Free-market thinking characterized the US-based Washington Consensus, which was then exported to low-income countries in the form of structural adjustment programmes, offering IMF loans in return for market reforms.[51]

As the importance of markets has risen, they have become an organizing principle within and between nations rather than simply a tool that *can* serve society's needs. Engineer and entrepreneur Amory Lovins argues that 'markets make a good servant but a bad master, and a worse religion.'[52] It's true that markets are neither immoral nor moral – they are simply amoral. They have no interest in whether or not a citizen has enough to eat, or in the well-being of a forest as a lifeline of biodiversity. Markets have to be told to care by regulators, or alerted to the potential of profit. This becomes a civilization threat if deregulation is paired with a sociopolitical framework that struggles to assert the public interest in directing markets. In the real world, the current economic system trends towards privatizing profits and socializing costs – especially via externalities.

Perhaps more competition and a multiplicity of market actors would help embody the public interest? Alas, as markets have opened up, globalized and grown, so too has the power of actors within them. It's argued that while free-market ideology aims to avoid monopolies (limited sellers in a market) and monopsonies (limited buyers), the paradigm of hands-off regulation ends up encouraging market concentration. This is because those that grow fastest and earliest can take the lion's share of the market and make it difficult for others to compete. Efforts to implement regulation to maintain competition are often seen as wildly ineffective. Of the hundred biggest economic organizations in the world, sixty-nine are now corporations and thirty-one are countries.[53] In the Food chapters, we saw how a handful of companies control a huge proportion of the agricultural sector, and this extends across all aspects of the economy. The US is an extreme case, where a few companies control well over 60% of many national markets. Three companies own 98% of the mobile contract market. Four companies own 75% of the beer market. Three own 81% of the arts and crafts stores. Three own 80% of the music market. A single funeral company owns 75% of the share of casket production and sales.[54] From the mundane to the sublime, markets have been monopolized. Concentration has been particularly dangerous in finance, where banks deemed 'too big to fail' were bailed out after the financial crash only to merge into even fewer banks with an even larger market share.[55] Since more market competition improves financial stability,[56] when (not if) we face the next financial crisis, the banks will be too big to bail out. This concentration of profits and power, along with political finance rules, goes some way towards explaining how, as we've heard, many US politicians ignore the wishes of their constituents in favour of donor interests.[57] Of course, some companies *can* act responsibly, but those details are for another chapter and, as the saying goes, 'power tends to corrupt'.[58]

We haven't touched on perhaps the most concerning concentration of new market power: information technology. Facebook owns 70% of the US social networking market, Google 91% of the *global* desktop search market, and Amazon 49% of all US e-commerce[59] – and these companies are now cornering other markets too, from Web services to home automation. Tech companies are also especially important to environmental and social problems in their role as algorithmic gatekeepers. They filter information and interpret the world for many, and research has shown that the outcome, which we'll return to in the Epilogue, is not good.

The rise of markets as an organizing principle and the increasingly monopolistic nature of these markets has helped diminish the government's role as a coordinator of economic activity. This is a big problem, as there are crucial roles that only governments can take on. Governments are the leading actors in early research and development, environmental regulation, managing social transitions and building infrastructure. Consider just one important role: while the costs of renewable energies have plummeted, due largely to market competition, this would have been inconceivable without huge amounts of government support. Governments in the UK, Denmark, US and China (among others) 'pushed' renewables into the marketplace via state-funded research of solar panels, batteries and wind power. Meanwhile, governments across the world 'pulled' renewables into the marketplace via renewable energy subsidies.[60] These 'push' and 'pull' policies enabled markets to do what they do best: drive down the cost of the nascent government-funded technologies. And while this process should have been faster (at one stage, only a handful of nations were supplying significant subsidies to 'pull' solar energy into the market), this is how governments and markets work together most effectively.

LABOUR, PRODUCTIVITY AND PRECARIOUS REALITIES

The reforms of the 1970s extended into labour markets too, with the idea that freer labour markets would improve productivity via easier hiring and firing and reduced worker entitlements (which were seen to be economically inefficient). The result of this belief has been the intentional weakening of labour regulations.[61] In a now familiar trend, this has reached extremes in countries like the US and UK, where many worker protections have been demolished, minimizing the social costs paid by companies of doing business. In the UK, symptoms range from zero-hour contracts to the appalling conditions in which gig economy workers are treated.[62] The result is a *precariat*: large communities with no economic or even time security. Psychologists and behavioural economists point out that such insecurity drives a level of scarcity that compromises decision-making faculties. It's difficult to break out of such a situation or engage fully in society when all thoughts are consumed by the day-to-day logistics of making ends meet.[63]

This precariousness was explicitly accepted from the start. It's a well-worn story – the promise was that free-market reforms would unleash such growth that the rising economic tide would lift all boats. The facts showed something very different. In the US, the hourly wage increased by 91% between 1948 and 1972 almost one-for-one with productivity growth at 97%; between 1973 and 2013, it only grew by 9%, against a 75% growth in productivity.[64] It's quite clear where the productivity gains went: between 1978 and 2013 average US salaries increased 12% in real terms, while those of CEOs increased 940%.[65] It's a similar but not quite as extreme story in many other countries. As wealth was deposited at the top, those earning most of their income from capital and land rather than labour reaped the dividends from productivity growth (increasingly, those with the largest capital assets today are also those with the highest income).[66] This reinforced a capitalist dynamic identified by economist Thomas Piketty, whereby returns from capital grow much faster than returns from labour over time.[67] When many of the capital assets of the rich (in the form of company stocks) were under threat during the financial crisis, it was these companies that were bailed out, not the poorer communities.[68]

The consequence for those not at the top? A massive increase in private debt. As we've seen, insecure working conditions and status anxiety caused by inequality degrade self-worth. The result has been a dynamic of materialism ('retail therapy') fuelled by cheap imports as compensation for the lost ability to work towards security (home ownership, for example), along with, increasingly, drugs and depression. Sociologist Wolfgang Streeck calls this 'coping, hoping, doping and shopping'.[69] We live in a system where large companies increasingly dictate how people spend their working lives (think of the restrictions on toilet breaks in warehouse jobs), and can manufacture people's material desires too, through advertising and marketing which, in 2017, made up an estimated 19% of all spending in the US.[70] All of this comes at a tremendous environmental, social and human cost.

This becomes a serious challenge when we consider how individual data is increasingly commodified via large tech companies. These data offer unprecedented access to our lived experience in a period of astonishing levels of corporate concentration. The fastest growing markets over the last decade have been those based around profiting from citizens' personal data and, by extension, their attention and emotions, through advertising and selling data to third parties. Governments and the public might eventually get a handle on these monopoly issues but, as innovation accelerates, the time it takes to get a handle on them is outpaced by the changes society is undergoing. Some thinkers are worried that it will become increasingly possible to have our interests rewritten without us knowing.[71] Indeed, they suggest that this is already happening at some level – for instance, by concerted Russian information warfare campaigns conducted on large internet platforms such as Facebook and Twitter.[72]

In sum, under-regulated capitalism can create some forms of abundance in resources, for example in 'fast fashion', but only when environmental and social values are ignored. Without strong restrictions, it also commodifies values in such a way as to render scarce many of the things that matter most to us: time, security, privacy, and relationships unmediated by the anxiety and mental-health problems of struggling to get by and assessing our status relative to others.

ECONOMIC MODELS LITERALLY DISCOUNT THE FUTURE

With massive upheaval around the corner, it would be good to have models to explore what 'around the corner' looks like. Unfortunately, many economic models, including those used in finance and by central banks, don't include the environment or natural resources... at all. These models are computer representations of what economist Kenneth Boulding called the 'cowboy economy': modelled economies unconcerned with environmental impacts because, much like cowboys in the American West, they assume there are always vast new markets into which to expand. Leave your waste behind and ride on. However, there are some models that do include environmental factors, the most influential of which are from climate-change economics. These are called *integrated assessment models* (IAMs) and results from their scenarios are often reported in IPCC assessments. Their structure, assumptions and policy impacts are so important that, as obscure as some of the terminology is, we really need to spend some time on them.

If you had the power to pull the levers of the global economy, how might you conduct and pay for the zero-carbon transition? Perhaps you'd ask experts to lay out your options, explaining how effective they are, how much they cost, and how quickly they could be implemented. IAMs represent a selection of levers in computer code that optimize how much to pull each lever and when to do so based on cost or feasibility. The choice of the lever, the assumptions behind the calculations, and the order in which they are pulled dictate the outcome. There are two main types of IAM: simple and complex. In *simple IAMs*, the lowest transition cost is found by balancing the estimated costs of mitigating climate change against the estimated costs of climate change. In *complex IAMs*, storylines about how society might progress in the future are combined with different emission scenarios, information which then optimizes the pulling of levers to best meet climate targets. For instance: should there be more wind energy now or more carbon capture in the future?

Has a shiver run down your spine? If so, you've spotted a problem already: all IAMs optimize or estimate costs based on GDP and market values. These policy-informing models inherit all the problems we've already discussed. While IAMs include carbon prices or climate costs, they don't include the cost of air pollution, noise pollution, energy security, water quality, inequality and so on – essentially, they are unable to 'see' the value of a better world ahead of us (*if* we do manage to mitigate fast enough), so they cannot even suggest better/sustainable/ liveable options!

A more technical problem is how models value the future. Economists don't ignore the future but they literally discount the future against today. Given the commonly used 5% annual discount rate, avoiding $100 of flood damage in 2100 is only worth paying $3 today. This is something akin to suggesting that the safety of a human in 2100 is only worth 3% of a person's safety today. There are reasons for thinking this way – though, in each case, there are significant objections.

- People generally prefer to have things today rather than at some point in the future. In one oft-cited study, participants preferred $50 immediately instead of waiting six months for $100 (an annualized interest rate of 400%!).[73] This implies that we (should) discount the future. However: 1) This thinking is not universal across cultures. 2) Preferences change if people learn about the mathematics behind their options. 3) It is a big leap to assume that individual, instinctual preference is the same as society's informed preference. Societies may, on aggregate, be happy to forgo today to avoid catastrophic climate change.
- As technology improves, IAMs assume that humanity's ability to both mitigate and adapt will improve. Future people should be better placed to fix the problems people today are creating, so why not leave it up to them? The thinking is that investing now to save future (richer) humans is a bit like a poorer country (i.e. Mali) bailing out a richer one (i.e. Germany) today. This is an inequality argument – this time between the present and the future – but it turns out that the poorest stand to suffer so much more from climate change that, under any set of inequality assumptions, it is best to mitigate now, and quickly.[74]
- We don't know for certain what will happen in the future. If you're open to risk, then you'll discount the future more and hope that climate-change impacts are on the lower end of what the science projects. (If you're rich, this probably means gambling with other people's lives – it's a privilege to have sufficient resources to be open to risk.)

This discounting of the future is not inconsequential. In the mid-2000s, British economist Nicholas Stern produced a model which, while it had the same structure, totally disagreed with the standard simple IAM of the time (designed by the economist we met earlier, William Nordhaus). Stern's model recommended immediate climate action,[75] while Nordhaus's advocated a very leisurely pace. The difference? The discount rate. Nordhaus used a 5% discount rate, while Stern used 1.4%.[76] (Now Stern advocates using different models entirely.) [77] Some ethical philosophers and environmentalists suggest that that rate should be zero or perhaps even negative – the value of an investment today should increase over time, since the damages avoided from climate change are so massive.[78] Note that complex IAMs also use higher discount rates of 5% (some researchers have attempted to investigate the influence of this on model results and find that reducing this rate to 2% doubles the carbon price and halves the need for speculative negative emissions[79]).

Unfortunately for realists everywhere, the way in which simple IAMs attempt to include climate damages doesn't agree with what climate scientists are saying. Simple IAMs suggest there would be a tiny 10% reduction in GDP by 2100 for a six-degree Celsius rise in global temperature – a level of catastrophic warming.[80] It gets worse. Climate damages are 'fat-tailed', meaning that there's a small (non-negligible) risk of a total cataclysm – what economists call 'unlimited downside exposure'. It turns out that fat tails are mathematically incompatible with the type of cost–benefit analyses most economists love, including those used in simple IAMs.[81] These crucial technicalities accumulate so disastrously that the economist Robert Pindyck answers the title of his 2013 paper 'Climate Change Policy: What Do the Models Tell Us?' with the first two words of the abstract: 'very little'.[82] In just one example of how far away some models are from the real world, some simple IAM studies recommend an 'optimal' warming of 3.5 degrees Celsius – a level that would result in inconceivable suffering, which would pass several tipping points, and would make it very difficult for organized civilization to continue.[83]

Let's give complex IAMs a round in the reality ring. Their high discount rates and other modelling choices mean that the majority of models reported in the 2015 IPCC assessment involve scenarios that drag carbon out of the atmosphere at a massive scale, using BECCS – an approach we've seen to be highly implausible.[84] Perhaps the mad scientist cliché has some basis in reality if the average model reported in the premier assessment on the science of climate change assumes that, by 2030, humanity will be 'compressing, transporting and burying an amount of CO_2 by volume – two to four times the amount of fluids that the global oil and gas industry deals with today'![85]

Even if we could assume there is a way to fix the flaws of these models, IAMs would still fail to represent the real world. Most models are based on concepts of balance or equilibrium. For example: a carbon tax is introduced, and models find a new equilibrium as its impact spreads throughout the economy. But we're not in Kansas any more. Nature is in a state of violent disequilibrium. Since humans have appropriated so many resources from the environment and the economy is in fact a subsystem of nature, it's hard to see how equilibrium modelling can work from here on. We know that complex behaviours like lock-ins (social and technological), feedbacks, path dependence (choices now have an impact on the availability of choices far into the future), and tipping points are absolutely critical. Alas, IAMs struggle to represent them at all. For instance, complex IAMs can't model new city designs that encourage environmentally friendly behaviour, nor can they represent a world in which climate-driven extremes provoke food shortages. In short, the ways in which humans can fail are multifarious and complex,[86] but the ways we can model our failure aren't.

Are these models just academic curiosities, or do they matter for policy? Modellers are the first to say that these are just tools to explore options; that 'all models are wrong, but some are useful',[87] and that they aren't intended to directly inform policy. But in practice they are used for just that. Since BECCS have been on the table, it's been all the rage in government net-zero emission reports. In the UK government's response to the Committee on Climate Change's 2019 Progress Report, it mentions carbon capture and sequestration around five times as much as solar power.[88] This is exactly what we might have expected, given concerns about the overextension of economic models more generally.[89] Such models play an important political role in hiding difficult ethical, social and political decisions behind opaque, technocratic and highly questionable assumptions. Like the visible parts of icebergs, these models lie on 'vast bodies of submerged ideology'.[90] The veil of quantitative objectivity gives political actors the 'evidence-based' cover to justify slower action (in the case of heavily discounted damages), and to avoid crucial public debates about the kind of world we want to live in; in other words, what trade-offs we can live with.

With such models, which optimize both the status quo and GDP, politically divisive decisions can be delayed until the next election cycle. Energy transitions can be delayed until oil companies make that extra barrel of cash. As with GDP, models and measurements can become self-fulfilling realities. In this case, the longer action is delayed (seemingly justified by economic models), the more negative emissions will be necessary. Perhaps, in the end, the best action a modeller could take to help humanity would be to *not* present the results of these models to policymakers, keeping them in the academic background in protest.

Much of the harm might be caused by overconfidence that we can optimize interactions between the environment and the economy. In making them so complicated, our IAMs may fall foul of the maxim that 'everything should be made as simple as possible, but no simpler'. Simpler, less opaque (and easier to criticize and ignore) models suggest that we continue to follow a scenario where the physical limits to growth on Earth will eventually assert themselves on human society and lead to a 'sudden and uncontrollable decline in both population and industrial capacity' in the second half of the twenty-first century.[91] The reason for this is not that resources and other factors run out; rather, that the Earth system as a whole runs out of resiliency against Earth-agnostic economies and their wastes.

GLOBALIZATION AND AUTOMATION PRESENT NEW CHALLENGES

These global environmental–economic models are run for several large regions, implying that the changes they recommend (the levers they suggest pulling) would be broadly adopted by many (if not all) international governments. This is necessary because local economies and environments have been increasingly connected to the global economy and environment via globalization. Dysfunctions in these systems become global dysfunctions – perhaps intractably so. (Climate change affects the whole of the Earth's system, if at different rates.) To give just one example of the globalized economy and its environmental impacts: buying a smartphone is the culminating act of a vast interplay of global market connections, most of which place no value on the environmental and social resiliency needed to safeguard the future. Manufacturing a smartphone requires about seventy-five of the 118 elements in the periodic table (the human body needs only thirty), with metals and plastics sourced from so many places in such complex supply chains that it's difficult for companies themselves to know where their components come from. These materials are then refined into microcomponents in a network of factories in dozens of countries. The microcomponents travel to other countries for manufacture – more than thirty countries for the iPhone, from Singapore to Italy, Taiwan to Brazil. Many companies then send those components to China, where the phones are constructed, to be re-exported around the world – in Apple's case, to the seventy countries that have Apple stores.[92] When we buy products today, we throw pebbles into a global pond with impacts rippling far out of sight and out of mind: forests felled to make way for mines in sub-Saharan Africa; Canadian topsoil sacrificed for oil… the list could go on, but surely Earth's resilience can't.

These stories have been made possible by the movement of capital around the world. Investments can freely move from nation to nation. This makes any attempt at regulating production much harder, especially since the entities with a lot of power in the system are the companies themselves. Sometimes companies do attempt to address working and environmental conditions in their supply chains, but often they have limited visibility of what is going on, or are in any case incentivized to look the other way until they become a public-relations issue.

The same globalized complexity is mirrored socially, where individual wealth made from social investments and environmental damage is offshored using international tax havens, precluding its reinvestment into public services. We've already heard that the amount of anonymous private wealth in tax havens is now somewhere between $9 and $36 trillion globally.[93] Companies as well as individuals shelter massive amounts of wealth abroad while they benefit from huge public investments in the form of research by universities, government labs and the military. Economist Mariana Mazzucato provides another good example in Apple's iPhone, a product that almost entirely depends on technologies that were initially developed by the public sector (including GPS, touchscreens, the internet, etc.).[94] The US government also shelters Apple from competition nationally and overseas. After all this, Apple is able to avoid 'paying forward' into this innovation system via a corporation tax regime that allows them to move wealth abroad. It's not that corporation tax is the best way to earn returns from public investment – public shareholding or other private–public structures might be better options[95] – but, with no skin in the game, these parties are insulated from the socioenvironmental impacts of their decisions. At this point, just when we most need concerted action nationally and internationally among *all* actors of society, especially in investments for the future, we see wealth siphoned away to private, sheltered enclaves, eroding the communal drive to confront and fix our problems.

We can't fully address all the ways in which globalization has altered societies, economics and the environment, but we can reflect that this represents a historically unprecedented increase in complexity in the global civilization in supply chains, wealth, national sovereignty and more. There are many aspects of the energy and food transition that only governments can organize, especially in areas of market failure such as energy efficiency, public transport, and large-scale infrastructure like long-distance, high-capacity transmission lines. In many cases, governments have access to cheap capital: they can fund solutions at very low interest rates. What's more, governments are one of the few institutions that can directly place international pressure on other large institutions, including other governments. However, a large amount of climate denial has been driven by a fear of government intervention. The irony is that now, due to the delay climate denial has caused, we'll see far more government intervention than would have been needed. After all, it won't be Nike saving people from hurricanes, or Amazon rescuing flood victims. Of course, governments have serious flaws – especially when internal expertise has been eroded or compromised – but they are critical actors in environmental transitions and are theoretically more responsive to broader public concerns than corporate interests, in democracies at least.

IT'S THE PLANET, STUPID!

If there were sufficient efforts to address many of the deep problems in this chapter – if we *could* rid ourselves of the dynamics of inequality, materialism and consumerism (dynamics that seem very stable), and if markets priced environmental and social damage – society would likely still privilege itself over natural functions. The knowledge of how to fully value the resilience of the ecosystem or the climate as a whole will elude us for a long time to come. We don't know exactly where tipping points lie, let alone how to produce a system that would avoid stepping on them.

The potential for damage and our hubris to think we can grasp the full extent of the threats we face was summed up in a smaller context by the reply the British Academy gave to the Queen when she asked why no one had predicted the financial crisis: 'In summary, Your Majesty, the failure to foresee the timing, extent and severity of the crisis... was principally the failure of the collective imagination of many bright people... to understand the risks to the systems as a whole.'[96]

CHAPTER TEN

◡ HOPE ◡

Valuing the Future

Economic problems have no sharp edges; they shade off imperceptibly into politics, sociology, and ethics. Indeed, it is hardly an exaggeration to say that the ultimate answer to every economic problem lies in some other field.

KENNETH BOULDING, *The Economics of Peace*, 1945

Everything depends on what the people are capable of wanting.

ERRICO MALATESTA, *Life and Ideas*, 1965

Pursuing growth at the expense of the environment is no longer an option. Marginal efficiency gains are not enough.

HANS BRUYNINCKX, Director of the European Environment Agency, commenting on 'The European Environment – State and Outlook 2020' report, 2020

We live on the cusp of astonishing changes in the global economy. A revolution in information and automation is arriving with discombobulating speed. The upheaval will be so large that it's hard to consider future economic settings before addressing this digital elephant in the room. Fortunately, this revolution *might*, in part, facilitate a socially and environmentally sustainable future. However, we should note up front that while the potential futures for the energy, food and climate systems, although very uncertain, are at least partially bounded by biophysical realities; thinking about the messy world of society and economics is fraught. Many writers have fallen foul of trying to see into the economic future, from the physicist who, in 1954, enthused that nuclear power would make electricity 'too cheap to meter', to Albert Einstein who, in 1934, was quoted as saying that 'there is not the slightest indication that [nuclear energy] will ever be obtainable... the atom would have to be shattered at will' only to be proved wrong just eight years later.[1] The best we can do is examine some trends and highlight potential interactions, knowing full well we will be very likely to get things very wrong. A good start is to look at historic revolutions to see if there are general identifiable trends to expect.

For most of history, human wealth came from the land. Even after the restructuring of society through the agricultural revolution, wealth rested on the productivity of land and physical labour – of human or animal. But with the advent of the scientific revolution followed by the First Industrial Revolution, everything changed. Physical labour could be mechanized and wealth could be created from other sources. To give one example of the rapid change, by the end of the eighteenth century, automated looms had increased weaving productivity almost fourteen times compared to hand looms.[2] Translating a similar productivity increase today (while maintaining the volume of production) would render the five-day working week done and dusted by Monday morning! Mechanization, combined with improved transportation and an insatiable demand from new national markets and overseas colonies, kept factories expanding and the fires of the revolution growing.

In the long term, this was a process of liberation from physical hardship. In the short term, it spawned chaos. The combination of agricultural mechanization, the opening of markets, improved transportation and accelerating land enclosures (in which communally held land was privatized) pushed workers off farmland and country dwellers into cities to work in factories, prompting rapid urbanization. Jobs disappeared faster than they could be replaced. Labour costs and wages plummeted. In the absence of a welfare state, unlucky factory workers, stripped of their rural safety net, starved on the streets.[3] The Industrial Revolution transformed the top of the class system too, as the means of production began to shift further away from the landowners (the gentry) to those who owned the machines (the capitalists). Sociopolitical systems across Britain and industrializing Europe were subjected to huge pressure. Inequality soared, Dickens gained his setting for *Oliver Twist*, and newly displaced skilled workers responded with civil unrest to the changes they saw as deeply unfair. The Luddite riots – aimed mostly at smashing weaving machinery – so threatened the ruling classes that, at one point, more British soldiers were fighting on the streets of England than fighting Napoleon on the Iberian Peninsula.[4] This unhappy story was retold by jobless agricultural workers during the Swing Riots of the 1830s.

Rather than a single Industrial Revolution which spread from developments in Britain to elsewhere, historians consider this story the First Industrial Revolution, and it was already over a century old before it translated to an improvement in life for most people. Health indicators such as average height actually fell in the 1700s, mirroring a trend that followed the first agricultural revolution (foragers were on average several centimetres taller than the first agriculturalists).[5] Eventually, welfare began to improve dramatically in the latter half of the nineteenth century: the years between 1870 and 1890 saw the start of the Second Industrial Revolution, with urban centres benefitting from technologies like telegraphs, railways and gas heating, and public health infrastructure like sewer systems and civic water supplies.[6] The Second Industrial Revolution increased the goods travelling around, thanks to steam-powered rail and new shipping trade routes like the Suez Canal, which significantly reduced transportation costs. Given these two factors – imperial competition via this supercharging of transport, along with increasing inequality driven by the impacts of new technologies on workforces – economists are increasingly making the argument that these structural forces played a significant role in the First World War and the upheaval that followed.[7]

As the economist Joseph Schumpeter noted, 'Economic progress, in capitalist society, means turmoil.'[8] After the turmoil of the 1940s, across Europe and further afield, states developed ways to partly control the dislocations resulting from technological revolutions. While the wars had reduced inequality in a malign way by destroying capital and making everyone poorer,[9] the welfare states that rose from the ashes reduced inequality in a benign manner: through universal education, high tax rates and access to healthcare. Losing a job was no longer a life-threatening event, and universal education gave people at least a fighting chance of obtaining more skills and better-paying jobs.

From the late 1970s onwards, international policies and institutions encouraged further integration of national economies into a global economy. By 1990, globalization began to soar, with complex supply chains exploiting resource and labour cost differences between nations.[10] Lucrative mechanical jobs were 'offshored' or automated with new information and robotic technologies, and the power of unions declined. Physical manufacturing in high-income nations no longer commanded high wages, and information-based services in management, administration and creative industries took off. The disappearing manufacturing jobs were in some ways similar to the skilled weaving jobs pre-industrialization in that they paid well and offered security and their loss fractured the communities throughout industrial towns in high-income countries. Across the UK, US and Germany, the share of manufacturing work in the economy dropped by around half between 1970 and 2009. More recently, the information revolution combined with globalization to drive service-sector jobs offshore, including call centre and online administration jobs. The result was that many areas of the world saw their incomes increase, as did the richer groups in high-income nations (as they collected the profits from this offshoring). The low- and middle-income groups across high-income nations, however, lost out – in a big way; between 1988 and 2008 they saw almost no increase in income.[11] The political and social fallout from the wage stagnation of the lower and middle classes in richer countries is still playing out today.

It took around a century for the sociopolitical upheaval of the First Industrial Revolution to filter through society. The upheaval during the Second Industrial Revolution was compressed into roughly seventy years. Only thirty to forty years have passed since the first information revolution began, yet we are already experiencing the start of another, perhaps deeper, revolution. Jobs once thought absolutely safe from automation are now threatened – and it will seem even more unfair to those who lose out. Are we in a better place to weather the sociopolitical storms of this rapid revolution? And how might avoiding these storms help in the ongoing environmental crises we face?

AUTOMATION: MACHINE LEARNING AND ROBOTICS

Algorithms for artificial intelligence have been around for decades. What's new is the amount of processing power and data available for these algorithms to learn on. Processing power doubles every two to two and a half years, meaning that today's household video game consoles are significantly faster than building-sized supercomputers of twenty years ago.[12] The amount of digitally stored information is increasing at approximately the same rate. In 2014, the total digital storage capacity was roughly 500 times more than all the information stored by the DNA of the 7.2 billion people on the planet.[13]

Massive data flows, increasing processing power and energy to drive these computations have allowed for explosive innovation in many existing technologies. Think of a structural beam in a building: machine learning algorithms can now optimize the amount of material used while maintaining structural integrity (i.e. modelled as part of a larger building), which can save a significant percentage of the materials used, thus reducing environmental impacts.[14] This is crucial, given the level of expected future urbanization. Take another example: automated picking and packaging in warehouses.[15] Each bit of technology involved, from cameras to laser scanning, has been around for years if not decades, but only recently have computers begun to process their environments quickly enough to operate these vehicles without human intervention.[16] This one combination of existing technologies has the potential to displace millions of jobs in warehouses globally. Accelerating information flows and processing power have catalysed the automation revolution as steam power did for the machines of the Industrial Revolution. The speed of this information revolution is the speed of electrons and light, rather than metal and steel.

Warehouse automation is a handy example, but many automatable tasks don't require physical applications to displace human work. The office chair can be emptied too. This overhaul in work has been under way for some time. We don't notice because the moment technologies become 'normal' they are no longer thought of as automation. Online translation services, though taken for granted, are all machine-learned processes, as are the optical character-recognition algorithms used in scanners and cameras. Automated services are at various stages of maturity across broad swathes of the service economy, from retail to legal services, from journalism to finance. Automation and machine learning won't steal jobs outright but will replace tasks within jobs, probably reducing the total number of people needed.

Extrapolating these trends across the economy, experts estimate that tens of millions of jobs may disappear in the US alone. A highly cited Oxford University report from 2013 found that 47% of all current work in the US is susceptible to computerization.[17] The sectors with the highest probability of automation were services, sales, office and administration, and manufacturing. Similar estimates have been reported by the Bank of England, an EU report, and a presidential report to the US Congress.[18] The unanimous findings of governments, researchers and businesses demonstrate that this new revolution is not technohyperbole and that it will arrive more quickly than any previous job displacement in history. It will be 'inhumanely fast'.[19] Like previous revolutions, this may remove more drudgery from work... eventually. But the potential for sociopolitical upheaval is massive and will combine with increasing pressures from environmental crises.

HOW TO MAKE WORK WORK

These developments are coinciding with serious problems in the world of work. In two much-discussed surveys of work from the UK and the Netherlands, as many as 20% to 40% of people believed their jobs make no meaningful contribution to the world.[20] As anthropologist David Graeber has explored in his book of the same name, these are self-confessed 'bullshit jobs'. This figure is probably higher than reported, since most jobs, even if not completely bullshit, often involve bullshit tasks. We all have personal examples of pointless meetings, task forces and other 'paper-pushing' exercises (complete with glitchy electronic forms). Factored in, the total amount of meaningless work might be as high as 40% or 50%.[21] Other polling shows that general motivation and

job engagement are shockingly low, with 66% of all employees claiming
no active interest in the work they're doing.[22] In a 2013 survey (by the
polling company Gallup) of 230,000 employees across 142 countries,
only 13% of people reported being engaged by their jobs![23]

In his 1930 'Essay on The Economic Possibilities for Our Grandchildren',
economist John Maynard Keynes guessed that technologically advanced
countries would continue to enjoy productivity improvements such that
the working week would be just fifteen hours long by the end of the
century. He reckoned that the returns of innovation would translate
to more leisure, 'which science and compound interest will have won
for [each citizen], to live wisely and agreeably and well'.[24] As Graeber
suggests, this partly came true, but instead of working half the week,
we have somehow filled the remaining days with meaningless tasks.
The very time and freedom we need to engage in meaningful activities
and to imagine a better future is being squeezed by nonsense tasks.
Historically, we needed more labourers to produce goods and services
for increasing demand. Today, we have to keep needlessly increasing
production just to keep people employed! It could be that the sheer
quantity of meaninglessness in today's high-income work indicates that
we are already producing enough; productivity is now so high – thanks to
various information and automation technologies – that extra human
work is no longer needed (and so is meaningless). It could be that this
superfluous work has arisen out of social hierarchies and macroeconomic
requirements to keep people employed rather than any material need.

So what are our options? One is to opt out and fulfil Keynes's prediction. There are already moves by some businesses to shift to a four-day
working week. For example, Microsoft moved Japanese workers to a four-
day week and found a productivity increase of 40%.[25] Some governments
are also experimenting with making the shift to a shorter working week.[26]
The ultimate outcome of these policies is that the working week would be
as short as possible while maintaining livelihoods and the infrastructure
that drives knowledge, healthcare, etc. As automation continues, it is
perfectly feasible for the average forty-hour working week to shorten to
twenty hours or less. The remaining 40% to 50% of the working week
could be available for leisure, caring, volunteering – all the activities not

counted by GDP. It might also allow us to start practising the behaviours we will need for the future.

Although it would undoubtedly be a social benefit, evidence is mixed on the possible environmental impacts of shortening the working week. One study in Sweden found that a decrease in 1% of work time could result in a 0.7% reduction in carbon, but this was mostly due to reduced income and consumption and not, as would have been hoped, an increase in more, low-carbon leisure activities.[27] Another study of twenty-nine high-income OECD countries over the period 1970–2007 found that (even while controlling for GDP) lower working hours resulted in lower ecological footprints.[28] A consensus view hasn't been reached in this new area of study. However, even with the somewhat limited evidence, there are other ways in which cutting the working week can help environmental goals, which we'll revisit later in this chapter.

Besides opting out, another response would be to transition to better-paid work in sectors sheltered from computer intelligence trends. Over the long term, jobs have transitioned from physical labour to mental labour. AI is likely to take over certain types of mental labour such as repetitive 'thinking tasks' of many a sprawling spreadsheet, or pattern-recognition tasks like those often used in healthcare, such as the identification of a malignant mole or breast cancer.[29] The remaining tasks will be the ones in which humans have the edge. That edge could be social intelligence: humans' unique ability to 'get along' and cooperate. It will remain exceptionally hard for AI to grasp the intricacies of human interaction and empathy for some time – especially in large groups, where the combinations of emotions between different people make such calculations intractable for computer intelligence. It also happens that jobs with a human touch are often the ones we see as most socially important and meaningful: ones generally underpaid today, including caring and nursing. As wages decrease across automatable jobs, workers could be 'pushed' away from bullshit jobs and 'pulled' towards new meaningful work that will begin to command higher wages.

Where increasing globalization of production hit industrial towns hard, the automation revolution will hit the service-oriented companies of central business districts around the world, not to mention Indian call centres, Chinese factories and German car plants. Dystopian scenarios could play out, in which control over the automation software and robotics becomes increasingly concentrated in the hands of the wealthy, rendering any small- or medium-sized business economically moot. If we don't respond rapidly, we will see further, dramatic sociopolitical upheaval. Since this is a hope chapter, let's focus on how we can head off the malign sociopolitical pressures of this upheaval.

Benign policies would be needed both to ease the pressure points of a rapidly transitioning economic structure, and to control the concentration of wealth that often occurs during technological revolutions. One such policy response is a combination of Universal Basic Income (UBI), along with some level of Universal Basic Services (UBS). UBI would mean paying each citizen a monthly sum, regardless of their income, whereas UBS would provide a selection of basic services (which usually go beyond healthcare to include food, shelter and transport). These concepts have been waiting in the wings for a very long time, and it may well be that their time has come.

Since these policies are often disparaged as too idealistic, it's important to think about how they might work in practice. In their book *Scarcity*, behavioural economist Sendhil Mullainathan and psychologist Eldar Shafir show how managing time and money is cognitively complex, and that perceived or real scarcity can lead to *tunnelling* – a focus on short-term problems to the detriment of long-term goals (we could perhaps think of broader 'life goals' here). The anxiety prompted by time and money limitations can rewire the mind, narrowing *bandwidth* – the cognitive freedom to think about new solutions.[30] This can lead to negative feedbacks, like the critical transition of an eviction we explored earlier. Remember that when these critical transitions occur they are often hard to reverse, and that it's often better to prevent problems rather than treat their consequences. While it couldn't remove all scarcity people experience, an effective UBI/UBS *might* help break these feedbacks and provide increased cognitive freedom for considering the long term – including our individual roles in sustainable transitions.

It is often the scarcity preceding a deadline or a bill that precludes the ability of many to socialize, to participate in communities and to develop relationships – the things we think of as being the most important aspects of life. As psychologist Daniel Kahneman points out from decades of work on human decision-making, if you want to be happy, pick the right goals and spend a lot of time with friends.[31] Both of these require an alleviation of perceived scarcity, and can often be activities with low environmental impacts. Relatedly, happiness researchers Elizabeth Dunn and Michael Norton suggest that the best way to spend your money for happiness is, among other things, to invest in others and buy experiences (both of which can also have low environmental impact).[32]

Another important breakthrough would be severing the link between work and consumption. We are currently stuck in a consumerist dynamic that requires work in order to consume, sapping us of the time and energy to imagine alternatives. Behavioural economists would classify this as a move from *extrinsic* to *intrinsic motivation*.[33] If UBI/UBS provides the economic freedom and safety nets for self-actualization (the fulfilment of one's talent or potential), we might assume that this would be accompanied by higher levels of intrinsic motivation. We would be left to pursue the values we claim matter to us; we would rid ourselves of status anxiety and addictive material consumption; we would start to sever the vicious circle of work–consumption–work (sometimes called a 'hedonic treadmill' for the way that consumption can drive status-signalling to others that drives yet more consumption). This may seem idealistic or naive, but it is what research suggests most people want.[34]

Fortunately, UBI/UBS are broadly supported across society, from techno-utopians to environmentalists, and from the left side of the political spectrum to the right. They also hold the potential to cleverly elide the existing power structures: if everyone gets it, who can complain? Disagreement arises when we begin to discuss the balance, reach and depth of the policies, and who pays for them: who wins and who loses. Often, those on the left worry that the UBI could be used to do away with the other parts of the social safety net.[35] Those on the right worry that, if funded through progressive taxation (as they would have to be), UBI and UBS would remove the entrepreneurial incentive to innovate.[36] Others worry that work is important to concepts of self-worth, but this may not be true if people are able to pursue their own interests. It's not necessarily the ideas themselves that cause disagreement, but their implementation within existing socioeconomic dynamics.

KEY CONCEPTS FOR FUTURE ECONOMICS

As these large technosocial shifts unfold messily, what other economic policies are needed in the Anthropocene? While many people can agree on what's going wrong in the economy, disagreements arise when it comes to why, and it's even harder to be definitive about solutions. In fact, it would be foolish to be definitive, as Voltaire noted: 'To be uncertain is unpleasant; to be certain is absurd.'[37] What we can do is attempt to highlight the most useful concepts. With those caveats out of the way, I will now discuss some economic solutions that *might* be equivalent to renewables in the energy transition or dietary shifts in the food transition.

The first policy that the vast majority of experts agree on is a price on carbon. There are two main forms of carbon price: a carbon tax or a cap-and-trade scheme. A tax prices each tonne of carbon emitted but doesn't control the exact amount emitted (emissions are flexible; price is not); cap-and-trade sets a total limit and then distributes permits which can be traded (prices are flexible; emissions are not).[38] Already 20% of the world is covered by some sort of carbon price, but this price is often far too low to make a serious dent on emissions.[39] In some places, though, this is starting to change. For instance, the European Trading Scheme (ETS) has been struggling to have an impact for decades, but from 2018 to the close of 2021, the price steadily increased from €9 to around €86 per tonne. This, along with plant closures and cheap renewables, drove more electricity production from renewables than fossil fuels in 2021.[40] This price, though, is *still* too low. While the conservative estimates of the IMF suggest around $75 to $100 per tonne experts think this is still *three times smaller* than it should be, and that the price should probably ramp up quickly and steeply before tailing off slowly.[41] To encourage faster decarbonization, some EU countries have raised additional direct taxes on some emissions, and several more EU countries are considering further taxes.[42]

A problem with carbon prices arises when businesses move overseas, where they might not apply: a phenomenon called *carbon leakage*. This is not the end of the world... unlike the absence of a carbon tax. As a countermeasure, governments could tax imports based on their embodied carbon (the European Commission is already proposing a carbon border adjustment mechanism that would start operating in 2026).[43] Secondly, research suggests that, in the long run, it would be in a nation's self-interest to link into a global carbon price scheme.[44] A strong international carbon price would dramatically speed up the energy transition discussed before, since sticking with the status quo would become too expensive. The reliance on negative emissions in the second half of this century means a carbon price in one form or another is essential. Both natural climate solutions such as forestry and technological solutions like direct air capture need to be paid for somehow, and a carbon price would be one way of making them happen – remember that to pay a company to physically draw carbon out of the atmosphere on your behalf is around €950 per tonne as of 2019 (though again, this could come down very quickly and estimates already place costs in the near future at around $100 per tonne). As we heard earlier, the income generated by the tax would have to be redistributed if the public is to be kept onside. One suggestion is to progressively ease income taxes – as we heard earlier, to tax what we burn, not what we earn – which might align with broader equality goals too.

A functioning carbon price would be revolutionary but is still only a short-term step towards long-term sustainability. What of the other environmental crises? Biodiversity loss, microplastics pollution, nutrient pollution (via fertilizers and livestock), air pollution, material depletion and so on? We could raise other taxes to address each one, but we are back to that whack-a-mole reactionary situation, where we are attempting to overcome the sociopolitical resistance to each problem fast enough and with a tax large enough to prevent catastrophic damage (and avoid exceeding safe boundaries on the road to that damage).

So how do we play three-dimensional chess rather than whack-a-mole? In the long run, it's hard to see how the incremental commodification of natural and social values can prevent exceeding planetary boundaries. It's equally difficult to see how new economic ideas like the circular economy (moving from a model of make–use–dispose to circular resource use) can get off the drawing board even with a carbon price. Ultimately, we need to change how we measure value.

There are no straightforward solutions to this challenge, but the *best* suggestion is a widening of the conception of value using multiple measures, which many economists call a *dashboard approach*.[45] The single measurement of GDP would be replaced with a plurality of standardized national and global indicators reflecting different values in order to better represent progress. Countries already keep non-standardized *shadow accounts* for internal policymaking, some international institutions (such as the OECD) use diverse indicators in some reports, and further countries, such as Iceland, are starting to prioritize well-being over GDP.[46] Yet, as ecological economist Rutger Hoekstra suggests, this dashboard has to become both more formalized (so that it can be standardized and calculated within and between countries) and more prominent with decision-makers and the public.[47] Any new dashboard would eventually have to dominate decision-making as GDP does today.

A dashboard might include:

- environmental indicators, like those used in the planetary boundaries we heard about in the previous chapter;
- energy indicators like EROI, or thermodynamic concepts like useful energy;
- social indicators like incarceration rates, social mobility, corruption and equality;
- life evaluation indicators like health, education and time-use surveys (for example, to identify and measure 'bullshit work').

Harder to measure but crucial is the resiliency of networks in nature and society. These sorts of measures might include:

- ecological network indicators – for example, various biodiversity measures or climate-resilience indices;

- economic network indicators – for example, levels of market concentration, patent trolling or risk of financial contagion;
- community resilience indicators – for example, reported loneliness, atomization and social-network density;
- innovation indicators – for example, the density of patent production networks, levels of business and government investment in R&D, measures of communication between innovators.

Some of these indicators may seem controversial; you might think of better ones to include. But notice that the answer isn't to stop measuring. What can't be measured can't be managed, so by both broadening measurements and paying greater attention to them, we could begin to understand and manage them. To facilitate this change, we'll have to embrace diversity in the field of economics, which needs a drastic overhaul and new textbooks, to say the least. Economics will have to be combined with research from ecology, climatology, geology, biology, demography, sociology and political science.[48] This shouldn't be seen as an insurrection in economics, since, as economist Dani Rodrik argues, progress in economics doesn't originate vertically from more maths and models, but horizontally by drawing in knowledge from other disciplines.[49]

This dashboard approach would begin to restructure the economic system, giving us a better sense of progress... or lack thereof. We would still need to make decisions when indicators come into conflict. Ultimately, we would aim to avoid repeating past (and sadly current) mistakes, whereby models attempt to optimize consumption within limits. A bit like approaching the planetary boundaries with the accelerator to the floor, easing off only if we sense danger or enter recession. (Recall that our inability to fully apprehend all the complexities of feedbacks means that this could easily end in disaster.) While the resiliency indicators developed above might help, new ways of making decisions on trade-offs will be important. We can only guess how this might be done. Perhaps, instead of least-cost accounting over the short term, the organizing principle could become least-regrets over the long term. Some really optimistic experts, such as James Lovelock, engineer, scientist and inventor of the Gaia concept, suggest that new computer intelligences might help humanity deal with these quandaries; that they would understand the complexity of the world better than we ever could, or admit its moral compromises.[50] Since both human and machine intelligence would need a habitable planet (silicon-based computers would fail in a dramatically altered climate), we might work together to address environmental crises. In a strange way, this very centralized planning, combined with UBI/UBS, looks more like the communism Marx intended rather than the Soviet system the world saw.

This is highly speculative. If, in the short term at least, the organizing principle is to become least-regrets, it's hard to see how decisions wouldn't be bent towards the interest of the powerful. Decisions might follow the same inbuilt tendencies of today's capitalism to privatize profits while socializing costs – just on a more limited scale. Like so many responses to our problems, we will need virtuous feedbacks among government, society and business to constantly reinforce environmental and social policies. One potential solution would be to improve national democracies along with democratizing markets and corporations: in essence, to spread out power and make society more inclusive to all.[51]

Research suggests that intergenerational problems like environmental degradation can be addressed by *more* – not less – democracy. In some models, a 'resource is almost always destroyed if extraction decisions are made individually... In contrast, when extractions are democratically decided by vote the resource is consistently sustained.' The reason this works is that environmental cooperators can hold back non-cooperators, and it reassures the conditional cooperators (the people who only cooperate if others do).[52] On the basis of this, and the fact that democracies are on the whole preferable to alternatives, we could assume that further voting rights and democratization of production is valuable in limiting resource use and environmental damage.

If we make this (perhaps large) assumption, there is evidence that business would be better off too. A recent review of more than a hundred studies shows that employee-owned companies perform better and reduce employee stress. Other reviews come to similar conclusions.[53] Employee-owned companies raise productivity while outperforming others in terms of innovation, ability to withstand recession (resilience), sickness absence, employee satisfaction and equality.

The roles businesses see for themselves in society are broadening. In 2019, the influential US lobby group, The Business Roundtable, changed their position from 'the purpose of a corporation is to serve shareholders' to the suggestion that they should 'create value for all our stakeholders'. This change appears to be driven by an expansion of what society expects of corporations, the realization that the one metric of profit is not the only estimate of the health of a business (finally!), and an internal desire both to contribute more to society and to address the systemic problems society faces. It is yet to be seen whether this is a genuine attempt to update the role businesses play in society (in the US at least), or if it is empty rhetoric intended to ease the increasing pressure felt from politicians and society worried about the monopolistic tendencies of major corporations.[54]

Even before this statement, some shareholders have been filing more environmental and social resolutions (in revolt of management decisions), and the support for those resolutions is going up.[55] Equally, the reverse tactic of divestment – the pulling of money from environmentally damaging companies – also appears to be working (although it is very hard to conduct research to confirm this). Divestment has potentially raised the cost of loans for fossil fuel developments and could be choking off political and institutional support.[56] The result of these business pressures might be an increase in patient capital: investments that don't discount the future (as much) when compared to share buy-backs and other short-term boosts to shareholder returns.

The ultimate result of these changes might be that the economy as measured by GDP stagnates or shrinks. (With a broader conception of value, society might not even care – growth agnosticism might come to pass.[57]) With population growth starting to stabilize and even decline in some regions, the move to 'steady state' or de-growth models of economy is increasingly plausible. This shouldn't come as a surprise. Many economists have assumed that 'steady-state' economics would be the eventual end point of our economic development. Adam Smith, one of the first economists and the developer of the concept of the 'invisible hand' (which describes how information and goods can move through society in mutually beneficial ways via interactions between responsible individuals who don't know one another, rather than the regular misunderstanding that individualistic greed is good for the economy), wrote in 1776: 'In a country which had acquired that full complement of riches which the nature of its soil and climate [...] allowed it to acquire, [it] could, therefore, advance no further.'[58] If economies slow one way or another, if they become 'smaller than design rather than smaller by disaster',[59] this will facilitate other economic forms – for example circular economics – which have significant community benefits. Think repair cafés, recycling and upcycling projects and larger amounts of interpersonal trade. If we think of today's growing economic systems as waging a war with the biosphere, then these policies, combined with UBI/UBS and 'opting out' of meaningless work, might constitute an ecological ceasefire... or at least a reduction in hostilities.

REDISCOVERING VALUE IN THE LONG TERM

Societies have been arranged in many different modes over the last tens of thousands of years. The differences between foragers, agriculturalists and industrial capitalists are obviously vast. Each mode had very different impacts on the environment and on the humans living in it. Each mode has involved more energy and information use. Do the technological trends of cheap renewable energy, an explosion of machine intelligence augur a new mode of living? We have to hope so. Existing economic and social structures are simply inimical to the long-term stability of a comfortable planet. They also seem incapable of seeing societies through the technological revolution that's under way.

We can't definitively know what economic solutions we need for the future, but to quote John Maynard Keynes again, 'The real difficulty in changing any enterprise lies not in developing new ideas, but in escaping from the old ones.'[60] It might just be that the new revolution can provide the spark to escape these old ideas and transition to something new. As we've heard, the Industrial Revolution eventually eased concerns for physiological needs, allowing a shift from hard physical labour to more fine-skilled motor work. Next, the development of the welfare state and modern institutions provided both social and physical safety. This progress looks a little like progress along the hierarchy of needs proposed by psychologist Abraham Maslow in 1943.[61] In this hierarchy the physiological needs of food and shelter sit at the base, safety needs above that, then love and belonging, esteem, and finally self-actualization (the realization of talent and potential). In recent decades, we have seen a continued expansion in enfranchisement, driven by changes in the way society views women, LGBTQ+ rights, race, ethnicity and other humans living in different countries. This has the effect of broadening love and belonging in the hierarchy. There is a lot more to do, of course, and these developments have been geographically very uneven. Yet at the top of the hierarchy, among the higher global income earners (including those at every level in high-income nations), happiness surveys, workplace surveys and epidemiological data show that self-esteem and self-actualization are stagnant, perhaps even retrogressing.

The next mode for society will likely harness the social intelligence of humans – the creativity and empathy through which we can work towards intrinsic meaning, either by taking empathic work (which would eventually be better paid than other service work) or by choosing to opt out. At the root of this is the easing of time and money scarcity in order to engage in long-term thinking. Time-use surveys of what people say they enjoy most include time with family and friends, time in the outdoors, holidays (hopefully excluding flying minibreaks!), reading, playing music, gardening, building furniture, writing, inventing, cooking for one another, exercise and sex (perhaps driven by an exercise-fuelled libido).[62] Most of these things provide far more meaning in life than 'bullshit jobs'. And do any of these things sound like they need smokestacks and retail therapy?

Stronger social bonds and increased social engagement may result in a greater consensus on sustainable efforts, better democratic participation and increased community resiliency. We start to live as we mean to go on, values change in the meantime, and we have time to appreciate life. We become more productive, not less. We live healthier lives on a healthier planet with lower consumption and less psychological stress. This rediscovery of value would rewrite both the way individuals think, and the societies in which they live. Perhaps this is pure idealism, but this is the hopeful chapter. We know the consequences if we don't transform our economy (and therefore workforce), and we know we can't go somewhere without having even a vague idea of where it is. Our actions and votes have never been more important, as they are the most direct way to have a say in where we're headed.

EPILOGUE

✍ PESSIMISM ✍

Are We Almost at the End?

Had we but world enough, and time...

ANDREW MARVELL, *To His Coy Mistress*, 1681

Collapse, if and when it comes again, will this time be
global... World civilisation will disintegrate as a whole.

JOSEPH TAINTER, *The Collapse of Complex Societies*, 1988

[E]xtreme grief may ultimately vent itself in violence
– but more generally takes the form of apathy.

JOSEPH CONRAD, *Heart of Darkness*, 1902

One day in 1950, nuclear physicist Enrico Fermi was walking to lunch
when he wondered aloud, 'Where are they?' Fermi was puzzled as
to why we hadn't yet heard from extraterrestrial civilizations. How was
it that, with billions of stars in the galaxy, not one species was getting
in touch? It wasn't the first time someone had asked this question, but
it's becoming harder to answer with time.[1] Astronomers think that there
are perhaps 40 billion Earth-like planets orbiting in the Goldilocks zone:
neither too close nor too far away from the host star(s) to sustain life.[2]
Many years later, economist Robin Hanson responded to Fermi's question
by hypothesizing that civilizations might become systematically 'filtered
out'; that all life on any planet would inevitably meet some crucible or
fatal confluence of critical problems before it managed to make contact
with distant life forms.[3]

There may be other explanations of Fermi's paradox, but the climate crisis is clearly such a potential filter. It's big enough, bad enough, and complex enough. Biodiversity collapse and nuclear war could be others. There could be as yet unknown proto-filters waiting on the sidelines; for example, some worry deeply about the impact of general artificial intelligence (an AI with human levels of sentience, creativity and adaptability).[4]

Filtering, like the collapse of ancient civilizations, doesn't need to be driven by a single crisis alone. We can cast climate change as a catalyst for other filters, or, as the US military calls it, a *threat multiplier*.[5] Environmental pressures and resource constraints could escalate into conventional or thermonuclear war. Even if they don't, it could be death by a million cuts: increasing water shortages,[6] crop failures, extreme weather events and ongoing social decay. Filtration is not the same as human extinction – not for a long time at least. It is simply the collapse of technologies sufficient to be detected by other life outside our solar system.

Ecological research shows that biodiversity collapse is generally fast, and recovery is slow. The same seems to be true of anthropological studies of past civilizations and some researchers suggest that the speed of civilizational collapse is increasing over time – potentially in line with the increasing complexity of civilizations.[7] It's easy to see how any potential collapse today would be extremely fast, with its integrated supply chains and internationalized food production. Humans wouldn't go back to a mode of pre-industrial living, but the reduction in complexity that would result from a collapse would irrevocably change the way we live, and might deprive people of many of the resources essential to live in the first place.

We have almost all the answers needed to save ecosystems, both technological and social. Yet humanity has barely implemented the first steps needed to avoid just the one calamity of climate change. Taken completely, perhaps this 'Great Filter' isn't just a failure to adapt to one, or even several problems, but rather a process whereby the strengths that let species succeed become weaknesses once they have expanded too far and consumed too much. It is horrifying that we are talking of these concepts as seriously applying to humanity within the next few decades, but this is unquestionably the case. The apocryphal curse 'may you live in interesting times' has never been so appropriate.

ATTENTION AND AMNESIA

Extreme sports instructors say that bungee jumping is much scarier than skydiving. Despite being far higher in the air and detached from any anchor point, the skydiving human brain simply can't process the concept of being so high up. This might explain why we're not more scared now. Our brains have never felt the fear of ecological bungee jumping... because we've gone straight for the skydive. We are now in free fall, hurtling towards the ground. Only now we're feeling around for a non-existent parachute ripcord (though models reassuringly suggest they will be commercialized and rolled out on a planetary scale by the time we're a few feet from the ground).

You would imagine that we'd be starting to panic, but environmental crises still barely make the evening news. In 2020, across four major US TV networks, there was a total of 4.5 hours of news programming on climate change.[8] The greatest threats humanity has ever faced apparently deserve less than a minute a day on the evening news. Although the US media are particularly guilty of climate silence, other countries don't do much better. The word 'cake' was mentioned ten times more frequently than 'climate change' on UK TV in 2020 and 'banana bread' mentioned more than 'wind power' and 'solar power' combined.[9] Media across the world still struggle to connect extreme weather events with climate change to explain the ecological issues we face, or to describe the findings of scientific research in representative ways. News articles on heatwaves are often accompanied by happy-go-lucky images of ice creams, beaches and children playing in water fountains, rather than crops wilting, rivers parching and the newly vacant homes of the elderly, where many heatwave deaths occur.[10] These stories rarely mention climate change. Some research suggests that stories are many times more memorable than facts alone, but how can the story or the facts of climate change be memorable if they're not being told?[11]

The media's silence was neatly exposed in a thought experiment devised by the former editor of *Fortune* magazine. Suppose a unanimous body of the world's foremost astrophysicists had discovered that a massive meteor was going to hit Earth in ten years' time: 'The media would throw teams of reporters at it... After all, the race to stop the meteor would be the story of the century.'[12] While elegant, this analogy is an understatement. Climate change is the story of the next few thousand years. It also gives no agency to individuals or communities, both of which can act to limit emissions and adapt to climate change. But only if everyone is aware of the issues we face. By the time environmental crises occupy news cycles, it may be too late to save much of what we value.

Even if we were paying attention, researchers have found that the baseline for what is 'normal' in society changes so quickly that we forget how abnormal the present really is. One of the first documented examples of this was among fishery scientists. When asked to identify the size of the fish population before depletion, scientists would place the baseline fish population close to where it had been at the start of their careers, instead of the size it had been before human activity and fishing. This way, each generation of scientists estimates a lower baseline than the previous, and the expectations of the population are lowered, generation by generation. Scientists call this *shifting baseline syndrome*.[13] And this failure of reasoning is among *scientists*, who might be expected to calibrate their estimates correctly. Analyses have shown that baselines shift faster among the public. A research paper based on more than 2 billion social media messages found that the typical time horizon for what is considered 'normal weather' is between two and eight years.[14] This means that the horrifying climate-driven events of 2018 could appear to be a 'normal' year by 2020. If you were born after 1976, you have never experienced Holocene-like ocean temperatures. If you were born after 1985, you have never experienced a month in which the global average surface temperature was below the Holocene average for that same month.[15] It's not that we don't appreciate what we've got till it's gone; it's that we forget what we had in the first place. The same dynamics contributed to the disastrous anti-vaxxer movement: fuelled by misinformation, anti-vaxxers have seemingly overlooked the horror of measles until children have become ill or died, preferring to avoid vaccination, with its imagined risks. We have seen this repeated with Covid-19. It's only when people see these diseases in reality that they realize the gravity of the mistake. If people have no conception of how abnormal even recent weather events have been, how can we possibly expect society to apply the necessary political pressure to transform the way we live? We are the proverbial frogs in the pot of boiling water.

To the satisfaction of well-resourced disinformation campaigns run by fossil fuel interests, environmental action has become increasingly connected to identity across many countries – partly as a result of the lobbying and advertising from dark-money interests. This has facilitated a division between different groups and makes it easy for people to engage in *motivated reasoning*, a process by which we seek out information that conforms to pre-existing beliefs and filter information through a social, sometimes tribal lens.[16] You might think that education could be the key here; that if people learn more they will understand the evidence. Worryingly, there is some research showing that scientific literacy makes the matter worse; that as scientific literacy improves, it can make people better at motivated reasoning.[17] Imagine! Due to your scientific literacy, you are presented with more technical information, so you are increasingly likely to bend over backwards to make this consistent with your pre-existing world view. Max Planck once suggested that '[a] new scientific truth does not triumph by convincing its opponents and making them see the light, but rather because its opponents eventually die, and a new generation grows up that is familiar with it.'[18] This could be true more broadly, but what it doesn't factor in to the problems we face today is that we have no time to wait for generations to die. If we wait for that, new generations may not have the opportunity to live.

Our failures in reasoning are compounded by social media. As many as 68% of people in the US get some news from social media websites like Twitter, Facebook, Reddit and others.[19] In the UK, the most popular news source among 35% of the public is Facebook.[20] Most people now know that social media acts as an echo chamber and many expect the information shared to be inaccurate (57% of people in the US, according to polls[21]). Unfortunately, repetition (including of lies) is worryingly effective, and social media has a tendency to lure people towards more extreme, outlandish, clickbait content in order to accrue advertising revenue.[22] As research has shown, most of the most highly viewed YouTube videos discussing climate change *oppose* the scientific consensus.[23] It's not that these companies have this intention by default; it's how their algorithms optimize for attention. Since YouTube (owned by Alphabet, Google's parent company) and other social-media outlets have special exemption from being legally held responsible for the content disseminated on their platforms, there is little redress in sight. In 2019, while Google claimed to have stopped supporting climate-denial lobby groups several years prior, reporters found it was still making large contributions – because these are the same libertarian lobby groups defending Google's content exemptions.[24]

As the old media fails, so too is new media failing, on a calamitous scale. Mirroring the failure of environmental systems as a result of the economy not sufficiently valuing nature, Facebook, Google, Amazon and others are allowed to operate with scant regulatory oversight. They have availed of the implicit green light to 'move fast and break things': from facilitating the breaking of electoral-campaign finance laws to disseminating unfiltered, hate-inciting content (including Holocaust and climate denial). Now that they're so large, they are 'locked-in'. Regulators in the US and UK are investigating their actions, but whatever (if any) regulation may be introduced, it will likely be too little, far too late.[25]

THE NEXT PROGRESS TRAP: GEOENGINEERING?

When complex systems signal the need for change, any delay can be catastrophic. It's been a long delay indeed. As baselines shifted, the media looked the other way, governments argued, individuals struggled to grasp the sheer scale of environmental problems, and powerful interests blocked change. What was once a set of difficult but manageable problems has now become seemingly insurmountable. The whole edifice of today's complex and globalized society has been built on cheap energy and the externalization of environmental costs. This has allowed humans to get *locked-in* to a lifestyle that is reliant on many polluting networks to live – intricate supply lines, roads through suburban sprawl, air-conditioned megalopolises in the desert – networks that are heavily exposed to climate disruption. It's now a race between both social adaptation and technical innovations, and the globalized by-product of carbon emissions. It is the largest progress trap humans have ever faced. It is for this reason that new progress traps have been necessary to avoid previous traps, requiring even faster innovation. Indeed, accelerating waves of innovation could be a necessary feature of growing civilizations.[26]

Perhaps the next progress trap, in an attempt to manage the previous one, is the deliberate anthropogenic management of the global climate system itself: global *geoengineering*. It might offer a glimmer of hope to think that this is just another 'amazing' innovation which will give people the time to fix previous traps. The concept of geoengineering is evocative and a little confusing, because humans have been inadvertently engineering the climate system for some time.[27] Perhaps it's now our responsibility to take on this role actively, given we've cut into natural ecosystems so deeply? Should we acknowledge our complicity by engaging technologies however we can?

Of the several geoengineering technologies on the table, the two most discussed are Negative Emissions Technologies (NETs), which we talked about in Chapter Eight, and the bureaucratic-sounding Solar Radiation Management (SRM),[28] which would involve injecting a screen of aerosols – the current favourite being sulphur particulates – into the upper atmosphere to dim the sunlight. (They would be spread as evenly as possible, to limit chaotic regional weather fluctuations.[29]) The planet already experiences cooling from anthropogenic aerosols. Air pollution from fossil fuel infrastructure along with metal smelters already emits around 97 million tonnes of sulphur dioxide each year – a similar order of magnitude to the tens of million tonnes that would need to be injected into the upper atmosphere to dim the sun proactively (aerosols are more effective up high). Global temperatures would have been roughly 0.2

degrees Celsius higher between 1951 and 2010 without these current emissions – a substantial amount, considering the vast difference in impacts between 1.5 and two degrees Celsius.[30] The existing low-level screen is already starting to disappear, though, as countries do a better job of controlling pollutants, along with the expansion of sulphur-free renewables. Research finds that it is still preferable to get rid of these emissions, since the reduction in carbon from phasing out this infrastructure is larger than the temperature reduction from the shading.[31] Still, by injecting aerosols up high, where they can be most effective, some suggest that SRM might buy humanity time to sort itself out underneath the shade. Similar to how the 'Green Revolution' and industrial fertilizers fed a growing population (with huge environmental consequences), the pitch is to buy ourselves time until technology improves enough to fix all the problems we face.

It's not so simple. Like any progress trap, SRM would further damage the environment. Sulphur aerosols would increase depletion of the ozone layer and wouldn't prevent ocean acidification via continuing increases in carbon-dioxide levels. (On its own, acidification can cause mass extinction.)[32] While SRM would cool the planet relative to no SRM, it wouldn't wind back the climate clock to 'safe' pre-industrial conditions. Again, once you put a system under too much pressure, it can flip into a different state, and it is incredibly difficult, if not impossible, to flip it back. Imagine SRM was used to recover Arctic ice – it would have to dim the sunlight enough to 'overcorrect' the temperature – cooling the poles far below pre-industrial temperatures to encourage ice to reform.[33] Researchers don't know how much cooling would be needed, but it's likely that such extreme polar cooling would wreak havoc on weather systems, with massive consequences for agriculture and food supplies. That is a devastatingly steep price to pay for a 'quick' fix.

Instead, say SRM was used to take 'the edge off' climate change as a stopgap measure to stall tipping points. We'd have to know exactly when to use SRM and by exactly how much. But that presents another problem. Anything pre-emptive would have to be used *before* things get too far out of control, but how can we determine when the climate is 'out of control'? We slip into dynamics of signal and delay. By the time humanity realizes the urgency to act in unison, it could be too late to prevent long-term climate changes from unfolding, mirroring the dynamics in trying to decarbonize the economy today.

Even if SRM were to work flawlessly, who gets to decide what level of SRM is used, and when it's politically and socially palatable? Would richer nations take the lead? These are the same nations which, having caused most of the emissions, stand to lose less on average than poorer nations. This international decision is fraught. Whereas the extreme events of climate change appear as unwanted side effects of 'development', there is – at present – little suggestion that any country is actively harming another in the way that trade wars are explicitly harmful. For example, after the climate-change-exacerbated Indian dust storms that killed 143 in 2018, India didn't declare war on the US and EU for killing their citizens through the callous emission of two-thirds of all historic anthropogenic carbon.[34] Yet the moment SRM is used, extreme weather events become identifiable with an active decision by governments. It could easily be used as an excuse to ramp up international tensions. Indeed, many researchers find it hard to see how SRM could be implemented in such a way as to avoid large-scale war.[35]

Even discussing SRM constitutes a moral hazard.[36] By presenting it as an option, it may reduce the drive for mitigation and adaptation today. We've already seen how many models discount the future, suggesting to decision makers that we can rely on yet-to-be-developed tech solutions. The SRM trap requires us to have already sorted out the problems that have created previous traps. Some researchers suggest that SRM is essentially impossible to govern.[37] Nonetheless, despite all the downsides and precisely *because* it's a moral hazard, it's possible we'll see SRM this century. We could even see several geoengineering attempts at once – perhaps SRM and large-scale Direct Air Capture (DAC) – all the while powerful polluters continue to emit carbon.

In one sense, if SRM is used, it would be a constant reminder of the burden of past actions, because the high-altitude particulates would scatter light, turning blue skies slightly whiter, casting the world in a different hue.[38] We will simply look up to be reminded of the overwhelming impact of billions of individual actions and ancestors. In practice, though, this is no technological wizardry to rely on. As with other progress traps, when there is light at the end of the tunnel, that light is usually another train coming directly at you.

COVID-19 AS PRACTICE IN COLLECTIVE ACTION

Early in 2019, there was a hope that the Covid-19 pandemic would bring about a fundamental shift in humanity's trajectory. There was hope that the pandemic would necessitate and facilitate an important development in collective international action – the very action needed to address our climate and biodiversity crises. The similarities were pointed out in the earliest weeks of the pandemic, with a viral cartoon of three increasingly large waves, named 'Covid-19', 'Climate Change' and 'Biodiversity Crisis', which were about to crash on a small town, whose inhabitant commented: 'Be sure to wash your hands and all will be well.'[39]

There were early signs of hope on community and national levels. There was a flurry of reporting about cleaner and bluer skies, of purer water and lower emissions.[40] There was an appreciation that the most important things in life, such as spending time with family and friends, should never be taken for granted. On the national level, organizations and policymakers called for huge investments to stabilize economies in a 'green recovery' and to 'Build Back Better'.[41] It was argued that these policies would help to accelerate the energy transition and address ine-qualities. (Interestingly, the need for a Great Food Transition – in diets, food waste and food production – was largely overlooked, even though slaughterhouses were suffering from record infection rates and pandemics themselves often emerge from animal agriculture.) Rich countries would, apparently, provide aid to help poorer nations deal with the pandemic and procure vaccines.[42]

In reality, emissions reductions were fleeting, the 'green recovery' was paltry, and international cooperation was disastrous. Lockdown-related travel restrictions and lower consumption in 2020 helped emissions drop by 6.3%, but this rebounded to close to 2019's historic record by 2021.[43] So we cannot cite emissions reductions as being a positive outcome of the pandemic. Perhaps this is unfair; like an oil tanker, the global system driving carbon emissions simply cannot change direction overnight, which is to say within two or three years. But it should be shocking that promises of a 'green recovery', which might have meant a course correction, have largely evaporated.

The Global Recovery Observatory, based at the University of Oxford, found that government spending to prevent economic recession, while 'unprecedented and extraordinary', was 'inconsistent with the world's low carbon transition'. While total spending on cash transfers, investments and other Covid-19 relief reached $19.8 trillion between March 2020 and May 2021, overall spending in most countries encouraged more fossil fuel use rather than shifts to low-carbon energy (75% of this spending was in high-income nations). Specific investments, rather than cash transfers or tax breaks, made up around 15% of this total spending, and saw more spending on fossil fuels (20%) than low-carbon energy (19%). These estimates are similar to the International Energy Agency's assessment that only 3% of overall economic recovery spending was directed into green energy.[44]

The sense of international solidarity and collective action also evaporated quickly. After early announcements that rich nations would help to procure 1.8 billion vaccine doses for poorer nations, only 14% of that target had been delivered by October 2021. Only 12% of the doses promised have come from Western pharmaceutical companies, while several countries and institutions (including the Bill and Melinda Gates Foundation) initially refused to waive vaccine patents.[45] The international response is best exemplified by the International Monetary Fund's making $650 billion available to nations in pandemic recovery, with only $33.6 billion going to Africa, as the fund was divvied up in proportion to economic size.[46] Addressing climate change requires deep international cooperation and trust, and the addressing of historic inequities. Covid-19 is just one more grievance to add to the tally of international injustices.

The pandemic, while devastating, offered a real chance to spur the clean energy transition and to help address national and global inequalities. It's now clear that we blew that chance.[47] The new spending on fossil fuel infrastructure represents a further 'lock-in', whereby countries and companies will want to operate these assets for as long as possible. The huge sums of money spent by governments are likely to be a one-off, and we probably won't see the same amount of spending in the future on environmental aims. Meanwhile, while high-income nations refused to acknowledge their failure in meeting funding pledges at COP26, poorer nations were once more reminded (as if they needed any reminder) that we are not in it together.

THESE GLOBAL SYSTEMS

One way or another, humanity will be undergoing many of the largest transitions it has ever experienced on a scale and depth no human has ever seen before. Either environmental crises will unfold faster and faster, eventually outrunning the ability of many societies to adapt, or we are to make dramatic social and technical changes to how we live that will have significant results in the next few years. Our problems are so urgent and severe that it *must be* these societies, these institutions, these humans that deal with them. Time's up. It's 'us' or no one.

These societies. Unfortunately, it's the same societies in thrall to endless, poorly measured growth that have to do the job. Societies that have followed an economic model which overtly focuses on under-regulated markets, placing no value on nature and, by extension, the future. This lock is so tight because every level of society is engaged in this dynamic. Overturning it would require unprecedented national and international cooperation, along with social buy-in. It might require ideas and cooperation that we are currently unable to even consider, our imaginations limited by the overarching focus on money and quotidian concerns. This could be why 'it's easier to imagine an end to the world than an end to capitalism'.[48]

The same society in which powerful communities of people that may lose from the necessary changes have resisted any efforts to make a transition, in which insidious existing violent, extreme philosophies metastasize with the growing climate crisis. Philosophies like eco-fascism, a brand of thinking that identifies authoritarian, fascist governments as best placed to protect the planet and their indigenous people, the '[if only] Africans would stop cutting down trees and would stop making babies when it gets dark'[49] type of thinking, thinking which is always as dishonest as it is racist. Eco-fascists have already been responsible for two shootings – one in El Paso, Texas, and another in Christchurch, New Zealand – and their ideology appears to be becoming more popular.[50]

These institutions. The same institutions that have to fix this are the institutions that have failed to manage many social and technological transitions over the years. Institutions that have allowed for a fantastic level of wealth accumulation – the wealthiest 500 people gained $1.2 trillion in 2019, increasing their net worth 25% to $5.9 trillion (remember that the necessary global climate action would cost in the order of $1 trillion a year).[51] In 2020, it is estimated billionares became a further 20% richer.[52] The same institutions that starve communities of the funds they need to build resiliency for the future, that explicitly protect the neoliberal social dynamics that allow the rich to 'go it alone'. Combined with increasing tear-gassing of environmental activists,[53] physical abuse and increased rates of violence and murder (depending on the country), it's easy to see how disillusionment in and fear of national and global institutions make cultures ripe for more demagoguery. On an international level, there are very few ways for institutions to punish or incentivize environmental free-riders in a globalized, interconnected production and consumption system dominated by large corporate interests. The international response to massive deforestation of the Brazilian Amazon – deforestation that heralds its potentially irreversible transformation from rainforest to savannah – is ineffectual and half-hearted.[54] Similarly, the international pressure to halt the destruction of Canadian boreal forests for tar sands is non-existent. We are in a perfect storm, with the highest winds and largest waves ever recorded, the ship is leaking and the metaphorical captain is drunk.

These humans. The same humans who have to fix this are those who think that the 'middle path' is the only politically feasible avenue for change; they don't realize that incrementalism is catastrophism. The same humans who appear to be living in a pre-traumatic state, that 'dull cataleptic trance',[55] who are encouraged to discount the future and, pressed for time either by meaningless jobs or just trying to get by, can't consider alternatives or plan for the long term. (Propagandist Edward L. Bernays, adept at 'engineering the consent' of the public by any means possible, once said that the great enemy of any attempt to change human habits is inertia: 'Civilization is limited by inertia.'[56]) The same humans who suffer from a cognitive bias where losers will always fight harder than winners. In fact, humans suffer from so many cognitive biases that Wikipedia lists more than 150 of them.[57] One of these biases is optimism bias – the irrational belief that 'it'll be fine'. Optimism bias has been shown in stock traders who think they are less exposed to losses in the market, and in smokers who believe that their chances of lung cancer won't change with smoking.[58] Wishful thinking might be why so many grim books end with an optimistic chapter – which, by design, this book does. It's nicer to imagine that things can unfold well; that our children will outsmart us and won't resent us. Perhaps what follows in the next chapter is, itself, simply placation.

The Grass Is Greener

Eternity is not something that begins after you are dead.
It is going on all the time. We are in it now.

CHARLOTTE PERKINS GILMAN, *The Forerunner*, 1909

Just because we cannot see clearly the end of the road that
is no reason for not setting out on the essential journey. On
the contrary, great change dominates the world, and unless
we move with the change we will become its victims.

ROBERT F. KENNEDY, Warsaw, Poland, 1964

Finding acceptance may be just having more good days than
bad ones.... We can never replace what has been lost, but we
can make new connections, new meaningful relationships, new
interdependencies. Instead of denying our feelings, we listen
to our needs; we move, we change, we grow, we evolve.

ELISABETH KÜBLER-ROSS AND DAVID
KESSLER, *On Grief and Grieving*, 2005

Every shift in the way humans have organized themselves has presaged tumultuous environmental changes. Hunter-gatherers drove
megafauna extinction. Agriculturalists stabilized the global climate by
clearing forests, increasing atmospheric CO_2 from glacial to interglacial
levels. Mercantilism homogenized species by transporting animals and
plants across the planet. Industrial capitalism pressed pause on the next
glacial cycle completely. And the current hypercommercial capitalism
is now breaching multiple planetary boundaries at once.[1]

These shifts were made possible by technological and social innovations, facilitated by increasing information and energy flows through society. These innovations bred further innovations. For example, the more efficient and powerful Newcomen steam engine (1712) was able to pump water from coal mines for the first time, opening up vast seams of energy that became cheaper as its use increased (we heard of this sometimes counterintuitive phenomenon in Chapter Nine with Jevons's paradox). New energy supplies, increasing amounts of scientific information, and new institutions for protecting patents and property encouraged further innovation.[2] Feedbacks like these push societies into new stable states at higher organizational complexities.

During each phase of this planetary influence, the prevailing philosophy has been that the solution to pollution is dilution; that, given infinite space into which to expand and dilute its wastes, societies could continue to increase the amount of energy, resources and information flowing through them. Eventually, perhaps the only limits would be the laws of thermodynamics. Believing in unbounded human creativity, today's eco-modernists assert a *techno-cornucopianist* view of the future: that modern planetary environmental limits are categorically no different to the more localized environmental limits of the past. They suggest that, even in the worst case, we can manage with artificial ecosystems or building colonies on Mars. However, there *is* a category difference in these limits: represented by a vast chasm of knowledge between understanding ecosystem processes and artificially replicating entire ecosystems technologically. (Every attempt to do this has failed.)[3] Equally, there's a horrendous penalty for doing anything in space, both in the energy needed to accelerate away from Earth and in its intrinsic inhabitability. Personally, I think the idea of visiting Mars is wonderful, but to think that Mars could be a replacement for Earth is not only immoral (given the implications for people that billionaires would leave behind), but thermodynamically illiterate. Even in the worst case, escaping environmental collapse by way of Antarctic bunkers would be far more habitable and thermodynamically preferable than Martian survival.[4]

Given enough time, these options *may* become technologically possible – even though they may never be *ethically* acceptable. But the environmental issues we face are escalating at such scale and speed that technology is not enough. It's simply not possible to continue passively in the systems that many societies are accustomed to. If we want to avoid a catastrophic collapse, huge structural changes are unavoidable – requiring the overcoming of massive societal inertia. The extent to which we can overcome this inertia may determine what level of collapse humanity might undergo: a scale that runs from a modest reduction in complexity to a cascading failure of many of the systems that provide food, energy, water and other services. It is essential to remember that this outcome is a scale rather than a binary future of utopia or cataclysm. The efforts made now, even if insufficient to avoid tipping points, will slow their unfolding. Even in the very unlikely case of least hope – that we have already committed a fundamental transformation of the Earth's climate system – we still have some control over the accelerator (even if we've lost control of the brake). That is, the transformation can unfold over centuries or millennia, depending on our actions. Slowing this process will give breathing space to develop new institutions, new philosophies and new adaptations. In any eventuality you care to imagine, there is no excuse for inaction. Rapid, deep and sustained action is the best chance humanity has to transition to a species with long-term global potential.

THE GREAT TRANSITION

We have seen the desperate need for a global energy transition, a global food transition and a global economic transition. As problems like climate change are so all-encompassing, they are *everything problems*. The response is to offer system solutions. This would require globally rewiring the connections between societies, humans and the environment, along with a transformation of how we value human experience. In many ways, our current system seems too big and too hard to change. This is partly by design, since the structures of power through society are more likely to persist if it all seems too much and individuals passively accept the way the world looks. Many necessary solutions can seem flatly impossible: can we really embrace shorter working weeks when we would be threatened by the economies of other countries eating our lunch? How can we rein in inequality, corruption, environmental damage and social harm when we live in a globalized capitalism in which it is easy to move such harm out of sight (into different jurisdictions) and out of mind? How is it possible that any of these energy, food or economic transitions – each one utterly transformative on its own – can happen fast enough to avoid transgressing boundaries? If these difficulties are going to make us give up, then we should start getting the escape plan ready, along with developing the psychopathies that would be needed to live out a lonely, reprehensively misanthropic end-of-days, watching the suffering from afar (as some seem willing to do[5]). Giving up on system reform puts the survival of the current Earth system and the lives of billions in the 'too-hard' basket. This is clearly not an option.

We don't have to have a perfect image of the future right now. And no new mode of living will emerge fully formed. Industrial capitalism could barely be intuited by those living through the Industrial Revolution. Philosophers suggest that we can only begin to understand the present system fully at the very moment it fades out of view. As Hegel put it: 'the owl of Minerva [wisdom] spreads its wings only with the falling of the dusk.'[6] Optimistically, it seems as if we are starting to grasp quite how destructive the current system is; we might be already moving towards something new. But if it's hard to even understand the present, then it's nigh on impossible to build projections of the future (as evidenced by the attempts of Integrated Assessment Models demonstrate). What we have seen in this book is a selection of the various changes we will *probably* need if we are to avoid short-term catastrophe. These include:

- The continual enfranchisement of women globally; aspirations for higher standards of education and healthcare around the world as a priority; well-managed migration in society; investment in people.
- A deep and sustained shift to electricity as the main energy carrier, generated by as much renewable energy as possible; sophisticated electricity grids with more long- and short-term storage and transmission to help smooth out the variations in weather; implementation of energy efficiencies aided by machine learning and smarter grids.

- Three large changes in food systems: shifting diets, reducing waste, and improved food-production technologies that are environmentally regenerative rather than exploitative. The ultimate aim of all three would be to decrease human impacts and the amount of land needed *while* continuing to improve yields.

- The implementation of regulation and policy that *we know* would help reduce emissions but can be politically 'difficult': frequent-flyer levies, banning petrol vehicles in urban areas, large-scale building retrofit schemes, and much more. We know that we have to bring carbon levels in the atmosphere down in order to maintain a comfortable climate; we will need to implement as many natural climate solutions as possible, including protecting forests and expanding other forested areas, using agriculture to store carbon, protecting wetlands and peatlands, and potentially storing more carbon in the soils through techniques like biochar. We also know that even this is not enough. Fortunately, direct air capture of carbon is looking increasingly plausible as an aid, but it will be needed on a massive scale *along with* everything else mentioned. It may be particularly helpful for activities that are difficult to decarbonize, like remaining flights (after the number of flights have been slashed) or some industrial processes (although given our current inertia, it probably will be relied on for much more).

- To facilitate all this, a global carbon price will be needed, as will new ways of balancing inequality in an age of increasing automation and machine learning. A universal basic income and universal basic services (UBI/UBS) will be needed. Ultimately, the poor economic measure of GDP must be replaced, and the economy must develop a broader conception of value – to measure the importance of the environment and society as intrinsic values, rather than converted into commodified monetary amounts.

Some of these changes would set the scene for beneficial feedbacks which would cascade through social and natural systems, providing a further push towards a new, sustainable stable state. For instance, a universal basic income or universal basic services might give many people the freedom from extreme scarcity to plan for the future, improving equality and revising the meaning ascribed to work and the environmental impacts that arise from consumption. Although many of the Covid-19 supplements were subsequently cut, they provided a vision of how direct cash transfers can provide a path out of poverty and helped reduce prejudice around welfare support.[7] It's important to remember that many of these solutions make the world a better place to live in, in ways we are only just beginning to understand. For instance, there is evidence that cleaner air from the decarbonization transition would initiate a feedback in which human cognition could improve.[8] Other feedbacks would be seen through policies like a carbon tax, which would encourage more technological and social innovation for low-carbon activities, opening up more opportunities for people to engage in and enjoy a low-impact lifestyle.

Many of these changes will herald the development of new world views. Indeed, there will have to be big changes in personal philosophies to enable sufficient social licence for these great transitions to continue as deeply as they need to go. Societies have laboured under misapprehensions of fundamental human behaviour for too long – be it the *rational actor* assumption which supposes that humans can make perfectly optimized decisions, or the idea that humans prefer high levels of discounting used to shrug off the future. We will need new ways of thinking in broad systems, and a shift away from the overwhelming focus on flows of money to flows of energy, matter, information and environmental harm.

THE GREAT REALIZATION

Had this book been written in the mid 2010s, there would have been much less reason for hope. By the early 2020s, Extinction Rebellion and Greta Thunberg, among other activists and organizations, had become household names. One year later, the world was paying attention. That these are Western examples makes them no less important, since these movements are increasingly global. Global climate marches in September 2019 saw between six and eight million people out on the streets in 140 countries and 4,500 different locations. Historically, as we have heard, once a movement has achieved the active and sustained participation of very roughly 3.5% of the population (equivalent to around 237 million globally), it has never failed to achieve change – and some movements have succeeded with far fewer.

The window of what's possible has been thrown wide open, and that window did not open of its own accord.[9] Previously radical or unthinkable ideas now get mainstream traction for the first time. Polls suggest that most people in the UK support limits on flying,[10] plant-based diets are exploding, and many people worldwide say they are willing to sacrifice economic growth for environmental sustainability. In fact, the marketplace of ideas has expanded so much that some are openly advocating policies that would have a dramatic impact on how they make a living. In short, people are starting to display self-sacrifice in public. Sports journalists have suggested that Formula One is 'a very expensive luxury at a time of climate crisis'.[11] Politicians, civil servants and business leaders have written openly that economic growth must end.[12] Travel journalists have rejected flying, and are taking trains instead.[13] These discussions will continue to shift and expand in scope with increasing atmospheric concentrations of carbon and species losses.

It's safe to say that social tipping points are very close. We can expect more action across society, both as environmental impacts intensify and as demonstrators demand more. We have seen how the dichotomy between individual action and system change is false – it is used to delay action. The broad societal changes may even begin to engage cynics who think nothing can ever change and that people are too self-interested to 'save the planet'. For example, research carried out before and after the 2017 Washington DC climate marches asked consumers of conservative media (which represented the marches negatively) whether they agreed with the statement that 'people are too selfish to cooperate and to fix problems'. The numbers agreeing with this statement dropped from 60% before the march to 45% afterwards.[14]

We are beginning to see these social shifts boil over into politics. Green parties have seen unprecedented gains in the EU, Switzerland and New Zealand. Austrian voters have turned away from the right wing, with Greens winning 14% of the vote in 2019.[15] The salience of environmental issues in the US has increased dramatically, with a democratic primary campaign that saw a seven-hour climate town hall discussion – just one TV show saw double the climate coverage of four entire news networks in all of 2018.[16] There is an increasing broiling anxiety among many people that something is not right. The philosopher Glenn Albrecht suggests that we should already be thinking of this mental and existential stress as a distinct form of mental harm, which he calls *solastalgia*.[17] There is evidence that poor people experience this distress much more than the rich, who have the resources to adapt to such damage.[18] Leading medical journal *The Lancet* has included solastalgia as one of the many impacts of climate change on human health and well-being, and several countries have declared a mental-health crisis as ecological and climate impacts become increasingly damaging and noticeable.[19] Surveys show that 85% of people are worried about warming in the UK, with similar levels of concern in other countries.[20]

One way to deal with these mental-health developments is to draw from theories of grief and accept that nothing can be the same again, but that finding new ways of living that are socially and environmentally enriching is both possible and necessary. Others suggest that approaches like Alcoholics Anonymous would be effective in dealing with the quick changes needed in mindsets and behaviours. What these ideas have in common is an acceptance of the deep trouble humanity is facing, an awareness that our way of life will fundamentally change, and 'a reclaiming of self-mastery to higher ends'[21] (in this case including the involvement in community movements and political action). Environmental Studies scholar David Orr suggests that the Alcoholics Anonymous framework in particular is powerful because substance addiction is similar to consumption addiction, in that overconsumption ultimately has destructive effects on 'our places, ourselves and our children'.[22] All of this grief, rehabilitation and acceptance of change will ultimately have to include a greater acceptance that many of the environmental impacts caused by the industrialized world are now firmly on the shoulders of the poorest and youngest in the world.

Things will not just be 'okay'. But even in the grip of a potential collapse in the coming decades, all the changes mentioned will brighten this admittedly bleak picture. They will all help societies and individuals to adapt to an increasingly damaging climate. Just as the Roman empire gave way to the Byzantine, and the Mayans devolved into smaller city states, humanity will continue – even under collapse. Some think that collapse will be total, a near resetting of human existence to a pre-industrial level of living.[23] This is possible but not quite plausible. We could perhaps see some regression, but most scientists don't expect the tape to rewind to the start. There will be no complete relief from the pressures many people are already experiencing and which more are yet to experience, but nihilism is just as objectionable as blind optimism.

George Eliot said that what we call our despair is often only 'the painful eagerness of unfed hope'. This book's structure aims to clarify how much work is involved in feeding this hope. A hope that lies in acting, and seeing other people act. A hope for a cleaner, healthier, more equitable and fulfilling world. A hope that needs all our engagement, our energy and our will.

NOTES

Prologue: 'Do you think we're going to be okay?'

1   This certainly needs clarifying. Who are 'we' talking about when we say 'we'? It is difficult not to refer to 'we' as a human whole, yet 'we' are not a homogeneous group. Is 'we' the same person living in a New York town house with a car, two children and a place in the Hamptons? Or is the 'we' a Guatemalan immigrant living in Brooklyn trying to feed a family and coping with soaring heat in a non-air-conditioned apartment? While 'we' can never be a homogeneous group, it's very hard to avoid using it in a larger sense referring to all humans. I will be as specific as possible, but when I do use 'we' in the book, it is in this sense of a human whole.

2   T.S. Eliot, *Four Quartets*, Harcourt: New York, 1943, p. 14.

3   Equally, any book about the future needs to be a book about our environment. So many techno-optimistic books gush with exciting new possibilities for humanity, from automation to abundant energy, without facing up to the possibility that they all rely on an organized society which may not survive environmental breakdown.

4   Many different examples could be chosen, but for instance see changes in land: K.K. Goldewijk et al., 'Anthropogenic Land Use Estimates for the Holocene – HYDE 3.2'. *Earth System Science Data*, 9/2 (2017), 927–53. doi: 10.5194/essd-9-927-2017; for oceans: K. Mostofa et al., 'Ocean Acidification and Potential Impacts on Marine Ecosystems', *Biogeosciences*, 13/6 (2016), 1767–86, doi: 10.5194/bg-13-1767-2016; and for air: D. Archer and V. Brovkin, 'The Millennial Atmospheric Lifetime of Anthropogenic CO_2', *Climatic Change*, 90/3 (2008), 283–97. doi: 10.1007/s10584-008-9413-1. The latter also describes the impacts over thousands of years. A nice popular overview of what the disappearing of humanity might look like can be found in the book by Alan Weisman, *The World Without Us*, Thomas Dunne Books (St Martin's Press): New York, 2007.

5   For the figures on fresh water see C.J. Vörösmarty, C. Lévêque and C. Revenga, 'Chapter 7: Fresh Water', in R. Bos et al., eds, *Ecosystems and Human Well-Being: Current State and Trends: Findings of the Condition and Trends Working Group of the Millennium Ecosystem Assessment*, Island Press: Washington, DC, 2005, 165–207. For the ice-free land figures, see IPCC, 'Climate Change and Land: An IPCC Special Report on Climate Change, Desertification, Land Degradation, Sustainable Land Management, Food Security, and Greenhouse Gas Fluxes in Terrestrial Ecosystems' (summary for policymakers approved draft) (7 Aug. 2019). doi: 10.4337/9781784710644. For the figures on the planet's biological productivity, see F. Krausmann et al., 'Global Human Appropriation of Net Primary Production Doubled in the 20th Century', in *Proceedings of the National Academy of Sciences of the United States of America*, 110/25 (2013), 10324–9. doi: 10.1073/pnas.1211349110

6   Increases in hunger: Food and Agriculture Organization of the United Nations, 'The State of Food Security and Nutrition in the World' (2021), https://www.fao.org/hunger/en/, accessed 4 Jan. 2022. Life expectancy decreasing in the UK: Danny

Dorling and Stuart Gietel-Basten, 'Life Expectancy in Britain Has Fallen So Much That a Million Years of Life Could Disappear by 2058 – Why?', *The Conversation* (29 Nov. 2017), https://theconversation.com/life-expectancy-in-britain-has-fallen-so-much-that-a-million-years-of-life-could-disappear-by-2058-why-88063, accessed 4 Dec. 2021.

7   J. Bendell, 'Deep Adaptation: A Map for Navigating Climate Tragedy', *Institute for Leadership and Sustainability (IFLAS) Occasional Papers Volume 2* (Carlisle: IFLAS, 2018), http://www.lifeworth.com/deepadaptation.pdf, accessed 10 Dec. 2019.

8   Andrew McAfee, *More from Less: The Surprising Story of How We Learned to Prosper Using Fewer Resources – and What Happens Next*, Scribner: New York, 2019.

9   This is a very rough estimate, and some large percentage of this information is duplicated – just think of the number of copies of the same film on different hard drives around the world. However, it is still an astonishing amount of storage. See Melvin M. Vopson, 'The World's Data Explained: How Much We're Producing and Where It's All Stored', *The Conversation* (4 May 2021), https://theconversation.com/the-worlds-data-explained-how-much-were-producing-and-where-its-all-stored-159964, accessed 4 Jan. 2022.

10   Erik Brynjolfsson and Andrew McAfee, *The Second Machine Age: Work, Progress, and Prosperity in a Time of Brilliant Technologies*, W.W. Norton & Company: New York, 2014.

11   There are many books that tell this tale, such as Aaron Bastani's *Fully Automated Luxury Communism: A Manifesto*, Verso: London, 2018, or Peter Diamandis and Steven Kotler's *Abundance: The Future Is Better than You Think*, Free Press: New York, 2012.

12   Although not quite as strong a sentiment as that found in this book's text, for an overview of this sort of philosophy and history see D. Grinspoon, *Earth in Human Hands: Shaping our Planet's Future*, Grand Central Publishing: New York, 2016.

13   In his book *The Wizard and The Prophet: Two Remarkable Scientists and Their Dueling Visions to Shape Tomorrow's World*, Picador: London, 2017, author Charles C. Mann draws a divide between two types of environmental thinker. Wizards think technology and innovation will ultimately be able to provide the revolutions needed to overcome potential human population collapse. Prophets think the only recourse for society is to fundamentally revolutionize the way we live, cutting back our consumption and reintegrating ourselves with the ecosystems that provide for us. It would be possible to imagine that pessimism and hope for this book cuts this same divide. But in my view parts of both types of thinking are necessary. The profundity of the problems we face dictates that we must do much of both, and at once.

14   Papers by both Chapman et al. and Reser and Bradley are good examples of this 'fervent discussion'; see D.A. Chapman, B. Lickel and E. M. Markowitz, 'Reassessing Emotion in Climate Change Communication', in *Nature Climate Change*, 7 (2017), 850–2. doi:10.1038/s41558-017-0021-9, and J.P. Reser and G.L. Bradley, 'Fear Appeals in Climate Change Communication', in *Oxford Research Encyclopaedia of Climate Science* (2017), doi: 10.1093/acrefore/9780190228620.013.386. David Roberts also gives an excellent overview here: 'Does Hope Inspire More Action on Climate Change than Fear? We Don't Know', *Vox* (5 Dec. 2017), https://www.vox.com/energy-and-environment/2017/12/5/16732772/emotion-climate-change-communication, accessed 10 Dec. 2019.

15   It is actually a misconception that the effects take several decades to appear. The reality is closer to a decade: K.L. Ricke and K. Caldeira, 'Maximum Warming Occurs

About One Decade After a Carbon Dioxide Emission', *Environmental Research Letters*, 9 (2014). doi: 10.1088/1748-9326/9/12/124002

16 E. Staff and D. Orr, 'Conversation with David Orr', *Earth Island Journal* (26 Oct. 2008), http://www.earthisland.org/journal/index.php/magazine/entry/conversation/, accessed 10 Dec. 2019.

17 A perennial share on social media sites is a video of buses passing vertically placed wind turbines spinning in the middle of the road. 'This is the future of energy!' the video exhorts, without realizing that the whole process is losing energy as the wind used to run the turbine is from the passing buses which in turn experience drag and slow them down – it's the conversion of oil into moving a turbine!

18 Martin Wolf, 'The IMF Today and Tomorrow', *Finance & Development*, 56/2 (15 Jun. 2019), https://www.imf.org/external/pubs/ft/fandd/2019/06/the-future-of-the-imf-wolf. htm, accessed 10 Dec. 2019.

19 This is a quote attributed to many different people, as described here by Quote Investigator: https://quoteinvestigator.com/2013/10/20/no-predict/, accessed 10 Dec. 2019.

20 This comment was made by historian, author and commentator Timothy Garton Ash in a 2019 podcast for media company Intelligence Squared, https://play.acast.com/s/ intelligencesquared/decfa7a8-77ee-478c-a5e1-01507b0a4452, accessed 5 Jan. 2020. Ash's book about the revolution in 1989, *The Magic Lantern: The Revolution of '89 Witnessed in Warsaw, Budapest, Berlin and Prague*, was re-released in 2019 (Atlantic Books: London, 2019).

21 M. Scheffer et al., 'Early-Warning Signals for Critical Transitions'. *Nature*, 461/7260 (2009), 53–9. doi: 10.1038/nature08227

22 W. Steffen et al. 'Trajectories of the Earth System in the Anthropocene', *Proceedings of the National Academy of Sciences of the United States of America*, 115/33 (2018), 8252–9. doi/10.1073/pnas.1810141115

23 Katy Steinmetz, 'See Obama's 20-Year Evolution on LGBT Rights', *Time* (10 Apr. 2015), https://time.com/3816952/obama-gay-lesbian-transgender-lgbt-rights/, accessed 10 Dec. 2019.

24 S.L. Lewis and M.A. Maslin, 'Defining the Anthropocene', *Nature*, 519 (2015), 171–80. doi: 10.1038/nature14258.

25 Ibid.

26 Douglas Heaven, 'Nuclear Bomb Tests Reveal Brain Regeneration in Humans', *New Scientist* (7 Jun. 2013), https://www.newscientist.com/article/dn23665-nuclear-bomb-tests-reveal-brain-regeneration-in-humans/, accessed 10 Dec. 2019.

27 Sarah Gibbens, 'Plastic Proliferates at the Bottom of the World's Deepest Ocean Trench', *National Geographic* (13 May 2019), https://www.nationalgeographic.com/ news/2018/05/plastic-bag-mariana-trench-pollution-science-spd/, accessed 10 Dec. 2019; and Damian Carrington, 'Microplastics "Significantly Contaminating the Air", Scientists Warn' , *The Guardian* (14 Aug. 2019), https://www.theguardian.com/ environment/2019/aug/14/microplastics-found-at-profuse-levels-in-snow-from-arctic-to-alps-contamination, accessed 4 Jan. 2022.

28 An estimated 9.2 billion tonnes of plastic have been produced since the 1950s (Laura Parker, 'We Made Plastic. We Depend on It. Now We're Drowning in It', *National Geographic* (9 Apr. 2019), https://www.nationalgeographic.co.uk/2018/05/we-made-plastic-we-depend-it-now-were-drowning-it, accessed 20 Dec. 2019. Cling film weighs around 18.3g per m². The surface area of the planet is 510 million km². If you were to stretch the film just a little it would cover just fine.

29   Based on an emission of 652g of carbon per litre of petrol and Professor Ken Caldeira's calculations in 'How Much Ice Is Melted by Each Carbon Dioxide Emission' (24 Mar. 2018), https://kencaldeira.wordpress.com/2018/03/24/how-much-ice-is-melted-by-each-carbon-dioxide-emission/, accessed 10 Dec. 2019.

30   J. Syvitski, 'Anthropocene: An Epoch of Our Making', Global Change, 78 (2012).

31   There are around ten tonnes of air above each square metre of Earth's surface – with a shoulder width of 40 cm and a depth of 20 cm, each person has a surface area of 0.08 square metres. Average EU emissions are 7.2 tonnes per person per year.

32   C. N. Waters et al., 'The Anthropocene Is Functionally and Stratigraphically Distinct from the Holocene', Science, 351/6269 (2016), 137–47. doi: 10.1126/science.aad2622

33   As of 2012, LEGO had made around 5.5 billion LEGO minifigures; given the projection made by Randall Munroe and the expansion of the LEGO market: https://www.explainxkcd.com/wiki/index.php/1281:_Minifigs, accessed 10 Dec. 2019. This was based on a LEGO press release that announced 4 billion minifigures in 2003 (private communication). Given there are now eighty-six LEGO pieces for every person on Earth (private communication) it's very likely that minifigures outnumber the global human population.

Chapter One – Pessimism – Has the Bomb Exploded?

1   Though polar bears can no longer reach the island today due to declining sea ice.

2   F.L. Miller, S. J. Barry and W.A. Calvert, 'St. Matthew Island Reindeer Crash Revisited: Their Demise Was Not Nigh – But Then, Why Did They Die?', Rangifer, 16/4 (2005), 185–97. doi: 10.7557/2.25.4.1783

3   This is approximated using various estimates of the human population by 10,000, including K. Klein Goldewijk et al., 'The HYDE 3.1 Spatially Explicit Database of Human-Induced Global Land-Use Change Over the Past 12,000 Years', Global Ecology and Biogeography, 20 (2011), 73–86. doi: 10.1111/j.1466-8238.2010.00587.x; and Colin McEvedy and Richard Jones, Atlas of World Population History, Facts on File: New York, 1978. There were periods where human populations were substantially lower than this, potentially during a genetic bottleneck about 70,000 years ago; see for example J. Hawks et al., 'Population Bottlenecks and Pleistocene Human Evolution', Molecular Biology and Evolution, 17/1 (2000), 2–22. doi: 10.1093/oxfordjournals.molbev.a026233

4   T. Kobashi, J.P. Severinghaus and J.M. Barnola, '4 ± 1.5 ° C Abrupt Warming 11,270 Yr Ago Identified from Trapped Air in Greenland Ice', Earth and Planetary Science Letters, 268/3 (2008), 397–407. doi: 10.1016/j.epsl.2008.01.032

5   UK covered in ice: T.J. Crowley, 'Ice Age Terrestrial Carbon Changes Revisited', Global Biogeochemical Cycles, 9/3 (1995), 377–89. doi: 10.1029/95GB01107. With regard to trees growing in the Arctic, this was during the Eemian period 130,000 years ago, which was the last interglacial before the Holocene. The Eemian had slightly higher temperatures than today but lower carbon levels, implying that we will shoot right past this most recent analogue to today's temperature.

6   See for instance Jared Diamond, Collapse: How Societies Choose to Fail or Succeed (rev. ed.), Penguin: London, 2011; or Anthony McMichael, Climate Change and the Health of Nations, Oxford University Press: London, 2017.

7   P.U. Clark et al., 'Consequences of Twenty-First-Century Policy for Multi-Millennial Climate and Sea-Level Change', Nature Climate Change, 6 (2016), 360–9 (2016). doi: 10.1038/nclimate2923

8   M.P. Richards, 'A Brief Review of the Archaeological Evidence for Palaeolithic and Neolithic Subsistence', *European Journal of Clinical Nutrition*, 56 (2002), 1270–8. doi: 10.1038/sj.ejcn.1601646

9   S.T. Kang, 'Irrigation in Ancient Mesopotamia', *Journal of the American Water Resources Association*, 8/3 (1972), 1289. doi: 10.1111/j.1752-1688.1972.tb05185.x

10   The development of sustained agriculture and inequality is thought to be very haphazard in practice. There were other forms of inequality and agriculture across human societies before, and contemporaneously with, the rise of these city states. It is hard to know definitively, but the hierarchies seen in regions such as Sumer do appear to represent a departure from the hierarchies seen in other societies of the time. For more see David Graeber and David Wengrow, *The Dawn of Everything: A New History of Humanity*, Farrar, Straus and Giroux: New York, 2022.

11   This is estimated using a total of 80,000 people for the major city of Uruk, adding other cities' populations and probably a high agricultural population.

12   Ronald Wright, *A Short History of Progress*, House of Anansi Press: Toronto, 2004.

13   Anthony McMichael, *Climate Change and the Health of Nations*, Oxford University Press: New York, 2017.

14   Thomas Homer-Dixon outlines the process of these calculations in his book *The Upside of Down: Catastrophe, Creativity, and the Renewal of Civilization*, Island Press: Washington, DC, 2006. For further details, see www.theupsideofdown.com/rome/colosseum, last accessed 10 Dec. 2019.

15   D. Hughes and V. Thirgood, 'Deforestation, Erosion, and Forest Management in Ancient Greece and Rome.' *Journal of Forest History*, 26/2 (1982), 60–75. doi: 10.2307/4004530

16   Anthony McMichael, *Climate Change and the Health of Nations*, Oxford University Press: London, 2017.

17   B.L. Turner and J.A. Sabloff, 'Classic Period Collapse of the Central Maya Lowlands: Insights About Human-Environment Relationships for Sustainability'. *Proceedings of the National Academy of Sciences of the United States of America*, 109/35 (2012), 13908–14. doi: 10.1073/pnas.1210106109

18   D.G. Beresford-Jones et al., 'The Role of Prosopis in Ecological and Landscape Change in the Samaca Basin, Lower Ica Valley, South Coast Peru from the Early Horizon to the Late Intermediate Period', *Latin American Antiquity*, 4595/20 (2009), 303–32. doi: 10/1017/S1045663500002650

19   T.H. McGovern, 'Climate, Correlation, and Causation in Norse Greenland', *Arctic Anthropology*, 28/2 (1991), 77–100.

20   Robin Fleming, *Britain After Rome: The Fall and Rise, 400–1070*, Allen Lane: London, 2010, 27; and idem, *The Material Fall of Roman Britain, 300–525 CE*, University of Pennsylvania Press: Pennsylvania, 2021.

21   Leonard E. Read talks about the ability to make pencils in his essay 'I, Pencil', *The Freeman* (Dec. 1957). You can read about this in detail here: https://www.econlib.org/library/Essays/rdPncl.html?chapter_num=2#book-reader, accessed 10 Dec. 2019. As for toasters, in 2018, the designer Tom Thwaites attempted to build a toaster from scratch. He found that the toaster he had bought for £3.99 had 400 parts made from more than a hundred different materials. He cut the materials down to five essentials and, after several months sourcing, refining them and building the toaster, he had a 'working' appliance. The toaster functioned for five seconds before melting entirely. He called it a partial success. You can read more at http://www.thetoasterproject.

org/, accessed 10 Dec. 2019, or read Thwaites's book, *The Toaster Project*, Princeton Architectural Press: New York, 2011.

22   H. J. Zahid, E. Robinson and R.L. Kelly, 'Agriculture, Population Growth, and Statistical Analysis of the Radiocarbon Record', *Proceedings of the National Academy of Sciences of the United States of America*, 113/4 (2015), 931–95. doi: 10.1073/pnas.1517650112

23   A. Koch et al., 'Earth System Impacts of the European Arrival and Great Dying in the Americas after 1492', *Quaternary Science Reviews*, 207 (2019), 13–36. doi: 10.1016/j. quascirev.2018.12.004

24   N. Nunn and N. Quan, 'The Potato's Contribution to Population and Urbanization: Evidence from an Historical Experiment'. *Quarterly Journal of Economics*, 126/2 (2011), 593–650. doi: 10.1093/qje/qjr009

25   S.L. Lewis and M.A. Maslin, 'Defining the Anthropocene'.

26   Francis Bacon, *Novum Organum, Book I: Aphorisms Concerning the Interpretation of Nature and the Kingdom of Man*, Aphorism CXXIX, John Bill: London, 1620.

27   K.K. Goldewijk et al. 'Anthropogenic Land Use Estimates for the Holocene – HYDE 3.2'. *Earth System Science Data*, 9/2 (2017), 927–53 (2017). doi: 10.5194/essd-9-927-2017

28   Stephen M. Gardiner and David A. Weisbach, *Debating Climate Ethics*, Oxford University Press, Oxford, 2016, pp. 1–5.

29   For population growth see the World Bank, 'Population growth (annual %)' (2019), https://data.worldbank.org/indicator/SP.POP.GROW?view=chart, accessed 10 Dec. 2019.

30   Paul Ehrlich, *The Population Bomb*, Ballantine Books: New York, 1968.

31   This is a statement which is yet to be published in a scientific article (indeed, it would be hard to conclusively make such an assessment). However, expert elicitation – the process of seeing what experts think – is a completely valid way to get estimates when no formal analyses are forthcoming. In this case, the prominent environmental scientist Will Steffen makes the claim here: 'Climate Change 2017: The Nature of the Challenge: Prof Will Steffen [video], YouTube (uploaded 17 Aug. 2017), https://www.youtube.com/watch?v=yiHc8vbTOTI (at 36:56). Another prominent climate scientist, Hans Joachim Schellnhuber, Director of the Potsdam Institute for Climate Impact Research, made similar comments here: James Kanter, 'Scientist: Warming Could Cut Population to 1 billion', *New York Times: Dot Earth* (13 Mar. 2009), https://dotearth.blogs.nytimes.com/2009/03/13/scientist-warming-could-cut-population-to-1-billion, both sources accessed 10 Dec. 2019.

32   A. Maddison, *Contours of the World Economy, 1–2030 AD: Essays in Macro-Economic History*, Oxford University Press: Oxford, 2007.

33   This is a slightly thorny comparison between a flow (living humans) and stocks (the amount of material used in human civilization). Approximately though, an estimated 11 billion people lived in the twentieth century it is less than ten times). See F. Krausmann et al., 'Global Socioeconomic Material Stocks Rise 23-fold over the 20th Century and Require Half of Annual Resource Use'. *Proceedings of the National Academy of Sciences of the United States of America*, 114/8 (2017), 1880–5. doi: 10.1073/pnas.1613773114

34   World Bank, 'Decline of Global Extreme Poverty Continues But Has Slowed: World Bank' (19 Sep. 2018), https://www.worldbank.org/en/news/press-release/2018/09/19/decline-of-global-extreme-poverty-continues-but-has-slowed-world-bank, accessed 10 Dec. 2019.

35 These calculations come from Our World In Data, which compares extreme poverty in the world including and excluding China: https://ourworldindata.org/grapher/ projected-share-extreme-poverty?tab=chart&time=2015..2030&country=CHN, accessed 10 Dec. 2019.

36 We have explored this industrially intensive dynamic in several academic articles, including M. Jiang et al., 'Different Material Footprint Trends between China and the World in 2007–2012 Explained by Construction- and Manufacturing-Associated Investment', One Earth, 5, 1 (2022), 109–19. doi:10.1016/j.oneear.2021.12.011; and M. Jiang et al., 'Provincial and Sector-Level Material Footprints in China', Proceedings of the National Academy of Sciences, 116, 52 (2019), 26484–8490. doi:10.1073/ pnas.1903028116.

37 Niall McCarthy, 'China Used More Concrete in 3 Years than the U.S. Used in the Entire 20th Century', Forbes (5 Dec. 2014), https://www.forbes.com/sites/ niallmccarthy/2014/12/05/china-used-more-concrete-in-3-years-than-the-u-s-used-in-the-entire-20th-century-infographic/, accessed 10 Dec. 2019.

38 M.R. Raupach et al., 'Sharing a Quota on Cumulative Carbon Emissions', Nature Climate Change, 4 (2014), 873–9. doi: 10.1038/nclimate2384

39 These statistics derive from L. Chancel and T. Piketty, 'Carbon and Inequality: from Kyoto to Paris', Paris School of Economics (2015). doi: 10.13140/RG.2.1.3536.0082. The 'average European' statistic was computed by energy and climate scientist Kevin Anderson using these data.

40 This is a rough approximation and almost certainly an underestimate. To make this rough estimate I take Figure A.3 of Chancel and Piketty (2015), and scale those in the lowest 50% to the average European footprint of 7.2 tonnes.

41 See the theoretical model in S. Motesharrei, J. Rivas and E. Kalnay, 'Human and Nature Dynamics (HANDY): Modeling Inequality and Use of Resources in the Collapse or Sustainability of Societies', Ecological Economics, 101 (2014), 90–102. doi: 10.1016/j. ecolecon.2014.02.014; the broader discussion in Jared Diamond, Collapse: How Societies Choose to Fail or Succeed (rev. ed.), Penguin Books: London, 2011; and the mixed approach of historical analysis and theoretical modelling in Peter Turchin and Sergey A. Nefedov, Secular Cycles, Princeton University Press: Princeton, NJ, 2009.

42 Milanovic explains that Marcus Crassus had an estimated fortune of 200 million sesterces, which at an interest rate of 6% (the common rate at that time) equated to an annual income of 12 million sesterces. The average income of Roman citizens was thought to be around 380 sesterces. A similar calculation can be made for Carlos Slim today. Branko Milanovic, The Haves and the Have-Nots: A Brief and Idiosyncratic History of Global Inequality, Basic Books: New York, 2012.

43 For the figures on billionaires in China, see Tara Francis Chan, 'Communist China Has 104 Billionaires Leading the Country While Xi Jinping Promises to Lift Millions out of Poverty', Business Insider (2 Mar. 2018), https://www.businessinsider. my/billionaires-in-china-xi-jinping-parliament-income-inequality-2018-3/. This is based on the Hurun Global Rich List 2018, http://www.hurun.net/EN/Article/ Details?num=2B1B8F33F9C0, accessed 10 Dec. 2019. The figures on lawmakers in the US are calculated for 2016 using the methodology described at the Open Secrets Project: http://www.opensecrets.org/personal-finances/top-net-worth. This $6 billion is dominated by two wealthy examples: Donald Trump and Betsy de Vos (at $2.2 billion). However, the estimate would still be high even without these two billionaires included.

44 Note that, as with the inequality example above, percentages hide how large the differences are in absolute terms. The 25% evaded by the top 0.01% results in a larger absolute number than than the 2.8% avoided by other income brackets. See also A. Alstadsæter, N. Johannesen and G. Zucman, 'Who Owns the Wealth in Tax Havens? Macro Evidence and Implications for Global Inequality', *Journal of Public Economics*, 162 (2018), 89–100. doi: 10.1016/j.jpubeco.2018.01.008

45 David Leonhardt, 'The Rich Really Do Pay Lower Taxes than You', *New York Times* (6 Oct. 2019), https://www.nytimes.com/interactive/2019/10/06/opinion/income-tax-rate-wealthy.html, accessed 10 Dec. 2019.

46 This is summarized by the International Monetary Fund: Nicholas Shaxson, 'Tackling Tax Havens', Finance & Development, 56/3 (2019), https://www.imf.org/external/pubs/ft/fandd/2019/09/tackling-global-tax-havens-shaxon.htm. The lower estimate comes from Gabriel Zucman, 'How corporations and the wealthy evade taxes', *New York Times* (10 Nov. 2017), https://gabriel-zucman.eu/how-corporations-avoid-taxes/, and the higher estimate from James S. Henry, 'Taxing Tax Havens', *Foreign Affairs* (12 Apr. 2016), https://www.foreignaffairs.com/articles/panama/2016-04-12/taxing-tax-havens, accessed 20 Dec. 2019. An important caveat is that you couldn't simply liquidate these assets for the headline total. As you sold off all the gold, diamonds, art, etc., the market price for these goods would drop. Yet in the comparison it is also likely that the transition to a carbon-free economy would be cheaper than the headline figure.

47 Estimates hover around this sort of level. Bogdanov et al. found a result of $22.5 trillion globally from a country-by-country modelling of the energy system at an hour-by-hour resolution. The 1.5-degree IPCC special report gave an additional $830 billion per year globally to meet the 1.5-degree target (with quite a large range and based on modelling which is critiqued later on in the book). See D. Bogdanov et al., 'Radical Transformation Pathway Towards Sustainable Electricity via Evolutionary Steps', *Nature Communications* (2019). doi: 10.1038/s41467-019-08855-1; V. Masson-Delmotte et al., eds, 'Summary for Policymakers', in *Global Warming of 1.5°C. An IPCC Special Report on the Impacts of Global Warming of 1.5°C Above Pre-Industrial Levels and Related Global Greenhouse Gas Emission Pathways, in the Context of Strengthening the Global Response to the Threat of Climate Change, Sustainable Development, and Efforts to Eradicate Poverty*, IPCC, World Meteorological Organization: Geneva, 2018. doi: 10.1017/CBO9781107415324; and International Renewable Energy Agency, 'Renewable Energy Can Support Resilient and Equitable Recovery', https://www.irena.org/newsroom/pressreleases/2020/Apr/Renewable-energy-can-support-resilient-and-equitable-recovery, accessed 4 Jan. 2022.

48 Gabriel Zucman makes this point based on data of charitable giving for the top 400 wealthiest individuals in the US, available here: IRS, 'SOI Tax Stats – Top 400 Individual Income Tax Returns with the Largest Adjusted Gross Incomes' (19 Nov. 2019), https://www.irs.gov/statistics/soi-tax-stats-top-400-individual-income-tax-returns-with-the-largest-adjusted-gross-incomes, accessed 10 Dec. 2019.

49 A good description of how these dynamics operate in practice is available in Anand Giridharadas, *Winners Take All: The Elite Charade of Changing the World*, Allen Lane: London, 2019.

50 Mike Collins gives a good summary of different sectors and their returns from lobbying in his article 'Buying Government with Lobbying Money', *Forbes* (28 Mar. 2015), https://www.forbes.com/sites/mikecollins/2015/03/28/buying-government-with-lobbying-money-2/#4b81fadf6b93, accessed 10 Dec. 2019.

51 This draws from research in Branko Milanovic's book *Global Inequality: A New Approach for the Age of Globalization*, Harvard University Press: Cambridge, MA, 2016. The statistic for lawmakers being five to six times more likely to respond is from Larry M. Bartels, 'Economic Inequality and Political Representation', Papers (8 Nov. 2005), Princeton University, Research Program in Political Economy, available here: https://ideas.repec.org/p/ecl/prirpe/08-11-2005.html, accessed 20 Dec. 2019. The statement from Bartels originates from Larry M. Bartels, *Unequal Democracy: The Political Economy of the New Gilded Age - Second Edition*, Princeton University Press: Princeton, NJ, 2017. Other work showing similar findings is available in Martin Gilens, *Affluence and Influence: Economic Inequality and Political Power in America*, Princeton University Press: Princeton, NJ, 2012 and M. Gilens and B.I. Page, 'Testing Theories of American Politics: Elites, Interest Groups, and Average Citizens', *Perspectives on Politics*, 12/3 (2014) 564–81. doi: 10.1017/S1537592714001595

52 A.J. McMichael et al., 'Global Environmental Change and Health: Impacts, Inequalities, and the Health Sector, *British Medical Journal*, 336 (2008), 191–4. doi: 10/1136/bmj.39392.473727.AD

53 R.J. Brulle, 'The Climate Lobby: A Sectoral Analysis of Lobbying Spending on Climate Change in the USA, 2000 to 2016', *Climatic Change* 149 (2018), 289–303. doi: 10.1007/s10584-018-2241-z

54 J. Farrell, 'Network Structure and Influence of the Climate Change Counter-Movement', *Nature. Climate Change* 6/4 (2015), 370–374. doi: 10.1038/nclimate2875

55 For more information on these sorts of cognitive biases see Daniel Kahneman's *Thinking, Fast and Slow*, Penguin: London, 2012.

56 See for instance G. Hofstede, 'Dimensionalizing Cultures: The Hofstede Model in Context', *Online Readings in Psychology and Culture*, 2 (2011), 1–26. doi: 10.9707/2307-0919.1014; and M. Wang, M.O. Rieger and T Hens, 'The Impact of Culture on Loss Aversion', *Journal of Behavioral Decision Making*, 30/2 (2016), 270–81. doi: 10.1002/bdm.1941

57 Thanks to my colleague Dr David Zetland for this example.

58 This dynamic was described in D. Harvey, 'Globalization and the "Spatial Fix"', *Geografische Revue: Zeitschrift für Dichtung und Diskussion*, 2 (2001), 23–30.

59 Some of the history is given in 'Sustainability Criteria', European Commission, https://ec.europa.eu/energy/en/topics/renewable-energy/biofuels/sustainability-criteria, accessed 10 Dec. 2019.

60 Roger Harrabin, 'EU Rethinks Biofuels Guidelines', *BBC News* (14 Jan. 2018), http://news.bbc.co.uk/2/hi/europe/7186380.stm, accessed 10 Dec. 2019.

61 Arthur Neslen, 'MEPs Vote to Ban the Use of Palm Oil in Biofuels', *The Guardian* (4 Apr. 2017), https://www.theguardian.com/sustainable-business/2017/apr/04/palm-oil-biofuels-meps-eu-transport-deforestation-zsl-greenpeace-golden-agri-resources-oxfam, accessed 10 Dec. 2019.

Chapter Two – Hope – Better Placed than Ever

1 This term was coined by John Rawls in *A Theory of Justice*, Belknap Press: Cambridge,1971.

2 These data come from a composite of other datasets built by the Gapminder project. More details are available here: https://www.gapminder.org/data/documentation/gd005/, accessed 10 Dec. 2019.

3 World Bank, 'Mortality Rate, Under-5 (per 1,000 live births)', https://data.worldbank. org/indicator/SH.DYN.MORT, accessed 10 Dec. 2019.

4 M.I. Norton and D. Ariely, 'Building a Better America – One Wealth Quintile at a Time', *Perspectives on Psychological Science* 6/1 (2011), 9–12. doi: 10.1177/174569161039 3524

5 One wonders what the results would look like if done 'behind' the veil of ignorance.

6 For the US see J. Marlon et al., 'Yale Climate Opinion Maps 2018', Yale Program on Climate Change Communication (7 Aug. 2018): https://climatecommunication. yale.edu/visualizations-data/ycom-us-2018/?est=prienv&type=value&geo=national. For China: 57% of Chinese people believe that environmental protection should take precedence 'even at the risk of curbing economic growth'. Only 21% put more growth above the quality of the environment. Daniela Yu and Anita Pugliese, 'Majority of Chinese Prioritise Environment Over Economy: Yet China Faces Challenges with Rapid Urbanization', Gallup (8 Jun. 2012), https://news.gallup.com/poll/155102/ majority-chinese-prioritize-environment-economy.aspx. For Europe see B. Anderson, T. Böhmelt and H. Ward, 'Public Opinion and Environmental Policy Output: A Cross-National Analysis of Energy Policies in Europe', *Environmental Research Letters*, 12/11 (2017). doi: 10.1088/1748-9326/aa8f80. For Brazil: the same Gallup Poll cited above found that only 7% of Brazilians felt that economic growth should be given priority, even if the environment suffers to some extent.

7 For a description of these concerns see Darrell Bricker and John Ibbitson, 'By the End of This Century, the Global Population Will Start to Shrink', *Medium* (29 Jan. 2019), https://onezero.medium.com/by-the-end-of-this-century-the-global-population-will-start-to-shrink-2f606c1ef088, accessed 10 Dec. 2019. And their book: Darrell Bricker and John Ibbitson, *Empty Planet: The Shock of Global Population Decline*, Signal Books: Oxford, 2019.

8 United Nations Population Division, 'World Population Prospects: 2019'. Available at the World Bank website, headed 'Fertility Rate, Total', https://data.worldbank.org/ indicator/sp.dyn.tfrt.in, accessed 10 Dec. 2019.

9 This is if you don't take migration into account. Inward migration brings the total reduction in population down to 430,000 people, according to the *Financial Times* in Robin Harding, 'Japan's Population Decline Accelerates Despite Record Immigration' (12 Apr. 2019), available to subscribers at https://www.ft.com/content/29d594fa-5cf2-11e9-9dde-7aedca0a081a, accessed 10 Dec. 2019.

10 It's hard to say whether this decline will increase further with time. On the one hand there are movements such as the 'birthstrikers', a group of women who have pledged not to have babies while the issues of climate change and ecological collapse go unaddressed. Conversely, we may expect more initiatives from governments to encourage people to start families.

11 However, researchers believe that China's population and fertility rates are actually lower than state estimates; see Jane Lanhee Lee, 'Researcher Questions China's Population Data, Says It May Be Lower', Reuters (3 Dec. 2021), https://www. reuters.com/world/china/researcher-questions-chinas-population-data-says-it-may-be-lower-2021-12-03/, accessed 9 Jan. 2022.

12 This data is taken from World Bank figures for 2017. You can view all trends next to one another at 'Fertility Rates, Total (Births per Woman) – China, United Kingdom, India, Argentina', https://data.worldbank.org/indicator/SP.DYN.TFRT.IN?locations=CN-GB-IN-AR, accessed 10 Dec. 2019.

13 This data is taken from Gapminder, 'Children per Woman (Total Fertility Rate)', version 12, https://www.gapminder.org/data/documentation/gd008/, accessed 10 Dec. 2019.

14 Rwanda Ministry of Health. 'Accelerating Fertility Decline to Trigger the Demographic Dividend in Rwanda' (Oct. 2017), https://www.afidep.org/resource-centre/downloads/policy-briefs/accelerating-fertility-decline-to-trigger-the-demographic-dividend-in-rwanda/, accessed 10 Dec. 2019.

15 *The Economist*, 'The UN Revises Down its Population Forecasts' (22 Jun. 2019), https://www.economist.com/graphic-detail/2019/06/22/the-un-revises-down-its-population-forecasts, accessed 10 Dec. 2019.

16 The 'peak child' term was coined by Hans Rosling. For example, he used it in his TED talk 'Religions and Babies' in 2012. The full transcript and video of the talk can be found at https://www.ted.com/talks/hans_rosling_religions_and_babies/transcript, accessed 10 Dec. 2019.

17 Early academic analysis was available here: A. Aassve et al., 'Early Assessment of the Relationship Between the COVID-19 Pandemic and Births in High-Income Countries', *Proceedings of the National Academy of Sciences*, 118/36 (2021). doi: 10.1073/pnas.2105709118. A broader discussion of birth rates during and post-COVID is here: Darrell Bricker, 'Bye, Bye, Baby? Birthrates Are Declining Globally – Here's Why It Matters', *World Economic Forum* (15 Jun. 2021), https://www.weforum.org/agenda/2021/06/birthrates-declining-globally-why-matters/, accessed 9 Jan. 2022.

18 Darrell Bricker and John Ibbitson, *Empty Planet: The Shock of Global Population Decline*. Oxford: Signal Books, 2019.

19 Sarah Harper, *How Population Change Will Transform Our World*, Oxford University Press: Oxford, 2019.

20 E.O. Wilson, *The Diversity of Life*, Penguin Books: London, 2001, p. 314.

21 All these data come from various UN departments via the World Bank's data portal. Fertility rate: https://data.worldbank.org/indicator/sp.dyn.tfrt.in, and income per capita: https://data.worldbank.org/indicator/ny.gdp.pcap.cd, accessed 10 Dec. 2019.

22 K.C. Samir and W. Lutz, 'The Human Core of the Shared Socioeconomic Pathways: Population Scenarios by Age, Sex and Level of Education for All Countries to 2100', *Global Environmental Change*, 42 (2017), 181–92. doi: 10.1016/j.gloenvcha.2014.06.004

23 R.J. Barro and J.W. Lee, 'A New Data Set of Educational Attainment in the World, 1950–2010'. *Journal of Development Economics* 104 (2013), 184–98. doi: 10.1016/j.jdeveco.2012.10.001; and Pamela Jakiela and Susannah Hares, 'Mind the Gap: 5 Facts About the Gender Gap in Education', Center for Global Development (17 Jun. 2019), https://www.cgdev.org/blog/mind-gap-5-facts-about-gender-gap-education, accessed 20 Dec. 2019.

24 Jakiela and Hares, 'Mind the Gap'.

25 W. Lutz and K.C. Samir, 'Global Human Capital: Integrating Education and Population', *Science*, 333/6042 (2011), 587–92. doi: 10.1126/science.1206964

26 For instance see N. Ravishankar et al., 'Financing of Global Health: Tracking Development Assistance for Health from 1990 to 2007', *Lancet*, 373/9681 (2009), 2113–24. doi: 10.1016/S0140-6736(09)60881-3; and Sneha Barot, 'The Benefits of Investing in International Family Planning – and the Price of Slashing Funding', *The Guttmacher Institute*, Volume 20 (1 Aug. 2017), https://www.guttmacher.org/gpr/2017/08/benefits-investing-international-family-planning-and-price-slashing-funding, accessed 20 Dec. 2019.

27 G. Sedgh, S. Singh and R. Hussain, 'Intended and Unintended Pregnancies Worldwide in 2012 and Recent Trends', *Studies in Family Planning*, 45/3 (2014), 301–14. doi: 10.1111/j.1728-4465.2014.00393.x.

28 One interesting indirect public information campaign is the use of soap operas to discuss these themes. There are examples from around the world, including Brazil, India, Ethiopia and the UK. For more on this see Charles Kenny, *Getting Better: Why Global Development Is Succeeding*, Basic Books: New York, 2012.

29 For instance N.J. Cook, T. Grillos and K.P. Andersson, 'Gender Quotas Increase the Equality and Effectiveness of Climate Policy Interventions', *Nature Climate Change*, 9 (2019), 330–4. doi: 10.1038/s41558-019-0438-4; and B. Agarwal, 'Gender and Forest Conservation: The Impact of Women's Participation in Community Forest Governance', *Ecological Economics*, 68/11 (2009), 2785–99 (2009). doi: 10.1016/j.ecolecon.2009.04.025

30 For instance R. Forsythe et al., 'Fairness in Simple Bargaining Experiments', *Games and Economic Behavior*, 6/3 (1994), 347–69. doi: 10.1006/game.1994.1021; C.C. Eckel and P.J. Grossman, 'Are Women Less Selfish than Men?', *Economic Journal*, 108 (1998), 726–35. doi: 10.1111/1468-0297.00311; R. Selten and A. Ockenfels, 'An Experimental Solidarity Game', *Journal of Economic Behavior & Organization*, 34/4 (1998), 517–39. doi: 10.1016/S0167-2681(97)00107-8; C. Engel, 'Dictator Games: A Meta Study', *Experimental Economics*, 14 (2011), 583–610. doi: 10.1007/s10683-011-9283-7; and B. Agarwal, 'Participatory Exclusions, Community Forestry, and Gender: An Analysis for South Asia and a Conceptual Framework, *World Development*, 29/10 (2001), 1623–48. doi: 10.1016/S0305-750X(01)00066-3

31 More female decision makers are more likely to protect land: C. Nugent and J.M. Shandra, 'State Environmental Protection Efforts, Women's Status, and World Polity: A Cross-National Analysis', *Organization & Environment*, 22/2 (2009), 208–29. doi: 10.1177/1086026609338166. Greater likelihood of ratification of treaties: K. Norgaard and R. York, 'Gender Equality and State Environmentalists', *Gender & Society*, 19/4 (2005), 506–22. doi: 10.1177/0891243204273612. Stricter climate change policies: A. Mavisakalyan and Y, Tarverdi, 'Gender and Climate Change: Do Female Parliamentarians Make a Difference?', *European Journal of Political Economy*, 56 (2019), 151–64. doi: 10.1016/j.ejpoleco.2018.08.001

32 For example, a review of ninety-one countries drew this conclusion: C. Ergas and R. York, 'Women's Status and Carbon Dioxide Emissions: A Quantitative Cross-National Analysis', *Social Science Research*, 41/4 (2012), 965–76. doi: 10.1016/j.ssresearch.2012.03.008. Another cross-national analysis is by L.A. McKinney and G.M. Fulkerson, 'Gender Equality and Climate Justice: A Cross-National Analysis', *Social Justice Research*, 28 (2015), 293–317. doi: 10.1007/s11211-015-0241-y. And another: Z. Lv and C. Deng, 'Does Women's Political Empowerment Matter for Improving the Environment? A Heterogeneous Dynamic Panel Analysis', *Sustainable Development*, 1–10 (2019). doi: 10.1002/sd.1926

33 A.M. McCright, 'The Effects of Gender on Climate Change Knowledge and Concern in the American Public', *Population and Environment*, 32/1 (2010), 66–87. doi: 10.1007/s11111-010-0113-1

34 R.J. Bord and R.E. O'Connor, 'The Gender Gap in Environmental Attitudes: The Case of Perceived Vulnerability to Risk', *Social Science Quarterly*, 78/4 (1997), 830–40. doi: 10.1007/978-94-007-5518-5_12

35 Mary Halton, 'Climate Change "Impacts Women More than Men"', BBC News (8

Mar. 2018), https://www.bbc.com/news/science-environment-43294221, accessed 10 Dec. 2019.

36   A. M. McCright and R.E. Dunlap, 'The Politicization of Climate Change and Polarization in the American Public's Views of Global Warming, 2001–2010', *Sociological Quarterly*, 52/2 (2011), 155–94. doi: 10.1111/j.1533-8525.2011.01198.x

37   UN Women, 'Facts and Figures: Ending Violence Against Women' (2018), https://www.unwomen.org/en/what-we-do/ending-violence-against-women/facts-and-figures, accessed 10 Dec. 2019.

38   B. Gault et al., 'Paid Parental Leave in the United States', Institute for Women's Policy Research (Washington, DC: Institute for Women's Policy, 2014), https://www.dol.gov/wb/resources/paid_parental_leave_in_the_united_states.pdf, accessed 10 Dec. 2019.

39   C. Liu, 'Are Women Greener? Corporate Gender Diversity and Environmental Violations', *Journal of Corporate Finance*, 52 (2018), 118–42. doi: 10.1016/j.jcorpfin.2018.08.004

40   Julia Carpenter, 'Women in the Fortune 500: 64 CEOs in Half a Century', CNN (Aug. 2017), https://money.cnn.com/interactive/pf/female-ceos-timeline/, accessed 10 Dec. 2019.

41   There appear to be backlashes across many countries. One line of evidence is taken from statements by political figures, for example, Nigerian president Muhammadu Buhari's statement that his wife 'belongs to my kitchen and my living room' (BBC News, 'Nigeria's President Buhari: My Wife Belongs in Kitchen' (14 Oct. 2016), https://www.bbc.com/news/world-africa-37659863); or Turkish president Recep Tayyip Erdoğan's claim that 'a woman who rejects motherhood, who refrains from being around the house, however successful her working life is, is deficient, is incomplete' (BBC News, 'Turkey's Erdogan Says Women Who Reject Motherhood "Incomplete"'(5 Jun. 2016), https://www.bbc.com/news/world-europe-36456878); Trump's grabbing women 'by the pussy' or Nigel Farage's excuse that 'it's the kind of thing, if we are being honest, that men do' (BBC News, 'Obscene Donald Trump Comments "Alpha Male Boasting" – Farage' (Oct. 2016), https://www.bbc.com/news/uk-politics-37601422). Other, more systematic evidence comes from an Amnesty International report showing that women are exposed to high levels of threatening language online, especially women of colour. In all, 7.1% of tweets sent to women journalists and politicians were abusive. You can read more here: Amnesty International, 'Troll Patrol Findings: Using Crowdsourcing, Data Science & Machine Learning to Measure Violence and Abuse Against Women on Twitter', https://decoders.amnesty.org/projects/troll-patrol/findings. Institutional examples include the recent changes to laws in Russia where policymakers have decriminalized domestic violence, as long as there is no 'substantial bodily harm' and is 'only' once a year (Marianna Spring, 'Decriminalisation of Domestic Violence in Russia Leads to Fall in Reported Cases', *The Guardian* (16 Aug. 2018), https://www.theguardian.com/world/2018/aug/16/decriminalisation-of-domestic-violence-in-russia-leads-to-fall-in-reported-cases). In the US there are several state-level and, with the shifting of the supreme court, potentially federal-level efforts to restrict reproductive and family planning rights, along with bills to increasingly restrict access to abortion (BBC News, 'What's Going on in the Fight Over US Abortion Rights?' (14 Jun. 2019), https://www.bbc.com/news/world-us-canada-47940659). It's hard to definitively show a backlash, but for more information on the existing and emerging barriers for women, see Caroline Criado-Perez, *Invisible Women: Exposing Data Bias in a World Designed for Men*, Penguin: London, 2019.

42 Although there are (many) issues with the metrics used, this gives a broad estimate of the rate of change: Oliver Cann, '108 Years: Wait for Gender Equality Gets Longer as Women's Share of Workforce, Politics Drops', World Economic Forum (18 Dec. 2018), https://www.weforum.org/press/2018/12/108-years-wait-for-gender-equality-gets-longer-as-women-s-share-of-workforce-politics-drops/, accessed 10 Dec. 2019.

43 D.Z. O'Brien and J. Rickne, 'Gender Quotas and Women's Political Leadership', *American Political Science Review*, 110/1 (2016), 112–26. doi: 10.1017/S0003055415000611. For more details on these issues see Criado-Perez, *Invisible Women*.

44 In the original French: *L'extension des privilèges des femmes est le principe général de tous progrès sociaux*. Charles Fourier, *Théorie des Quatre Mouvements*, volume 2, chapter 4, 1808.

45 Paul Hawken, ed., *Drawdown: The Most Comprehensive Plan Ever Proposed to Reverse Global Warming*, Penguin: New York, 2017.

46 G. Terry, 'No Climate Justice Without Gender Justice: An Overview of the Issues', *Gender & Development*, 17/1 (2009), 5–18. doi: 10.1080/13552070802696839

47 These data come from a composite of other data sets built by the Gapminder project. More details are available here: https://www.gapminder.org/data/documentation/gd005/, accessed 10 Dec. 2019.

48 These statistics come from a summary of the research at the Our World in Data project: Max Roser, Cameron Appel and Hannah Ritchie, 'Human Height', *Our World in Data* (2019): https://ourworldindata.org/human-height, accessed 10 Dec. 2019.

49 On IQ increases in Britain, see J.R. Flynn, 'Requiem for Nutrition as the Cause of IQ Gains: Raven's Gains in Britain 1938–2008', *Economics & Human Biology*, 7/1 (2009), 18–27. doi: 10.1016/j.ehb.2009.01.009; on the 'top 98%' figure, see Mohamed Nagdy and Max Roser, 'Intelligence', *Our World In Data* (2019), https://ourworldindata.org/intelligence, accessed 10 Dec. 2019.

50 J. Pietschnig and M. Voracek, 'One Century of Global IQ Gains: A Formal Meta-Analysis of the Flynn Effect (1909–2013)', *Perspectives on Psychological Science*, 10/3 (2015). doi: 10.1177/1745691615577701

51 Ibid.

52 For the figures on people living in extreme poverty, see United Nations, 'Goal 1: End Poverty in All Its Forms Everywhere': https://www.un.org/sustainabledevelopment/poverty/, accessed 10 Dec. 2019. For those on increases in hunger, see Food and Agriculture Organization of the United Nations, 'The State of Food Security and Nutrition in the World' (2021), https://www.fao.org/hunger/en/, accessed 4 Jan. 2022.

53 $7.40 as an indicator of poverty: P. Edward, 'The Ethical Poverty Line: A Moral Quantification of Absolute Poverty', *Third World Quarterly*, 27 (2006), 377–93. doi: 10.1080/01436590500432739. Between $10 and $15 as an indicator of poverty: L. Pritchett, 'Who is Not Poor? Proposing a Higher International Standard for Poverty', *Social Science Research Network Electronic Journal* (2008). doi: 10.2139/ssrn.1111717. US poverty line: US Department of Health & Human Services, 'HHS Poverty Guidelines For 2019' (1 Feb. 2019), https://aspe.hhs.gov/poverty-guidelines, accessed 10 Dec. 2019.

54 S. Chen and M. Ravallion, 'More Relatively-Poor People in a Less Absolutely-Poor World', *Review of Income and Wealth*, 59/1 (2013), 1–28. doi: 10.1111/j.1475-4991.2012.00520.x

55 All data rely on World Bank statistics. All the following data were accessed 4 Jan. 2020. Health expenditure: https://data.worldbank.org/indicator/

SH.XPD.CHEX.PC.CD?end=2016&locations=CR-US&most_recent_year_ desc=true&start=2000&view=chart%29. GDP per capita: https://data.worldbank. org/indicator/NY.GDP.PCAP.CD?locations=US-CR. Life expectancy at birth: https:// data.worldbank.org/indicator/SP.DYN.LE00.IN?locations=CR-US&most_recent_ year_desc=true.

56 Life expectancy at birth: Oommen C. Kurian, 'India's New #1: J&K Surpasses Kerala's Life Expectancy, Except At Birth', India Spend (Oct. 2016), https://archive. indiaspend.com/cover-story/indias-new-1-jk-surpasses-keralas-life-expectancy-except-at-birth-64335, accessed 4 Jan. 2020. Income per capita: Government of Kerala State Planning Board 'Economic Review 2017', http://spb.kerala.gov.in/ER2017/web_e/ ch12.php?id=1&ch=12, accessed 4 Jan. 2020.

57 A. Kenny and C. Kenny, Life, Liberty and the Pursuit of Utility: Happiness in Philosophical and Economic Thought, Imprint Academic: London, 2006.

58 Data from the World Bank, 'Access to Electricity (% of Population)', https://data. worldbank.org/indicator/eg.elc.accs.zs, accessed 10 Dec. 2019.

59 S. Pachauri, 'Household Electricity Access a Trivial Contributor to CO_2 Emissions Growth in India', Nature Climate Change, 4/12 (2014), 1073–6. doi:10.1038/ NCLIMATE2414

60 See Saroja Koneru, 'The Changing Cost of Medicine in Developing Countries', The Borgen Magazine (24 Jul. 2016), https://www.borgenmagazine.com/cost-of-medicine-developing-countries/, accessed 10 Dec. 2019. For a full-length exploration of the subject, see Kenny, Getting Better.

61 A good, nuanced and data-driven look at this is available here: Will Koehrsen, 'Has Global Violence Declined? A Look at the Data', Medium (6 Jan. 2019), https://towardsdatascience.com/has-global-violence-declined-a-look-at-the-data-5af708f47fba

62 Max Roser, 'War and Peace', Our World in Data (2019), https://ourworldindata.org/ war-and-peace, accessed 10 Dec. 2019.

63 Ibid.

64 A. Clauset, 'Trends and Fluctuations in the Severity of Interstate Wars', Science Advances, 4/2 (2018), 26–8. doi: 10.1126/sciadv.aao3580

65 For example more frequent storms: A. Grinsted, J.C. Moore and S. Jevrejeva, 'Homogeneous Record of Atlantic Hurricane Surge Threat Since 1923', Proceedings of the. National Academy of Sciences of the United States of America, 109/48 (2012), 19601–5. doi: 10.1073/pnas.1209542109. More economic damage: M. Coronese et al., 'Evidence for Sharp Increase in the Economic Damages of Extreme Natural Disasters', Proceedings of the National Academy of Sciences of the United States of America, 116/43 (2019), 21450–5. doi: 10.1073/pnas.1907826116

66 EMDAT, 'Total Number of Deaths as a Result of Natural Disasters', OFDA/CRED International Disaster Database, Université catholique de Louvain: Brussels, 2019, via Hannah Ritchie and Max Roser, 'Natural Disasters', Our World in Data (2019), https://ourworldindata.org/natural-disasters, accessed 10 Dec. 2019.

67 A. Grinsted, P. Ditlevsen and J.H. Christensen, 'Normalized US Hurricane Damage Estimates Using Area of Total Destruction', 1900–2018. Proceedings of the National Academy of Sciences of the United States of America, 109/46, 1–5 (2019). doi:10.1073/ pnas.1912277116

68 International Organization for Migration, World Migration Report 2013: Migrant Well-Being and Development, United Nations Publications: New York, 2013.

69 Sarah Harper, *How Population Change Will Transform Our World*, Oxford University Press: Oxford, 2016.

70 Florence Jaumotte, Ksenia Koloskova and Sweta C. Saxena, 'Impact of Migration on Income Levels in Advanced Economies', IMF Spillover Task Force (International Monetary Fund, 2016), available at https://www.imf.org/en/Publications/Spillover-Notes/Issues/2016/12/31/Impact-of-Migration-on-Income-Levels-in-Advanced-Economies-44343, accessed 10 Dec. 2019. For a summary see idem, 'Migrants Bring Economic Benefits for Advanced Economies', IMFBlogs (Oct. 2016), https://blogs.imf.org/2016/10/24/migrants-bring-economic-benefits-for-advanced-economies/, accessed 10 Dec. 2019.

71 Jonathan Portes, 'The Economic Impacts of Immigration to the UK', VOX CEPR Policy Portal (6 Apr. 2018), https://voxeu.org/article/economic-impacts-immigration-uk, accessed 10 Dec. 2019.

72 W. Lutz and S Scherbov, 'The Contribution of Migration to Europe's Demographic Future: Projections for the EU25 to 2050', Interim Report IR07–024, International Institute for Applied Systems Analysis (2006).

73 United Nations, *Replacement Migration: Is It a Solution to Declining and Ageing Populations?* (United Nations: New York, 2000).

74 BBC News, 'Migrant Crisis: One Million Enter Europe in 2015' (22 Dec. 2015), https://www.bbc.com/news/world-europe-35158769, accessed 10 Dec. 2019.

75 *The Japan Times*, 'Japan Has the World's Lowest Proportion of Working-Age People, U.N. Report Says' (18 Jun. 2019), https://www.japantimes.co.jp/news/2019/06/18/national/social-issues/japan-worlds-lowest-proportion-working-age-people-u-n-report-says/, accessed 10 Dec. 2019.

76 United Nations, 'World Population Prospects 2019', https://population.un.org/wpp/DataQuery/, accessed 10 Dec. 2019.

77 This is true in the case of the US, for example: *The Economist*, 'Immigrants Boost America's Birth Rate' (30 Aug. 2017), https://www.economist.com/graphic-detail/2017/08/30/immigrants-boost-americas-birth-rate, accessed 10 Dec. 2019. But also, for instance, the case of Syrians (with a fertility rate of 2.92) migrating to Turkey (with a fertility rate of 2.05). For these fertility statistics see World Bank data: 'Fertility Rate, Total (Births per Woman) – Syrian Arab Republic, Turkey', https://data.worldbank.org/indicator/SP.DYN.TFRT.IN?end=2017&locations=SY-TR&start=1960&view=chart, accessed 4 Jan. 2020.

78 R.H. Adams and J. Page, 'Do International Migration and Remittances Reduce Poverty in Developing Countries?' *World Development*, 33/10 (2005), 1645–69. doi: 10.1016/j.worlddev.2005.05.004

79 Education: K. Gyimah-Brempong and E. Asiedu, 'Remittances and Investment in Education: Evidence from Ghana', *Journal of International Trade & Economic Development*, 24/2 (2015), 173–200, doi: 10.1080/09638199.2014.881907. Healthcare: M.L. Held, 'A Study of Remittances to Mexico and Central America: Characteristics and Perspectives of Immigrants', *International Journal of Social Welfare*, 26/1 (2016), 75–85. doi: 10.1111/ijsw.12225. Food shortages: R. Generoso, 'How Do Rainfall Variability, Food Security and Remittances Interact? The Case of Rural Mali', *Ecological Economics*, 114 (2015), 188–98. doi: 10.1016/j.ecolecon.2015.03.009

80 J. Bouoiyour and A. Miftah, 'The Impact of Remittances on Children's Human Capital Accumulation: Evidence from Morocco', *Journal of International Development*, 28/2 (2016), 266–80. doi: 10.1002/jid.3147

81   J. Lehne and F. Preston, 'Making Concrete Change: Innovation in Low-carbon Cement and Concrete', Chatham House Reports (Royal Institute of International Affairs: 2018).

82   In 2014, CO_2 emissions in Bangladesh were 0.474 metric tons per capita (https://data.worldbank.org/indicator/EN.ATM.CO2E.PC?locations=BD). CO_2 emissions in Rwanda were 0.076 metric tons per capita (https://data.worldbank.org/indicator/EN.ATM.CO2E.PC?locations=RW). The CO_2 emissions for a round-trip flight from London to New York are 0.986 tonnes (*The Guardian*, 'Carbon Calculator: Find Out How Much CO_2 Your Flight Will Emit', https://www.theguardian.com/travel/2019/jul/31/carbon-calculator-find-out-how-much-co2-your-flight-will-emit), all websites accessed 10 Dec. 2019. Although these estimates don't include the additional radiative forcing (and consequent warming) of contrails.

83   Global Financial Integrity, 'Illicit Financial Flows to and from 148 Developing Countries: 2006–2015' (2019).

84   As shown in several models of multi-level selection (and references contained) in Peter Turchin, *Ultrasociety: How 10,000 Years of War Made Humans the Greatest Cooperators on Earth*, Beresta Books: Chaplin, CT, 2015.

85   G.J. Abel et al., 'Climate, Conflict and Forced Migration', *Global Environmental Change*, 54 (2019), 239–49. doi: 10.1016/j.gloenvcha.2018.12.003

86   Peter Turchin, *Ultrasociety: How 10,000 Years of War Made Humans the Greatest Cooperators on Earth*, Beresta Books: Chaplin, CT, 2015, pp. 1-15.

87   For polio statistics see World Health Organization, 'Poliomyelitis' (Jul. 2019), https://www.who.int/news-room/fact-sheets/detail/poliomyelitis. For Guinea worm disease see World Health Organisation, 'Dracunculiasis (guinea-worm disease)' (Mar. 2019), https://www.who.int/news-room/fact-sheets/detail/dracunculiasis-(guinea-worm-disease). For statistics on vaccination rates see UNICEF, 'Immunization', https://www.unicef.org/immunization. For details on spearheading responses to HIV/Aids see United Nations, 'AIDS', https://www.un.org/en/sections/issues-depth/aids/index.html, all accessed 10 Dec. 2019.

88   UN, 'International Day for the Preservation of the Ozone Layer, 16 September', https://www.un.org/en/events/ozoneday/background.shtml, accessed 10 Dec. 2019.

89   Jack Doyle, 'DuPont's Disgraceful Deeds', *Multinational Monitor*, 12/10 (1991), https://www.multinationalmonitor.org/hyper/issues/1991/10/doyle.html, accessed 4 Jan. 2020. Wiliam Glaberson, 'Behind Du Pont's Shift On Loss of Ozone Layer', *New York Times* (26 Mar. 1988), https://www.nytimes.com/1988/03/26/business/behind-du-pont-s-shift-on-loss-of-ozone-layer.html, accessed 28 Feb. 2020.

90   For an overview see Sigal Samuel's article: 'Should Animals, Plants, and Robots Have the Same Rights As You?' *Vox* (4 Apr. 2019), https://www.vox.com/future-perfect/2019/4/4/18285986/robot-animal-nature-expanding-moral-circle-peter-singer, accessed 10 Dec. 2019.

Chapter Three – Pessimism – Slaves to Power

1   Georgescu-Roegen, N., *Southern Economic Journal*, 41:3, 1975, 381.

2   Vaclav Smil, *Energy in Nature and Society: General Energetics of Complex Systems*, MIT Press: Cambridge, MA, 2007.

3   J.V. Henderson, A. Storeygard and D.N. Weil, 'Measuring Economic Growth from Outer Space', *American Economic Review*, 102/2 (2012), 994–1028. doi: 10.1257/aer.102.2.994

4   A six-watt LED bulb has a light output of around 450 lumens. An approximate candlepower conversion is one candle = 12.57 lumens.

5   A. Usubiaga-Liaño, P. Behrens and V. Daioglou, 'Energy Use in the Global Food System', Journal of Industrial Ecology (2020), 1–11. doi: 10.1111/jiec.12982; and S. Wood and A. Cowie, A Review of Greenhouse Gas Emission Factors for Fertiliser Production, IEA Bionergy: Paris, 2004.

6   J. Hendrickson, 'Energy Use in the U.S. Food System: A Summary of Existing Research and Analysis', Center for Integrated Agricultural Systems, University of Wisconsin: Madison, WI, 1996.

7   Laura Cozzi and Apostolos Petropoulos, 'Growing Preference for SUV Challenges Emissions Reductions in Passenger Car Market', International Energy Agency (15 Oct. 2019), https://www.iea.org/commentaries/growing-preference-for-suvs-challenges-emissions-reductions-in-passenger-car-market, accessed 10 Dec. 2019.

8   Martin Ross, 'More Flights and Fuller Aircraft as UK Air Traffic Continues to Grow', Civil Aviation Authority (27 Jul. 2018), https://www.caa.co.uk/Blog-Posts/More-flights-and-fuller-aircraft-as-UK-air-traffic-continues-to-grow/, accessed 10 Dec. 2019.

9   For instance see the reporting from Michelle Nijhuis: 'Three Billion People Cook Over Open Fires – with Deadly Consequences', National Geographic (14 Aug. 2017), https://www.nationalgeographic.com/photography/proof/2017/07/guatemala-cook-stoves/, accessed 10 Dec. 2019.

10 A European or US fridge-freezer uses around 495 kWh per year, as described in Intelligent Energy Europe, 'Technical Guidebook Efficient Cold Products in Residential and Commercial Sectors', https://ec.europa.eu/energy/intelligent/projects/sites/iee-projects/files/projects/documents/proefficiency_technical_guide_efficinet_cold_products_en.pdf, and the average Ethiopian uses around 70 kWh per person per year, World Bank, 'Electric Power Consumption (kWh per Capita) – Ethiopia', https://data.worldbank.org/indicator/EG.USE.ELEC.KH.PC?locations=ET, both accessed 4 Jan. 2020.

11 Delphine Strauss, 'French "Gilets Jaunes" Show Pain of Macron's Tax Policy', Financial Times (3 Dec. 2018), available to subscribers at https://www.ft.com/content/b6297b3a-f4bd-11e8-9623-d7f9881e729f, accessed 10 Dec. 2019.

12 BBC, 'Why is There a Backlash Against Climate Policies?' [podcast] (2 Dec. 2019), The Inquiry, https://www.bbc.co.uk/programmes/w3csyth0, accessed 7 Jan. 2020.

13 Declan Walsh, 'Alienated and Angry, Coal Miners See Donald Trump as Their Only Choice', New York Times (19 Aug. 2016), https://www.nytimes.com/2016/08/20/world/americas/alienated-and-angry-coal-miners-see-donald-trump-as-their-only-choice.html, accessed 10 Dec. 2019.

14 The fatality figures for coal mining alone demonstrate how dangerous it is: see 'Fatality Reports', United States Department of Labor, https://www.msha.gov/data-reports/fatality-reports/search, accessed 10 Dec. 2019.

15 One of many examples was the notice of proposed rulemaking issued to revitalize the coal industry on the basis of grid resiliency. This was ultimately rejected, but many in the industry were highly critical of the attempt. See 'Notice of Proposed Rulemaking for the Grid Resiliency Pricing Rule' Department of Energy (2017), https://www.energy.gov/downloads/notice-proposed-rulemaking-grid-resiliency-pricing-rule, accessed 10 Dec. 2019.

16 Emily Holden, 'Fossil Fuel Firms Linked to Trump Get Millions in Coronavirus Small Business Aid', The Guardian (1 May 2020), https://www.theguardian.com/

environment/2020/may/01/fossil-fuel-firms-coronavirus-package-aid, accessed 15 Jan. 2022.

17 Irina Ivanova, 'America's Largest Miners Coal Union Asks Manchin to Reconsider Opposition to Build Back Better', CBS (22 Dec. 2021), https://www.cbsnews.com/news/manchin-united-mine-workers-build-back-better/, accessed 15 Jan. 2022.

18 On the UK's subsidies for fossil fuels, see Damian Carrington, 'UK Has Biggest Fossil Fuel Subsidies in the EU, Finds Commission', The Guardian (23 Jan. 2019), https://www.theguardian.com/environment/2019/jan/23/uk-has-biggest-fossil-fuel-subsidies-in-the-eu-finds-commission. On the slashing of subsidies for solar power, see Jillian Ambrose, 'Home Solar Panel Installations Fall by 94% As Subsidies Cut', The Guardian (5 Jun. 2019), https://www.theguardian.com/environment/2019/jun/05/home-solar-panel-installations-fall-by-94-as-subsidies-cut, accessed 10 Dec. 2019.

19 Sandra Laville, 'Top Oil Firms Spending Millions Lobbying to Block Climate Change Policies, Says Report', The Guardian (22 Mar. 2019), https://www.theguardian.com/business/2019/mar/22/top-oil-firms-spending-millions-lobbying-to-block-climate-change-policies-says-report, accessed 10 Dec. 2019.

20 'Private Jets Receive Ludicrous Tax Breaks That Hurt the Environment', The Economist (7 Mar. 2019), https://www.economist.com/leaders/2019/03/07/private-jets-receive-ludicrous-tax-breaks-that-hurt-the-environment, accessed 10 Dec. 2019.

21 This land would now be aggregated into larger parcels of land so the fields would be larger also.

22 For the 90% figure see R. C. Allen, 'Economic Structure and Agricultural Productivity in Europe, 1300–1800', European Review of Economic History, 4/1, (2000), 1–26. doi: 10.1017/s1361491600000125. For today's 1.2% figure see: World Bank, 'Employment in Agriculture (% of Total Employment)', https://data.worldbank.org/indicator/SL.AGR.EMPL.ZS, accessed 10 Dec. 2019.

23 Digital History, 'Housework in Late 19th Century America' (2019): http://www.digitalhistory.uh.edu/topic_display.cfm?tcid=93, accessed 10 Dec. 2019. For more information see: C.L. Shehan and A.B. Moras, 'Deconstructing Laundry: Gendered Technologies and the Reluctant Redesign of Household Labor', Michigan Family Review, 11 (2006), 39–54.

24 H. Rosling, Factfulness: Ten Reasons We're Wrong About the World – and Why Things Are Better Than You Think, Flatiron Books: New York, 2018.

25 This includes the embodied energy used in imports and is based on data from David Mackay, Sustainable Energy – Without the Hot Air, UIT Cambridge: Cambridge (2009).

26 For details on the minimum wage see: UK Government, 'National Minimum Wage and National Living Wage Rates' (2019), https://www.gov.uk/national-minimum-wage-rates, accessed 10 Dec. 2019. For rates of debt see: Amelia Hill, 'Seven in 10 UK Workers Are "Chronically Broke", Study Finds', The Guardian (25 Jan. 2018), https://www.theguardian.com/money/2018/jan/25/uk-workers-chronically-broke-study-economic-insecurity, accessed 10 Dec. 2019.

27 World Health Organization, '9 out of 10 People World-wide Breathe Polluted Air, but More Countries Are Taking Action' (2 May 2018), https://www.who.int/news-room/detail/02-05-2018-9-out-of-10-people-worldwide-breathe-polluted-air-but-more-countries-are-taking-action, accessed 8 Dec. 2019.

28 European Court of Auditors, Special Report no 23/2018: Air Pollution: Our Health Still Insufficiently Protected (11 Sep. 2018). doi:10.2865/363524

29   70,000 articles see Dr Maria Neira, WHO Director of Public and Environmental Health, as quoted in: Damian Carrington, 'Revealed: Air Pollution May Be Damaging "Every Organ in the Body"', *The Guardian* (17 May 2019), https://www.theguardian.com/environment/ng-interactive/2019/may/17/air-pollution-may-be-damaging-every-organ-and-cell-in-the-body-finds-global-review, accessed 10 Dec. 2019.

30   For aggravated asthma see D. E. Schraufnagel et al., 'Air Pollution and Noncommunicable Diseases: A Review by the Forum of International Respiratory Societies' Environmental Committee, Part 1: The Damaging Effects of Air Pollution', *Chest*, 155/2 (2019), 409–416. doi: 10.1016/j.chest.2018.10.042. For lung cancer: R. T. Burnett et al., 'An Integrated Risk Function for Estimating the Global Burden of Disease Attributable to Ambient Fine Particulate Matter Exposure', *Environmental Health Perspectives*, 122/9 (2014), 397-403. doi:10.1289/ehp.1307049. For Alzheimer's: J. Kilian and M. Kitazawa, 'The Emerging Risk of Exposure to Air Pollution on Cognitive Decline and Alzheimer's Disease – Evidence from Epidemiological and Animal Studies', *Biomedical Journal*, 41/3 (2018), 141-162. doi:10.1016/j.bj.2018.06.001. For miscarriage risk: L. Zhang et al., 'Air Pollution-Induced Missed Abortion Risk for Pregnancies', *Nature Sustainability*, 2 (2019), 1011–1017. doi:10.1038/s41893-019-0387-y. For diabetes: B. Bowe et al., 'The 2016 Global and National Burden of Diabetes Mellitus Attributable to PM 2·5 Air Pollution', *Lancet Planetary Health*, 2/7 (2018), 301-312. doi:10.1016/S2542-5196(18)30140-2. For low birth rates: R.B. Smith et al., 'Impact of London's Road Traffic Air and Noise Pollution on Birth Weight: Retrospective Population Based Cohort Study', *British Medical Journal*, 359 (2017), 1–13. doi:10.1136/bmj.j5299. For depression and suicide: see I. Braithwaite et al., 'Air Pollution (Particulate Matter) Exposure and Associations with Depression, Anxiety, Bipolar, Psychosis and Suicide Risk: A Systematic Review and Meta-Analysis', *Environmental Health Perspectives*, 127/12 (2019). doi: 10.1289/EHP4595

31   J. Heissel, C. Persico, and D. Simon, *Does Pollution Drive Achievement? The Effect of Traffic Pollution on Academic Performance*. NBER *Working Paper* (November 2019), IZA Institute of Labor Economics, http://ftp.iza.org/dp12745.pdf, accessed 10 Dec. 2019.

32   For example, see: C.T. Loftus et al., 'Prenatal Air Pollution and Childhood IQ: Preliminary Evidence of Effect Modification by Folate', *Environmental Research*, 176 (2019), 108505. doi:10.1016/j.envres.2019.05.036 or X. Zhang, X. Chen and X. Zhang, 'The Impact of Exposure to Air Pollution on Cognitive Performance', *Proceedings of the National Academy of Sciences of the United States of America*, 115/37 (2018), 9193–9197. doi: 10.1073/pnas.1809474115

33   A. Sharma and P. Kumar, 'A Review of Factors Surrounding the Air Pollution Exposure to In-Pram Babies and Mitigation Strategies'. *Environment International*, 120 (2018), 262–278. doi: 10.1016/j.envint.2018.07.038

34   L.P. Clark, D.B. Millet and J.D. Marshall, 'Changes in Transportation-Related Air Pollution Exposures by Race-Ethnicity and Socioeconomic Status: Outdoor Nitrogen Dioxide in the United States in 2000 and 2010', *Environmental Health Perspectives*, 125/9 (2017). doi: 10.1289/EHP959

35   For overall trends, see United States Environmental Protection Agency, 'Particulate Matter (PM2.5) Trends', https://www.epa.gov/air-trends/particulate-matter-pm25-trends, accessed 9 Jan. 2022. For reporting on the particulate matter consequences of wildfires, see Vivian Ho, 'West Coast Cities Face the World's Worst Air Quality

As Wildfires Rage', *The Guardian* (14 Sep. 2020), https://www.theguardian.com/world/2020/sep/14/west-coast-air-quality-wildfires-oregon-california-washington, accessed 9 Jan. 2022.

36 P.J. Landrigan et al., 'The Lancet Commission on Pollution and Health', *The Lancet*, 391/10119 (2017), 462–512. doi: 10.1016/S0140-6736(17)32345-0

37 J.S. Apte et al., 'Ambient PM 2.5 Reduces Global and Regional Life Expectancy', *Environmental Science & Technology Letters* (2018), 546–551. doi:10.1021/acs.estlett.8b00360

38 Damian Carrington and Matthew Taylor, 'Air pollution is the "new tobacco" warns WHO Head', *The Guardian* (27 Oct. 2018), https://www.theguardian.com/environment/2018/oct/27/air-pollution-is-the-new-tobacco-warns-who-head, accessed 4 Jan. 2019.

39 World Health Organization, 'COP24 Special Report: Health and Climate Change 38' (2018), https://www.who.int/globalchange/publications/COP24-report-health-climate-change/en/, accessed 10 Dec. 2019.

40 For the EU, see P. Behrens et al., 'Climate Change and the Vulnerability of Electricity Generation to Water Stress in the European Union', *Nature Energy*, 2 (2017). doi: 10.1038/nenergy.2017.114. For the US, see M.T.H. Van Vliet et al., 'Power-generation System Vulnerability and Adaptation to Changes in Climate and Water Resources', *Nature Climate Change*, 6 (2016), 375–380. doi:10.1038/nclimate2903

41 Akshat Rathi, 'Yet Another Reason for the Doom of Europe's Fossil-Fuel Power Plants: Not Enough Water', *Quartz* (26 Jul. 2017), https://qz.com/1038035/climate-change-driven-droughts-could-cause-power-shortages-in-europe-according-to-a-new-study/, accessed 10 Dec. 2019.

42 Christopher F. Schuetze, 'The Rhine, a Lifeline of Germany, Is Crippled by Drought', *New York Times* (4 Nov. 2018), https://www.nytimes.com/2018/11/04/world/europe/rhine-drought-water-level.html, accessed 10 Dec. 2019.

43 Bate Felix and Sybille de La Hamaide, 'Latest Hot Spell Set to Deepen Drought Pain in France', *Reuters* (17 Jul. 2019), https://www.reuters.com/article/us-france-drought/latest-hot-spell-set-to-deepen-drought-pain-in-france-idUSKCN1UC1V8, accessed 10 Dec. 2019.

44 Tom DiChristopher, 'Trump Says Breaking with Saudi Arabia Would Send Oil Prices "Through the Roof"', *CNBC* (20 Nov. 2018), https://www.cnbc.com/2018/11/20/trump-says-breaking-with-saudi-arabia-would-send-oil-prices-through-the-roof.html, accessed 10 Dec. 2019.

45 This includes direct and indirect subsidies, including environmental costs like climate change, air pollution, other vehicle externalities like congestion, along with direct subsidies for production and consumption. It should be noted that the carbon costs assumed in the study are conservative, at $40 per tonne, which many think should be much higher. For the research see D. Coady et al., 'Global Fossil Fuel Subsidies Remain Large: An Update Based on Country-Level Estimates', *IMF Working Papers*, 19/89 (2019), 1–36.

46 Simon English, 'Cheney Had Iraq in Sights Two Years Ago', *The Telegraph* (22 Jul. 2003), https://www.telegraph.co.uk/news/worldnews/northamerica/usa/1436785/Cheney-had-Iraq-in-sights-two-years-ago.html, accessed 10 Dec. 2019. Paola Totaro, 'Memos Show Oil Motive in Iraq War', *Sydney Morning Herald* (20 Apr. 2011), https://www.smh.com.au/world/memos-show-oil-motive-in-iraq-war-20110419-1dnkf.html, accessed 10 Dec. 2019.

47   Joe Stiglitz and Linda Bilmes, 'The Three Trillion Dollar War', *The Times* (23 Feb. 2008), https://www.thetimes.co.uk/article/the-three-trillion-dollar-war-kw5qcglpwgg, accessed 10 Dec. 2019.

48   Chloe Holden, 'The Price of a Fully Renewable US Grid: $4.5 Trillion', *Green Tech Media* (28 Jun. 2019), https://www.greentechmedia.com/articles/read/renewable-us-grid-for-4-5-trillion, accessed 4 Jan. 2020. This was based on the Wood Mackenzie consultancy report: 'Deep Decarbonisation: the Multi-Trillion-Dollar Question' (2019), https://www.woodmac.com/news/feature/deep-decarbonisation-the-multi-trillion-dollar-question/, accessed 4 Jan. 2020.

49   The BBC, 'Nord Stream 2: Go-ahead for Russian gas pipeline angers Ukraine', BBC (31 Oct. 2019), https://www.bbc.com/news/world-europe-50247793, accessed 10 Dec. 2019.

50   Michael T. Klare, *Rising Powers, Shrinking Planet: The New Geopolitics of Energy*, Metropolitan Books: New York, 2008.

51   F. Hill, C.G. Gaddy and I. Danchenko, event of the Brookings Institution, 'The Mystery of Vladimir Putin's Dissertation', 30 Mar. 2006. Slides and notes available here: https://www.brookings.edu/events/the-mystery-of-vladimir-putins-dissertation/, *Brookings Institute* (2006), accessed 10 Dec. 2019.

52   William J. Broad, 'In Taking Crimea, Putin Gains a Sea of Fuel Reserves', *New York Times* (17 May 2014), https://www.nytimes.com/2014/05/18/world/europe/in-taking-crimea-putin-gains-a-sea-of-fuel-reserves.html, accessed 10 Dec. 2019.

53   Steve Mollman, 'The US Says China Is Blocking $2.5 Trillion in South China Sea Oil and Gas', *Quartz* (25 Aug. 2019), https://qz.com/1694322/south-china-seas-oil-and-natural-gas-pretty-important-after-all/, accessed 10 Dec. 2019.

54   Niharika Mandhana, 'Vietnam Told China to Get Out of Its Waters. Beijing's Response: No, You Get Out', *The Wall Street Journal* (1 Nov. 2019), https://www.wsj.com/articles/vietnam-told-china-to-get-out-of-its-waters-beijings-response-no-you-get-out-11572625722, accessed 10 Dec. 2019.

55   C. Mora et al., 'Dredging in the Spratly Islands: Gaining Land but Losing Reefs', *PLoS Biology*, 14/6 (2016), 4–10. doi: 10.1371/journal.pbio.1002422

56   Rachael Bale, 'Critical Reefs Destroyed in Poachers' Quest for World's Biggest Clams', *National Geographic* (30 Aug. 2016), https://www.nationalgeographic.com/news/2016/08/wildlife-giant-clam-poaching-south-china-sea-destruction/, accessed 10 Dec. 2019.

57   Agence France-Presse, 'Chinese Warship Sails Within Yards of US destroyer in "Unsafe" Encounter', *The Guardian* (2 Oct. 2018), https://www.theguardian.com/world/2018/oct/01/chinese-warship-american-destroyer-uss-decatur-unsafe-encounter, accessed 10 Dec. 2019.

58   Andreas Malm, *The Progress of This Storm*, Verso: London, 2017.

59   R. Winkelmann et al., 'Combustion of Available Fossil Fuel Resources Sufficient to Eliminate the Antarctic Ice Sheet', *Science Advances*, 1/8 (2015), 1–6. doi: 10.1126/sciadv.1500589

60   See, for instance: G. Supran and N. Oreskes, 'Assessing ExxonMobil's Climate Change Communications (1977-2014)', *Environmental Research Letters*, 12/8 (2017). doi: 10.1088/1748-9326/aa815f

61   Amy Lieberman and Susanne Rust, 'Big Oil Companies Unite to Fight Regulations; but Spent Millions Bracing for Climate Change', *The Los Angeles Times* (31 Dec. 2015), https://www.latimes.com/nation/la-na-oil-operations-20151231-story.html, accessed 4 Jan. 2019.

62 Naomi Oreskes and Eric M. Conway, *Merchants of Doubt: How a Handful of Scientists Obscured the Truth on Issues from Tobacco Smoke to Global Warming*, Bloomsbury Publishing: London, 2010.

63 Ben Chapman, 'BP and Shell Planning for Catastrophic 5°C Global Warming Despite Publicly Backing Paris Climate Agreement', *The Independent* (27 Oct. 2017), https://www.independent.co.uk/news/business/news/bp-shell-oil-global-warming-5-degree-paris-climate-agreement-fossil-fuels-temperature-rise-a8022511.html, accessed 10 Dec. 2019.

64 You'd need an area three times the size of Belgium full of algal ponds to supply 10% of EU oil demand, see Kevin Flynn, 'Algal Biofuel Production Is Neither Environmentally Nor Commercially Sustainable', *The Conversation* (8 Aug. 2017), https://theconversation.com/algal-biofuel-production-is-neither-environmentally-nor-commercially-sustainable-82095, accessed 10 Dec. 2019.

65 A paid advertisement from Shell hosted at *New York Times*, 'A Net-Zero Emissions World by 2070?', https://www.nytimes.com/paidpost/shell/net-zero-emissions-by-2070.html, accessed 4 Jan. 2019.

66 V. Masson-Delmotte et al., eds, 'Summary for Policymakers' in *Global Warming of 1.5°C. An IPCC Special Report on the Impacts of Global Warming of 1.5°C Above Pre-Industrial Levels and Related Global Greenhouse Gas Emission Pathways, in the Context of Strengthening the Global Response to the Threat of Climate Change, Sustainable Development, and Efforts to Eradicate Poverty*, IPCC, World Meteorological Organization: Geneva, 2018. doi: 10.1017/CBO9781107415324

67 Frédéric Simon, 'Exxon Lobbyists Allowed to Keep EU Access Badges', *EURACTIV* (12 Apr. 2019), https://www.euractiv.com/section/climate-environment/news/exxon-lobbyists-allowed-to-keep-eu-access-badges/, accessed 10 Dec. 2019.

68 Auke Hoekstra, 'Photovoltaic Growth: Reality Versus Projections of the International Energy Agency', *Steinbuch: Blog page of Maarten Steinbuch* (12 Jun. 2017), https://steinbuch.wordpress.com/2017/06/12/photovoltaic-growth-reality-versus-projections-of-the-international-energy-agency/, accessed 4 Jan. 2020.

69 These estimates have a second damaging effect. They depress the financial interest in renewables. Institutional investors have the trillions of dollars needed to supercharge the energy transition but they prefer to invest in huge, one-off projects they can do the due diligence on. If major international entities are saying that it will still take some time for renewables to win then they are less likely to make that bet. Here the modular advantages of solar and wind become a weakness. Big investors don't generally deal with smaller projects, which are what renewable energy projects are in their multitudes. This locks out a large amount of funding for no other reason than institutional arrangements. The message also needs to get out. Partly this is due to the amount of money financiers are funnelling into the energy system and their obsolete understanding of the marketplace.

70 V. Masson-Delmotte et al., eds, 'Summary for Policymakers', in *Global Warming of 1.5°C. An IPCC Special Report on the Impacts of Global Warming of 1.5°C Above Pre-Industrial Levels and Related Global Greenhouse Gas Emission Pathways, in the Context of Strengthening the Global Response to the Threat of Climate Change, Sustainable Development, and Efforts to Eradicate Poverty*, IPCC, World Meteorological Organization: Geneva, 2018. doi: 10.1017/CBO9781107415324

71 *Rainforest Action Network*, 'Banking on Climate Change 2019' (20 Mar. 2019), https://www.ran.org/bankingonclimatechange2019/, accessed 10 Dec. 2019. This was

a multi-NGO report from organizations including the Rainforest Action Network, Sierra Club and Bank Track.

72 Steve Inskeep and Ashley Westerman, 'Why Is China Placing A Global Bet on Coal?', *National Public Radio* (29 Apr. 2019), https://www.npr.org/2019/04/29/716347646/why-is-china-placing-a-global-bet-on-coal, accessed 10 Dec. 2019

73 J.-F. Mercure et al., 'Reframing Incentives for Climate Policy Action', *Nature Energy*, 6 (2021), 1133–43. doi:10.1038/s41560-021-00934-2

74 United Kingdom Office for National Statistics, visualized here: S. O'Dea, 'UK Households: Ownership of Mobile Phones 1996–2018', *Statistica* (9 Aug. 2019), https://www.statista.com/statistics/289167/mobile-phone-penetration-in-the-uk/, accessed 10 Dec. 2019.

75 S.J. Davis and R.H. Socolow, 'Commitment Accounting of CO_2 Emissions', *Environmental Research Letters*, 9/8 (2014). doi: 10.1088/1748-9326/9/8/084018

76 Methane is much more potent as a warming gas than carbon dioxide. However, it does break down into carbon dioxide after twelve years. Because of this some sort of averaging of its impact must be made. The 3.2% figure given here is if you take the average over twenty years rather than the more common one hundred years. These choices in averaging are controversial but one thing that isn't is many companies are spending hundreds of billions of dollars on infrastructure which is likely to become stranded. For further information, see R.A. Alvarez et al., 'Greater Focus Needed on Methane Leakage from Natural Gas Infrastructure', *Proceedings of the National Academy of Sciences of the United States of America*, 109/17 (2012), 6435–6440. doi: 10.1073/pnas.1202407109

77 Warren Cornwall, 'Natural Gas Could Warm the Planet as Much as Coal in the Short Term', *Science* (21 Jun. 2018), https://www.sciencemag.org/news/2018/06/natural-gas-could-warm-planet-much-coal-short-term, accessed 10 Dec. 2019.

78 V. Smil, 'Examining Energy Transitions: A Dozen Insights Based on Performance', *Energy Research and Social Science*, 22 (2016), 194–197. doi: 10.1016/j.erss.2016.08.017

79 Jeff Tollefson, 'COVID Curbed Carbon Emissions in 2020 – but Not by Much', *Nature News* (15 Jan. 2021), https://www.nature.com/articles/d41586-021-00090-3, accessed 9 Jan. 2022.

80 Private communication with environmental economist Ranran Wang on econometric modelling in a forthcoming manuscript.

81 This is a (very) rough estimate from R. Yuan, P. Behrens et al., 'Carbon Overhead: The Impact of the Expansion in Low-Carbon Electricity in China 2015-2040', *Energy Policy*, 119 (2018), 97–104. doi: 10.1016/j.enpol.2018.04.027

82 Jeff Masters, 'The Top 10 Global Weather and Climate Change Events of 2021', Yale Climate Connections (11 Jan. 2022), https://yaleclimateconnections.org/2022/01/the-top-10-global-weather-and-climate-change-events-of-2021/, accessed 14 Jan. 2022.

83 Fiona Harvey, 'Biggest Food Brands Failing Goals to Banish Palm Oil Deforestation', *The Guardian* (17 Jan. 2020), https://www.theguardian.com/environment/2020/jan/17/biggest-food-brands-failing-goals-to-banish-palm-oil-deforestation, accessed 15 Jan. 2022.

84 Harvey, 'Biggest Food Brands'.

85 I found the physical space of COP26 to be stultifying. Mixed in between cafés selling largely animal-based meals (the huge fish and chip stand took centre stage in the main food hall) were drab meeting rooms. The one area with more life – the 'action

hub' – had corporate sponsorship prominently displayed on a continuous line of ticker tape, as if it were a sports event. Whereas the largely successful COP21 in Paris was held in an environment designed to put good decision-making first (with plentiful areas to de-stress and even have a rest), the Glasgow environment felt unnecessarily claustrophobic and stressful. For reporting of the policing at Glasgow, see Chris Green's 'COP26: UN Climate Change Conference in Glasgow Will Be Policed by 10,000 Officers Every Day', *iNews* (24 Jun. 2021), https://inews.co.uk/news/scotland/cop26-un-climate-change-conference-in-glasgow-will-be-policed-by-10000-officers-every-day-1069682, accessed 14 Jan. 2022; and Libby Page, 'COP26 Police Tactics Creating Atmosphere of Fear, Protesters Say', *The Guardian* (11 Nov. 2021), https://www.theguardian.com/environment/2021/nov/11/cop26-police-tactics-creating-atmosphere-of-fear-protesters-say, accessed 14 Jan. 2022.

86   Matt McGrath, 'COP26: Fossil Fuel Industry Has Largest Delegation at Climate Summit', *BBC News* (8 Nov. 2021), https://www.bbc.com/news/science-environment-59199484, accessed 14 Jan. 2022.

87   David Vetter, 'Top Climate Experts Blast Rich Nations For "Breakdown in Trust" at COP26', *Forbes* (26 Nov. 2021), https://www.forbes.com/sites/davidrvetter/2021/11/26/top-climate-experts-blast-rich-nations-for-breakdown-in-trust-at-cop26/, accessed 16 Jan. 2022.

88   For references and further discussion, see Paul Behrens, 'The Best- and Worst-Case Scenarios for the World as COP26 Ends', *Politico* (11 Nov. 2021), https://www.politico.com/news/magazine/2021/11/11/cop-26-climate-future-520817, accessed 14 Jan. 2022.

89   The rebound was reported on here: Jeff Tollefson, 'Carbon Emissions Rapidly Rebounded Following COVID Pandemic Dip', *Nature News* (4 Nov. 2021), https://www.nature.com/articles/d41586-021-03036-x, accessed 14 Jan. 2022; and expectations for increases in US emissions were reported on here: John Kemp, 'Column: U.S. Energy-Related Emissions Likely to Rise in 2022 and 2023: Kemp', *Reuters* (13 Jan. 2022), https://www.reuters.com/business/energy/us-energy-related-emissions-likely-rise-2022-2023-kemp-2022-01-12/, accessed 14 Jan. 2022.

Chapter Four   Hope – Power to the People

1   These details come from a mixture of sources: Andrew Nikiforuk, 'The Big Shift Last Time: From Horse Dung to Car Smog', *The Tyee* (6 Mar. 2013), https://thetyee.ca/News/2013/03/06/Horse-Dung-Big-Shift/; Jennifer Lee, 'When Horses Posed a Public Health Hazard', *New York Times, City Room Blogs* (9 Jun. 2008), https://cityroom.blogs.nytimes.com/2008/06/09/when-horses-posed-a-public-health-hazard/; and Brad Smith and Carol Ann Browne, 'The Day the Horse Lost Its Job', Microsoft Blogs (10 Dec. 2017), https://blogs.microsoft.com/today-in-tech/day-horse-lost-job/, and references therein, all accessed 10 Dec. 2019.

2   D. Bogdanov et al., 'Radical Transformation Pathway Towards Sustainable Electricity via Evolutionary Steps', *Nature Communications*, 10/1 (2019). doi: 10.1038/s41467-019-08855-1

3   Fred Lambert, 'Tesla is Working on New Battery that Lasts 1 Million Miles to Come Out Next Year, Says Elon Musk', *Electrek* (23 Apr. 2019), https://ww.electrek.co/2019/04/23/tesla-battery-million-miles-elon-musk/, accessed 10 Dec. 2019. Although Musk is prone to overestimation, this statement appears to be backed up by academic modelling: J.E. Harlow et al., 'A Wide Range of Testing Results

on an Excellent Lithium-Ion Cell Chemistry to Be Used as Benchmarks for New Battery Technologies', *Journal of the Electrochemical Society*, 166/13 (2019), A3031–44. doi:10.1149/2.0981913jes

4   Commercial flights: Guardian Staff, 'World's First Fully Electric Commercial Aircraft Takes Flight in Canada', *The Guardian* (11 Dec. 2019), https://www.theguardian. com/world/2019/dec/11/worlds-first-fully-electric-commercial-aircraft-takes-flight-in-canada. Commercial ferry: Fred Lambert, 'World's Largest All-Electric Ferry Completes Its Maiden Trip', *Electrek* (21 Aug. 2019), https://electrek.co/2019/08/21/ worlds-largest-electric-ferry/, both accessed 3 Jan. 2020.

5   Dr Christine Shearer, 'Analysis: The Global Coal Fleet Shrank for First Time on Record in 2020', *Carbon Brief* (3 Aug. 2020), https://www.carbonbrief.org/analysis-the-global-coal-fleet-shrank-for-first-time-on-record-in-2020, accessed 9 Jan. 2022.

6   S.J. Davis et al., 'Net-Zero Emissions Energy Systems', *Science* 360/6396 (2018). doi: 10.1126/science.aas9793

7   Wikipedia has a very solid round-up of the many different studies from different countries on the cost of electricity by different source: https://en.wikipedia.org/wiki/ Cost_of_electricity_by_source, accessed 10 Dec. 2019. It is important to be aware, though, that full 'system integration costs' for renewables might be higher when storage has to be taken into account.

8   McKinsey & Company, 'Global Energy Perspective 2019: Reference Case' (Jan. 2019).

9   R.F. Service, 'Solar Plus Batteries Is Now Cheaper than Fossil Power', *Science*, 365/6449 (2019), 108. doi: 10.1126/science.365.6449.108

10   International Energy Agency, 'Coal Consumption by Sector in China 2008–2014', International Energy Agency (16 Dec. 2019), https://www.iea.org/data-and-statistics/ charts/coal-consumption-by-sector-in-china-2008-2024, accessed 1 Mar. 2022.

11   Photovoltaic Power Systems Programme, 'Snapshot 2021', International Energy Agency (Apr. 2021), https://iea-pvps.org/snapshot-reports/snapshot-2021/, accessed 9 Jan. 2022.

12   Tim Buckley, Director of Energy Finance Studies at the Institute of Energy Economics and Financial Analysis (IEEFA), made these statements on the Energy Transition Show, 'Episode #91 – Energy Transition in India and Southeast Asia, Part 1' (20 Mar. 2019), which you can listen to here: https://xenetwork.org/ets/episodes/episode-91-energy-transition-in-india-and-southeast-asia-part-1/, accessed 10 Dec. 2019.

13   Solar panel prices: X. Wang and A. Barnett, 'The Evolving Value of Photovoltaic Module Efficiency'. *Applied Sciences*, 9/6 (2019), 1227. doi: 10.3390/app9061227. Lithium-ion batteries: C. Curry, 'Lithium-Ion Battery Costs and Market'. *Bloomberg New Energy Finance* (2015), 164–88, https://data.bloomberglp.com/bnef/sites/14/2017/07/ BNEF-Lithium-ion-battery-costs-and-market.pdf, accessed 10 Dec. 2019.

14   Ed Davey, 'What Is the Most Expensive Object on Earth?', BBC News (29 Apr. 2016), https://www.bbc.com/news/magazine-36160368, accessed 10 Dec. 2019.

15   James Temple, 'Why France Is Eyeing Nuclear Power Again', *MIT Technology Review* (16 Oct. 2019), https://www.technologyreview.com/s/614579/why-france-is-eyeing-nuclear-power-again/, accessed 10 Dec. 2019.

16   Anshul Joshi, 'India to Add 20,000MW of Nuclear Power Generation Capacity Over Next Decade: DAE Secy', *ETEnergyWorld* (18 Oct. 2019), https://energy. economictimes.indiatimes.com/news/power/india-to-add-20000-mw-of-nuclear-power-generation-capacity-over-next-decade-dae-secy/71652199, accessed 10 Dec. 2019.

17 Christopher Flavelle and Jeremy C.F. Lin, 'U.S. Nuclear Power Plants Weren't Built for Climate Change', *Bloomberg Businessweek* (18 Apr. 2019), https://www.bloomberg.com/graphics/2019-nuclear-power-plants-climate-change/, accessed 12 Dec. 2019.

18 There have been many (and increasing) examples over recent years, for instance *Reuters*, 'Hot Weather Cuts French, German Nuclear Output' (25 Jul. 2019), https://www.reuters.com/article/us-france-electricity-heatwave/hot-weather-cuts-french-german-nuclear-power-output-idUSKCN1UK0HR, accessed 10 Dec. 2019.

19 Y. Jin et al., 'Water Use of Electricity Technologies : A Global Meta-Analysis', *Renewable and Sustainable Energy Reviews*, 115 (2019), 109391. doi: 10.1016/j.rser.2019.109391

20 X. He et al., 'Solar and Wind Energy Enhances Drought Resilience and Groundwater Sustainability', *Nature Communications*, 10 (2019), 1–8. doi: 10.1038/s41467-019-12810-5

21 Large turbine arrays can diminish peak near-surface hurricane wind speeds by between twenty-five and forty-one metres per second and storm surge by between 6 and 79%: M.Z. Jacobson, C.L. Archer and W. Kempton, 'Taming Hurricanes with Arrays of Offshore Wind Turbines'. *Nature Climate Change*, 4 (2014), 195–200. doi: 10.1038/NCLIMATE2120 and Y. Pan, C. Yan and C.L. Archer, 'Precipitation Reduction During Hurricane Harvey with Simulated Offshore Wind Farms', *Environmental Research Letters*, 13/8 (2018). doi: 10.1088/1748-9326/aad245

22 Vaclav Smil, *Energy in Nature and Society: General Energetics of Complex Systems*, MIT Press: Cambridge, MA, 2007, p. 316.

23 My own rough estimate given an average annual growth rate for the last five years and an exponential calculation of the form $x = x_0(1+r)^t$. Where x is the future size, x_0 is the total size today, r is the growth rate, and t is the number of years in the future the calculation is made.

24 Tailing off of growth: Benjamin Wehrmann, 'Germany's Renewables Investments Falls on Lower Prices, Slower Capacity Growth', *Clean Energy Wire* (5 Sep. 2019), https://www.cleanenergywire.org/news/germanys-renewables-investments-falls-lower-prices-slower-capacity-growth, accessed 12 Dec. 2019. Growth tailing off after a fast expansion: G.J. Kramer and M. Haigh, 'No Quick Switch to Low-Carbon Energy', *Nature*, 462/7273 (2009), 568–9. doi: 10.1038/462568a

25 R. Wiser et al., 'Expert Elicitation Survey on Future Wind Energy Costs', *Nature Energy*, 1/10 (2016). doi: 10.1038/nenergy.2016.135

26 An overview of the potential and challenges faced by perovskites is available: Andy Extance, 'The Reality Behind Solar Power's Next Star Material', *Nature News Feature* (25 Jun. 2019), https://www.nature.com/articles/d41586-019-01985-y, accessed 12 Dec. 2019.

27 Charles A.S. Hall and Kent A. Klitgaard, *Energy and the Wealth of Nations: Understanding the Biophysical Economy*, Springer-Verlag: New York, 2012, p. 218.

28 P.E. Brockway et al., 'Estimation of Global Final-Stage Energy-Return-on-Investment for Fossil Fuels with Comparison to Renewable Energy Sources', *Nature Energy*, 4 (2019), 612–21. doi: doi.org/10.1038/s41560-019-0425-z

29 K. Wang et al., 'Energy Return on Investment of Canadian Oil Sands Extraction from 2009 to 2015', *Energies*, 10/5 (2017). doi: 10.3390/en10050614

30 C.J. Cleveland and P.A. O'Connor, 'Energy Return on Investment (EROI) of Oil Shale' *Sustainability*, 3 (2011), 2307–22. doi: 10.3390/su3112307

31 C.A.S. Hall, J.G. Lambert and S. Balogh, 'EROI of Different Fuels and the Implications for Society', *Energy Policy*, 64 (2014), 141–52. doi: 10.1016/j.enpol.2013.05.049

32 For instance see C.A.S. Hall, S. Balogh and D.J.R. Murphy, 'What Is the Minimum EROI That a Sustainable Society Must Have?', *Energies*, 2/1 (2009), 25–47. doi: 10.3390/en20100025. Here they state, 'The calculation of this is beyond the scope of this paper but our guess is that we would need something like a 5:1 EROI from our main fuels to maintain anything like what we call civilization.'

33 Wind: Hall, Lambert and Balogh, 'EROI of Different Fuels'. Solar: K.P. Bhandari et al., 'Energy Payback Time (EPBT) and Energy Return on Energy Invested (EROI) of Solar Photovoltaic Systems: A Systematic Review and Meta-Analysis', *Renewable and Sustainable Energy Reviews*, 47 (2015), 133–41. doi: 10.1016/j.rser.2015.02.057

34 J. van Zalk and P. Behrens, 'The Spatial Extent of Renewable and Non-Renewable Power Generation: A Review and Meta-Analysis of Power Densities and Their Application in the U.S.', *Energy Policy*, 123 (2018), 83–91. doi: 10.1016/j.enpol.2018.08.023

35 C.K. Miskin et al., 'Sustainable Co-Production of Food and Solar Power to Relax Land-Use Constraints', *Nature Sustainability*, 2 (2019), 972–80. doi: 10.1038/s41893-019-0388-x

36 International Energy Agency, 'Offshore Wind Outlook 2019: Wind Energy Outlook Special Report' (25 Oct. 2019), https://webstore.iea.org/offshore-wind-outlook-2019-world-energy-outlook-special-report, accessed 13 Dec. 2019.

37 A. Grubler et al., 'A Low Energy Demand Scenario for Meeting the 1.5 °C Target and Sustainable Development Goals without Negative Emission Technologies', *Nature Energy*, 3/6 (2018), 515–27. doi: 10.1038/s41560-018-0172-6

38 This quote is from physicist Amory Lovins. See Logan Ward, 'Amory Lovins: Solving the Energy Crisis (and Bringing Wal-Mart)', *Popular Mechanics* (1 Oct. 2017), https://www.popularmechanics.com/science/green-tech/a2146/4224757/, accessed 13 Dec. 2019.

39 This is from a concept called exergy – the amount of useful energy available to society. This is the maximum amount of energy available to a society given the constraints of the second law of thermodynamics. This means there are limits to the efficient conversion of fossil fuels into useful energy, and exergy takes this into account. The estimate in the text is derived from I.S. Ertesvåg, 'Society Exergy Analysis: A Comparison of Different Societies', *Energy*, 26/3 (2001), 253–70 (2001). doi: 10.1016/S0360-5442(00)00070-0

40 McKinsey & Company, 'Pathways to a Low-Carbon Economy: Version 2 of the Global Greenhouse Gas Abatement Cost Curve', McKinsey and Company Sustainability (Sep. 2013), https://www.mckinsey.com/business-functions/sustainability/our-insights/pathways-to-a-low-carbon-economy, accessed 13 Dec. 2019.

41 International Energy Agency, 'Energy Efficiency Market Report 2016' (Oct. 2016), available to download at https://webstore.iea.org/energy-efficiency-market-report-2016, accessed 13 Dec. 2019.

42 International Energy Agency, 'Sustainable Development Goal 7: Data and Projections' (Nov. 2019), https://www.iea.org/sdg/efficiency/, accessed 13 Dec. 2019.

43 Simon Nicol, Mike Roys and Helen Garrett, 'The Cost of Poor Housing to the NHS', Building Research Establishment (Watford: BRE Trust, 2011), https://www.bre.co.uk/filelibrary/pdf/87741-Cost-of-Poor-Housing-Briefing-Paper-v3.pdf, accessed 13 Dec. 2019.

44 K. Klein Goldewijk et al., 'The HYDE 3.1 Spatially Explicit Database of Human-Induced Global Land-Use Change Over the Past 12,000 Years', *Global Ecology and Biogeography*, 20, (2011), 73–86. doi: 10.1111/j.1466-8238.2010.00587.x.

45 United Nations, 'World Urbanization Prospects: the 2018 Revision' (16 May 2018), https://www.un.org/en/development/desa/population/theme/urbanization/index.asp, accessed 13 Dec. 2019.

46 Hal Harvey with Robbie Orvis and Jeffrey Rissman, *Designing Climate Solutions: A Policy Guide for Low-Carbon Energy*, Island Press: 2018, 59.

47 H. Lohse-Busch et al., 'Ambient Temperature (20°F, 72°F and 95°F) Impact on Fuel and Energy Consumption for Several Conventional Vehicles, Hybrid and Plug-In Hybrid Electric Vehicles and Battery Electric Vehicle', SAE Technical Paper 2013-01-1462, Proc. 2013 SAE World Congress and Exhibition, Detroit, 8 Apr. 2013; A. Bandivadekar et al., 'On the Road in 2035 – Reducing Transportation's Petroleum Consumption and GHG Emissions', Massachusetts Institute of Technology, LFEE 2008-05 RP (Jul. 2008); M. Baglione, M. Duty and G. Pannone, 'Vehicle System Energy Analysis Methodology and Tool for Determining Vehicle Subsystem Energy Supply and Demand', SAE Technical Paper 2007-01-0398 (16 Apr. 2007).

48 For parking statistics see David Z. Morris, 'Today's Cars Are Parked 95% of the Time', *Fortune* (13 Mar. 2016), https://fortune.com/2016/03/13/cars-parked-95-percent-of-time/. For land use statistics, see https://transportgeography.org/?page_id=5721, both accessed 10 Dec. 2019.

49 Chris Ogden, 'Serious Growth in E-Bike Sales Predicted by 2050', *Environment Journal* (10 Jul. 2019), https://environmentjournal.online/articles/serious-growth-in-e-bike-sales-predicted-by-2050/ accessed 4 Jan. 2020.

50 This is my own calculation.

51 Kristoffer Tigue, 'U.S. Electric Bus Demand Outpaces Production as Cities Add to Their Fleets', *Inside Climate News* (14 Nov. 2019), https://insideclimatenews.org/news/14112019/electric-bus-cost-savings-health-fuel-charging, accessed 4 Jan. 2020.

52 Alaric Nightingale, 'Forget Tesla, It's China's E-Buses that are Denting Oil Demand' (19 Mar. 2019), based on an analysis from Bloomberg New Energy Finance: https://www.bloomberg.com/news/articles/2019-03-19/forget-tesla-it-s-china-s-e-buses-that-are-denting-oil-demand, accessed 13 Dec. 2019.

53 There are many websites online where air quality can be compared now (although there are questions about the legitimacy of some monitoring stations in China). The best comparison is perhaps the US embassy in Beijing: 'Air Pollution: Real-Time Air Quality Index', https://aqicn.org/city/beijing/us-embassy/ with somewhere like London: 'London Air Pollution: Real-Time Air Quality Index', http://aqicn.org/city/london/, accessed 13 Dec. 2019, although pollution can change with the time of the day and year.

54 D.E. Schraufnagel et al., 'Health Benefits of Air Pollution Reduction.' *Annals of the American Thoracic Society*, 16/12 (2019), 1478–87. doi: 10.1513/AnnalsATS.201907-538CME

55 There is growing academic evidence of harm; reviews include M. Basner et al., 'Auditory and Non-Auditory Effects of Noise on Health', *Lancet*, 383/9925 (2014), 1325–32. doi: 10.1016/S0140-6736(13)61613-X. The magnitude is similar to air pollution in that one million healthy life years are lost every year from traffic-related noise: R. Kim, 'Burden of Disease from Environmental Noise' [conference paper], European Union Joint Research Centre (2011), available at https://www.env-health.org/IMG/pdf/25052011_Conference_on_Noise_-_Rokho_Kim_WHO_Burden_of_disease_Presentation.pdf. For popular reporting see Richard Godwin, 'Sonic Doom: How Noise Pollution Kills Thousands Each Year', *The Guardian* (3 Jul. 2018), https://www.theguardian.com/lifeandstyle/2018/jul/03/sonic-doom-noise-

pollution-kills-heart-disease-diabetes, and David Owen, 'Is Noise Pollution The Next Big Public-Health Crisis', *The New Yorker* (13 May 2019), https://www.newyorker.com/magazine/2019/05/13/is-noise-pollution-the-next-big-public-health-crisis, all references accessed 13 Dec. 2019.

56   For impacts on health see C.A. Celis-Morales et al., 'Association Between Active Commuting and Incident Cardiovascular Disease, Cancer, and Mortality: Prospective Cohort Study', *BMJ (Clinical Research Ed.)*, 357 (2017), 1456. doi: 10.1136/bmj.j1456. For those on happiness: J. Zhu and Y. Fan, 'Daily Travel Behavior and Emotional Well-Being: Effects of Trip Mode, Duration, Purpose, and Companionship.' *Transportation Research Part A: Policy and Practice*, 118 (2018), 360–73. https://doi.org/10.1016/j.tra.2018.09.019.

57   Bloomberg New Energy Finance, 'How Hydrogen Could Solve Steel's Climate Test and Hobble Coal' (2 Sep. 2019), https://about.bnef.com/blog/hydrogen-solve-steels-climate-test-hobble-coal/, accessed 13 Dec. 2019. These sorts of breakthroughs are sorely needed. In 2021 we conducted some research showing that even if we were to implement all current technologies and social changes (such as moving to smaller buildings), the building material sector would still be two times higher than its current emissions share. See X. Zhong et al., 'Global Greenhouse Gas Emissions from Residential and Commercial Building Materials and Mitigation Strategies to 2060', *Nature Communications*, 12:6126 (2021), 1–10, doi: 10.1038/s41467-021-26212-z.

58   Flying as accounting for 2.5% of total carbon emissions in 2019 is reported here: Hiroko Tabuchi, '"Worse than Anyone Expected": Air Travel Emissions Vastly Outpace Predictions', *New York Times* (19 Sep. 2019), https://www.nytimes.com/2019/09/19/climate/air-travel-emissions.html, accessed 13 Dec. 2019. Although this just discusses the carbon emissions, if we take into account the effect that contrails have on the energy balance in the atmosphere, it could be as high as around 5%. For example see B. Owen, B, D.S. Lee and L. Lim, 'Flying into the Future: Aviation Emissions Scenarios to 2050', *Environmental Science and Technology* 44/7 (2010), 2255–60. doi: 10.1021/es902530z. Figures about growth of demand by mid-century: Roz Pidcock and Sophie Yeo, 'Analysis: Aviation Could Consume a Quarter of 1.5C Carbon Budget by 2050', *Carbon Brief* (8 Aug. 2016), https://www.carbonbrief.org/aviation-consume-quarter-carbon-budget, accessed 13 Dec. 2019.

59   Engineer Hugh Hunt first thought of this comparison here: Hugh Hunt, 'Engineering – Carbon Waste' [video], YouTube (uploaded 15 Nov. 2017), https://www.youtube.com/watch?v=TUbogGT3XjY, accessed 13 Dec. 2019.

60   A.W. Schäfer et al., 'Technological, Economic and Environmental Prospects of All-Electric Aircraft', *Nature Energy*, 4/2 (2019), 160–6. doi: 10.1038/s41560-018-0294-x

61   C. Van Der Giesen, R. Kleijn and G.J. Kramer, 'Energy and Climate Impacts of Producing Synthetic Hydrocarbon Fuels from CO_2,' *Environmental Science & Technology*, 48/12 (2014), 7111–21. doi: 10.1021/es500191g. The estimate depends heavily on the cost of energy needed to produce the fuel and the taxes levied on kerosene. For more see Andrew Murphy, 'Lufthansa Takes First Steps Towards Non-Fossil Kerosene' (26 Feb. 2019), https://www.transportenvironment.org/news/lufthansa-takes-first-steps-towards-non-fossil-kerosene. For the higher estimate see energy analyst Euan Mearns' article, 'LCOE and the Cost of Synthetic Jet Fuel', *Energy Matters* (26 Oct. 2016), http://euanmearns.com/lcoe-and-the-cost-of-synthetic-jet-fuel/, both articles accessed 4 Jan. 2020.

62 This may seem confusing, since clouds reflect light away from the planet. The answer is that it depends on what frequency of light is being reflected. If contrails are thin, they can let the sun's optical light through but reflect the predominantly infrared light from the surface back down to Earth. For reading on contrails see Laura Naranjo, 'On the Trail of Contrails', EarthData, NASA (11 Oct. 2013), https://earthdata.nasa.gov/learn/sensing-our-planet/on-the-trail-of-contrails. And for future impacts, see L. Bock and U. Burkhardt, 'Contrail Cirrus Radiative Forcing for Future Air Traffic', *Atmospheric Chemistry and Physics*, 19 (2019), 8163–74. doi: 10.5194/acp-19-8163-2019

63 15% of the population taking 70% of flights: these data come from government statistics found in 'Public Experiences of and Attitudes Towards Air Travel: 2014', https://www.gov.uk/government/statistics/public-experiences-of-and-attitudes-towards-air-travel-2014, and were analysed in Campaign for Better Transport, 'New Runway Would Add Hundreds of Pounds to Holidays in Extra Carbon Costs, Say Campaigners' (8 Aug. 2016), https://bettertransport.org.uk/new-runway-would-add-hundreds-pounds-holidays-extra-carbon-costs-say-campaigners. 1% of flyers taking 20% of flights: Niko Kommenda, '1% of English Residents Take One-Fifth of Overseas Flights, Survey Shows', *The Guardian* (25 Sep. 2019), https://www.theguardian.com/environment/2019/sep/25/1-of-english-residents-take-one-fifth-of-overseas-flights-survey-shows, all sites accessed 13 Dec. 2019.

64 Estimates vary, but most suggest between two and four times. See for example Tim Fernholz, 'Why Flying First Class Increases Your Carbon Footprint by Six Times', *Quartz* (13 Jun. 2013), https://qz.com/94268/first-class-airline-passengers-are-to-blame-for-global-warming/, accessed 13 Dec. 2019.

65 Hiroko Tabuchi and Nadja Popovich, 'How Guilty Should You Feel About Flying?', *New York Times* (17 Oct. 2019), https://www.nytimes.com/interactive/2019/10/17/climate/flying-shame-emissions.html, accessed 13 Dec. 2019.

66 S. Wynes et al., 'Academic Air Travel Has a Limited Influence on Professional Success', *Journal of Cleaner Production*, 226 (2019), 959–67. doi: 10.1016/j.jclepro.2019.04.109

67 L. Simon et al., 'Increased Shear in the North Atlantic Upper-Level Jet Stream over the Past Four Decades', *Nature*, 572/7771 (2019), 639–42. doi: 10.1038/s41586-019-1465-z.

68 Sam Morgan, 'Nine EU Countries Urge New Commission to Tax Aviation More', EURACTIV (7 Nov. 2019), https://www.euractiv.com/section/aviation/news/nine-eu-countries-urge-new-commission-to-tax-aviation-more/, accessed 4 Jan. 2020.

69 The cost of around €950 per tonne can be paid to start-up company Climeworks to extract CO_2 from the atmosphere and place it into rocks; for more information, see https://www.climeworks.com/. The likelihood of the price dropping fast is taken from M. Fasihi, O. Efimova and C. Breyer, 'Techno-Economic Assessment of CO_2 Direct Air Capture Plants', *Journal of Cleaner Production*, 224 (2019), 957–80. doi: doi.org/10.1016/j.jclepro.2019.03.086.

70 Fasihi, Efimova and Breyer, 'Techno-Economic Assessment'.

71 Owen Jones, 'How to Stop Climate Change? Nationalise the Oil Companies', *The Guardian* (25 Apr. 2019), https://www.theguardian.com/commentisfree/2019/apr/25/climate-change-oil-companies-extinction-rebellion, accessed 4 Jan. 2020.

72 For instance see Ben Smee, 'Leading Australian Engineers Turn Their Backs on New Fossil Fuel Projects', *The Guardian* (21 Oct. 2019), https://www.theguardian.com/environment/2019/oct/21/leading-engineers-turn-their-backs-on-new-fossil-fuel-projects, accessed 13 Dec. 2019.

73   A good review of fossil-fuel vehicle phase-outs is available on Wikipedia: 'Phase-Out of Fossil Fuel Vehicles', https://en.wikipedia.org/wiki/Phase-out_of_fossil_fuel_vehicles, accessed 10 Dec. 2019. VW's eight EV factories: Edward Taylor, Jan Schwartz and Joseph White, 'VW Ramps up China Electric Car Factories, Taking Aim at Tesla', *Reuters*, (28 Oct. 2019), https://www.reuters.com/article/us-volkswagen-electric-focus/vw-ramps-up-china-electric-car-factories-taking-aim-at-tesla-idUSKBN1X71RV, accessed 4 Jan. 2020. Mercedes stopping investment: Fred Lambert, 'Daimler Stops Developing Internal Combustion Engines to Focus on Electric Cars', *Electrek* (19 Sep. 2019), https://ww.electrek.co/2019/09/19/daimler-stops-developing-internal-combustion-engines-to-focus-on-electric-cars/, accessed 4 Jan. 2020.

74   Germany: Julian Wettengel, 'Eastern German Coal States Welcome Proposal for Phase-Out Support', Clean Energy Wire (22 Aug. 2019), https://www.cleanenergywire.org/news/eastern-german-coal-states-welcome-proposal-phase-out-support; Spain: Arthur Neslen, 'Spain to Close Most Coal Mines in €250m Transition Deal', *The Guardian* (26 Oct. 2019), https://www.theguardian.com/environment/2018/oct/26/spain-to-close-most-coal-mines-after-striking-250m-deal, both accessed 4 Jan. 2020.

75   While many materials don't present a problem, some rare earth elements (which are actually not that rare, but are just more limited than other materials like cement) could become a problem. For why many materials don't present a problem, see E.G. Hertwich et al., 'Integrated Life-Cycle Assessment of Electricity-Supply Scenarios Confirms Global Environmental Benefit of Low-Carbon Technologies', *Proceedings of the National Academy of Sciences of the United States of America*, 112/20 (2015), 6277–82. doi: 10.1073/pnas.1312753111. For an overview of important metals needed in low-carbon transition, see R. Kleijn et al., 'Metal Requirements of Low-Carbon Power Generation', *Energy*, 36/9 (2011), 5640–8. doi: 10.1016/j.energy.2011.07.003. And for projections of the requirements of rare-earth metals see E. Alonso et al., 'Evaluating Rare Earth Element Availability: A Case with Revolutionary Demand From Clean Technologies', *Environmental Science & Technology*, 46/6 (2012), 3406–14. doi: 10.1021/es203518d

76   J. Lee et al., 'Reversible Mn2+/Mn4+ Double Redox in Lithium-Excess Cathode Materials', *Nature*, 556 (2018), 185–90. doi: 10.1038/s41586-018-0015-4. For a more general overview see Prachi Patel, 'Could Cobalt Choke Our Electric Vehicle Future?', *Scientific American* (1 Jan. 2018), https://www.scientificamerican.com/article/could-cobalt-choke-our-electric-vehicle-future/, accessed 13 Dec. 2019.

Chapter Five – Pessimism – Eating the Earth

1   F. Krausmann et al., 'Global Human Appropriation of Net Primary Production Doubled in the 20th Century', in *Proceedings of the National Academy of Sciences of the United States of America*, 110/25 (2013), 10324–9. doi: 10.1073/pnas

2   IPCC, 'Climate Change and Land: An IPCC Special Report on Climate Change, Desertification, Land Degradation, Sustainable Land Management, Food Security, and Greenhouse Gas Fluxes in Terrestrial Ecosystems' (summary for policymakers approved draft) (7 Aug. 2019). doi: 10.4337/9781784710644

3   P.H. Gleick and M. Palaniappan, 'Peak Water Limits to Freshwater Withdrawal and Use', *Proceedings of the National Academy of Sciences of the United States of America*, 107/25 (2010), 11155–62. doi: 10.1073/pnas.1004812107

4   This is estimated by F.M. Sabatini et al., 'Where Are Europe's Last Primary Forests?', *Diversity and Distributions*, 24/10 (2018), 1426–39. doi: 10.1111/ddi.12778. 80% of Europe's land area was once covered by native forest: European Environment Agency, 'Forest Dynamics in Europe and Their Ecological Consequences' (27 Nov. 2018), https://www.eea.europa.eu/themes/biodiversity/forests/forest-dynamics-in-europe-and, accessed 21 Dec. 2019.

5   Y.M. Bar-On, R. Phillips and R. Milo, 'The Biomass Distribution on Earth', *Proceedings of the National Academy of Sciences of the United States of America*, 115/25 (2018), 6506–11. doi: 10.1073/pnas.1711842115

6   According to the International Union for Conservation of Nature (IUCN), as reported by James Katner, 'One in 4 Mammals Threatened with Extinction, Group Finds', *New York Times* (6 Oct. 2008), https://www.nytimes.com/2008/10/07/science/earth/07mammal.html, accessed 13 Dec. 2019.

7   C.A. Hallmann et al., 'More than 75 Percent Decline Over 27 Years in Total Flying Insect Biomass in Protected Areas', *PLOS ONE*, 12/10 (2017). doi: 10.1371/journal.pone.0185809

8   More specifically the land and energy use via human systems. For instances see the Intergovernmental Science-Policy Platform on Biodiversity and Ecosystem services, 'Models of Drivers of Biodiversity and Ecosystem Change', https://ipbes.net/models-drivers-biodiversity-ecosystem-change, accessed 13 Dec. 2019.

9   In the US: Lauren Bauer and Diane Witmore Schanzenbach, 'Children's Exposure to Food Insecurity Is Still Worse than It Was Before the Great Recession', Brookings Institution (29 Jun. 2018), https://www.brookings.edu/blog/up-front/2018/06/29/childrens-exposure-to-food-insecurity-is-still-worse-than-it-was-before-the-great-recession/. In the UK: 'Food Insecurity in UK Is Among Worst in Europe, Especially for Children, Says Committee', *British Medical Journal*, 364/l126 (2019). doi: 10.1136/bmj.l126

10   M. Di Cesare et al. 'Trends in Adult Body-Mass Index in 200 Countries from 1975 to 2014: A Pooled Analysis of 1698 Population-Based Measurement Studies with 19.2 Million Participants', *Lancet*, 387/10026 (2016), 1377–96. doi: 10.1016/S0140-6736(16)30054-X

11   On the percentage of adults who are overweight, see World Health Organization, 'Obesity and Overweight' (16. Feb. 2018), https://www.who.int/news-room/fact-sheets/detail/obesity-and-overweight, accessed 13 Dec. 2019. On the percentage of those who are eating too little, see Food and Agriculture Organization of the United Nations, 'The State of Food Security and Nutrition in the World' (2019), http://www.fao.org/state-of-food-security-nutrition/en/, accessed 13 Dec. 2019.

12   Attributed to Franklin D. Roosevelt, in Susan Ratcliffe, ed., *Oxford Essential Quotations* (5th edn.), Oxford University Press: Oxford, 2017, https://www.oxfordreference.com/view/10.1093/acref/9780191843730.001.0001/q-oro-ed5-00008907, accessed 13 Dec. 2019.

13   DOE/Lawrence Berkeley National Laboratory. 'Scientists Hit Pay Dirt with New Microbial Research Technique: A Better Method for Studying Microbes in the Soil Will Help Scientists Understand Large-Scale Environmental Cycles', *ScienceDaily* (24 Jun. 2019), https://www.sciencedaily.com/releases/2019/06/190624111537.htm, accessed 9 Jan. 2019.

14   D. Pimentel and M. Burgess, 'Soil Erosion Threatens Food Production', *Agriculture*, 3 (2013), 443–63. doi:10.3390.

15 The figure of a third of global soils is taken from D. Pimentel et al., 'Environmental and Economic Costs of Soil Erosion and Conservation Benefits', *Science*, 267/5201 (1995), 1117–23. For a more recent estimate, see Grantham Centre for Sustainable Futures, 'A Sustainable Model for Intensive Agriculture' [briefing note] (2 Dec. 2015), http://grantham.sheffield.ac.uk/soil-loss-an-unfolding-global-disaster/. The figure of 3.4 tonnes per person annually is taken from United Nations News, '24 Billion Tons of Fertile Land Lost Every Year, Warns UN Chief on World Day to Combat Desertification' (16 Jun. 2019), https://news.un.org/en/story/2019/06/1040561, all sources accessed 16 Dec. 2019.

16 P. Panagos et al., 'Cost of Agricultural Productivity Loss Due to Soil Erosion in the European Union: From Direct Cost Evaluation Approaches to the Use of Macroeconomic Models', *Land Degradation and Development*, 29/3 (2018), 471–84. doi: 10.1002/ldr.2879

17 This estimate comes from Qiu Qiwen, head of the soil environment department of China's Ministry of Environmental Protection, and appears in Amanda Little, *The Fate of Food: What We'll Eat in a Bigger, Hotter, Smarter World*, Harmony: New York, 2019, 122. Qiwen estimates clean-up to cost $20,000 per acre, and China has about 200,000 acres of heavily contaminated farmland.

18 C. Rumpel et al., 'Put More Carbon in Soils to Meet Paris Climate Pledges', *Nature*, 564/7734 (2018), 32–4. doi: 10.1038/d41586-018-07587-4

19 S.E. Page et al., 'The Amount of Carbon Released from Peat and Forest Fires in Indonesia During 1997', *Nature*, 420/6911 (2002), 61–5. doi: 10.1038/nature01131

20 David Gibbs, Nancy Harris and Frances Seymour, 'By the Numbers: The Value of Tropical Forests in the Climate Change Equation', World Resources Institute (4 Oct. 2018), https://www.wri.org/blog/2018/10/numbers-value-tropical-forests-climate-change-equation, accessed 4 Jan. 2020.

21 J. Fargione et al., 'Land Clearing and the Biofuel Carbon Debt', *Science*, 319/5867 (2008), 1235–8. doi: 10.1126/science.1152747

22 Ibid.

23 T. Searchinger et al., 'Use of U.S. Croplands for Biofuels Increases Greenhouse Gases Through Emissions From Land-Use Change', *Science*, 319/5867 (2008), 1238–40. doi: 10.1126/science.1151861

24 Although there are ways in which the wastes can be used, if done carefully. For instance, sawdust and crop residues can be burned for electricity.

25 X.P. Song et al., 'Global Land Change from 1982 to 2016', *Nature*, 560/7720 (2018), 639–43. doi: 10.1038/s41586-018-0411-9

26 Tropical regions can also be reforested with commodity products like palm oil.

27 Benjamin Franklin, Poor Richard to the 'Courteous Reader', *Poor Richard's Almanac*, 1746. As cited in Benjamin Franklin, *Poor Richard's Almanac*, edited by Lloyd E. Smith, Haldeman-Julius Company: Girard, Kansas, 1925, p. 22.

28 S. Siebert et al., 'Groundwater Use for Irrigation – a Global Inventory', *Hydrology and Earth System Sciences*, 14/10 (2010), 1863–80. doi: 10.5194/hess-14-1863-2010

29 T. Sterner et al., 'Policy Design for the Anthropocene', *Nature Sustainability*, 2/1 (2019), 14–21. doi: 10.1038/s41893-018-0194-x

30 B.D. Tapley et al., 'Contributions of GRACE to Understanding Climate Change', *Nature Climate Change*, 9 (2019), 358–69. doi: 10.1038/s41558-019-0456-2

31 Of the world's largest aquifers, twenty-one out of thirty-seven are draining faster than they can refill: A.S. Richey et al., 'Quantifying Renewable Groundwater Stress with GRACE', *Water Resources Research*, 51/7 (2015), 1–22. doi:10.1002/2015WR017349

32 R.C. Buchanan et al., 'The High Plains Aquifer', Kansas Geological Survey, Public Information Circular 18, http://www.kgs.ku.edu/Publications/pic18/index.html, accessed 16 Dec. 2019. The study expanded this to a hundred years if all farmers cut water use by 20%.

33 American Geophysical Union, 'Groundwater Resources Around the World Could Be Depleted by 2050s', Phys.org (15 Dec. 2016), https://phys.org/news/2016-12-groundwater-resources-world-depleted-2050s.html, accessed 16 Dec. 2019.

34 M.T. Niles and C.R. Hammond Wagner, 'The Carrot or the Stick? Drivers of California Farmer Support for Varying Groundwater Management Policies', Environmental Research Communications, 1/4 (2019). doi: 10.1088/2515-7620/ab1778

35 P. Wester et al., The Hindu Kush Himalaya Assessment: Mountains, Climate Change, Sustainability and People (Springer Open, 2019), https://link.springer.com/content/pdf/10.1007%2F978-3-319-92288-1.pdf.1

36 J.M. Maurer et al., 'Acceleration of Ice Loss Across the Himalayas Over the Past 40 Years', Science Advances, 5/6 (2019), doi: 10.1126/sciadv.aav7266

37 M.K. Roxy et al., 'A Threefold Rise in Widespread Extreme Rain Events over Central India', Nature Communications, 8/1 (2017), 1–11. doi: 10.1038/s41467-017-00744-9

38 UN Water, 'Water Quality and Wastewater', https://www.unwater.org/water-facts/quality-and-wastewater/, accessed 16 Dec. 2019.

39 E. Shochat et al., 'Invasion, Competition, and Biodiversity Loss in Urban Ecosystems', Bioscience, 60/3 (2010), 199–208. doi: 10.1525/bio.2010.60.3.6

40 As remarked by astrophysicist Carl Sagan and Ann Druyan in Comet, Random House: New York, 1985, p. 273.

41 For the figures from Germany, see C.A. Hallmann et al., 'More than 75 Percent Decline over 27 Years in Total Flying Insect Biomass in Protected Areas', PLOS ONE, 12/10 (2017). doi: 10.1371/journal.pone.0185809. For the figures from France, see Agence France-Presse, '"Catastrophe" as France's Bird Population Collapses Due to Pesticides', The Guardian (21 Mar. 2018), https://www.theguardian.com/world/2018/mar/21/catastrophe-as-frances-bird-population-collapses-due-to-pesticides, accessed 16 Dec. 2019.

42 Bee populations declining all over Europe: European Food Safety Authority, 'Bee Health' (2018), http://www.efsa.europa.eu/en/topics/topic/bee-health, accessed 16 Dec. 2019. On the increase of colonies in Berlin, see Kate Connolly, 'Berlin's Bumbling Beekeepers Leave Swarms without Homes', The Guardian (9 Aug. 2019), https://www.theguardian.com/environment/2019/aug/09/berlin-beekeepers-leave-swarms-without-homes-schwarmfangers, accessed 16 Dec. 2019.

43 Rachel Carson, Silent Spring, Penguin Modern Classics: New York, 2000.

44 K.V. Rosenberg et al., 'Decline of the North American Avifauna', Science, 366/6461 (2019), 120–4. doi: 10.1126/science.aaw1313

45 This draws on the concept of keystone species: species that have a very large effect on the functioning of an entire ecosystem. Often these are apex predators such as sea otters, which keep kelp forests healthy.

46 For the figure of 8.7 million species see C. Mora et al., 'How Many Species Are There on Earth and in the Ocean?', PLOS Biology, 9/8 (2011), 1–8. doi: 10.1371/journal.pbio.1001127. For the species loss figures, see J. De Vos et al., 'Estimating the Normal Background Rate of Species Extinction', Conservation Biology, 29/2 (2014), 452–62. doi: 10.1111/cobi.12380

47 Although difficult to estimate, one highly cited study suggested that 80% of biodiversity loss is driven by agricultural systems: B.M. Campbell et al., 'Agriculture Production

as a Major Driver of the Earth System Exceeding Planetary Boundaries', *Ecology and Society*, 22/4 (2017), 1-2. doi: 10.5751/ES-09595-220408, p. 2.

48   D. Jablonski and W.G. Chaloner, 'Extinctions in the Fossil Record [and Discussion]', *Philosophical Transactions: Biological Sciences*, 344/1307 (1994), 11–17. doi: 10.1098/rstb.1994.0045

49   C.M. Lowery and A.J. Fraass, 'Morphospace Expansion Paces Taxonomic Diversification After End Cretaceous Mass Extinction', *Nature Ecology and Evolution*, 3/6 (2019), 900–4. doi: 10.1038/s41559-019-0835-0; and S. Sahney and M.J. Benton, 'Recovery from the Most Profound Mass Extinction of All Time', *Proceedings of the Royal Society B: Biological Sciences*, 275/1636 (2008), 759–65. doi: 10.1098/rspb.2007.1370

50   M. Scheffer et al., 'Early-Warning Signals for Critical Transitions'. *Nature*, 461/7260 (2009), 53–9 doi: 10.1038/nature08227

51   Otto, I., et al, 'Social tipping dynamics for stabilizing Earth's climate by 2050', *Proceedings of the National Academy of Sciences of the United States of America*, 117/5 (2020) 2354-65. doi: 10.1073/pnas.1900577117.

52   C.J. Gobler et al., 'Ocean Warming Since 1982 Has Expanded the Niche of Toxic Algal Blooms in the North Atlantic and North Pacific Oceans', *Proceedings of the National Academy of Sciences of the United States of America*, 114/19 (2017), 4975–80. doi: 10.1073/pnas.1619575114

53   Susie Cagle, 'Slimy Lakes and Dead Dogs: Climate Crisis Has Brought the Season of Toxic Algae', *The Guardian* (19 Sep. 2019), https://www.theguardian.com/environment/2019/sep/18/toxic-algae-climate-change-slimy-lakes-dead-dogs, accessed 16 Dec. 2019.

54   FAO, 'The State of Food Security and Nutrition in the World' (2019), http://www.fao.org/state-of-food-security-nutrition/en/, accessed 13 Dec. 2019.

55   Ibid.

56   E. Holt-Giménez et al., 'We Already Grow Enough Food for 10 Billion People… and Still Can't End Hunger', *Journal of Sustainable Agriculture*, 36/6 (2012), 595–8. doi: 10.1080/10440046.2012.695331

57   Data on wealth: World Bank, 'GDP, PPP (Constant 2011 International $)', https://data.worldbank.org/indicator/NY.GDP.MKTP.PP.KD. Data on undernourishment: World Bank, 'Prevalence of Undernourishment (% of Population)', https://data.worldbank.org/indicator/SN.ITK.DEFC.ZS?view=chart. Data on total population: World Bank, 'Population, Total', https://data.worldbank.org/indicator/SP.POP.TOTL, all accessed 4 Jan. 2020.

58   Thomas Fuller, *Gnomologia: Adagies and Proverbs, Wise Sentences and Witty Sayings, Ancient and Modern, Foreign and British*, printed for B. Barker: London, 1732, p. 149.

59   W. You and M. Henneberg, 'Meat Consumption Providing a Surplus Energy in Modern Diet Contributes to Obesity Prevalence: An Ecological Analysis', *BMC Nutrition*, 2 (2016), 1–11. doi: 10.1186/s40795-016-0063-9

60   See the livestock data at the FAO, 'Live Animals', http://www.fao.org/faostat/en/#data/QA, accessed 4 Jan. 2020.

61   The figure of 80% is taken from the following article on the FAO's website: 'Animal Production', http://www.fao.org/animal-production/en/, accessed 16 Dec. 2019. The figure for abstracted water consumption is taken from P.W. Gerbens-Leenes, M.M. Mekonnen and A.Y. Hoekstra, 'The Water Footprint of Poultry, Pork and Beef: A Comparative Study in Different Countries and Production Systems', *Water Resources and Industry*, 1–2 (2013), 25–36. doi: 10.1016/j.wri.2013.03.001.

62 M. Karki, S.S. Sellamuttu et al., 'The Regional Assessment Report on Biodiversity and Ecosystem Services for Asia and the Pacific: Summary for Policymakers', IPBES (Bonn: IPBES, 2018), https://ipbes.net/system/tdf/spm_asia-pacific_2018_digital. pdf?file=1&type=node&id=28394, accessed 16 Dec. 2019. Details on manure run-off are available: M. Motew et al., 'The Synergistic Effect of Manure Supply and Extreme Precipitation on Surface Water Quality', *Environmental Research Letters*, 13/4 (2018). doi: 10.1088/1748-9326/aaade6

63 A. Marques et al., 'Increasing Impacts of Land Use on Biodiversity and Carbon Sequestration Driven by Population and Economic Growth', *Nature Ecology and Evolution*, 3/4 (2019), 628–37. doi: 10.1038/s41559-019-0824-3

64 These data derive from a set of papers as reported in Global Forest Atlas, 'Cattle Ranching in the Amazon Region', Yale School of Forestry & Environmental Studies, https://globalforestatlas.yale.edu/amazon/land-use/cattle-ranching, accessed 28 Feb. 2020 and, from the same group: 'Soy Agricultural in the Amazon Basin', https:// globalforestatlas.yale.edu/amazon/land-use/soy, accessed 28 Feb. 2020.

65 For the figure of a third of crops going to feed animals, see E.S. Cassidy et al., 'Redefining Agricultural Yields: From Tonnes to People Nourished per Hectare', *Environmental Research Letters*, 8/3 (2013). doi: 10.1088/1748-9326/8/3/034015. In terms of numbers of people eating too much meat, the figure is roughly billions: the average meat consumption globally is forty-two kilograms per person per year, with about 50% of this as red meat (pork and beef). The maximum suggested intake depends on the source, but almost always lower than 500 grams per week, resulting in a maximum of twenty-six kilograms a year. Given that this is probably an underestimate and doesn't include other meat intake which can contribute to problems, it's safe to say that billions eat too much meat for their health.

66 P.J. Gerber et al., eds, 'Tackling Climate through Livestock: A Global Assessment of Emissions and Mitigation Opportunities', FAO (2013).

67 For this prediction, see OECD and FAO, 'OECD-FAO Agricultural Outlook 2019–2028' (2019). doi: 10.1787/agr_outlook-2019-en

68 For a discussion of how governments may implement dietary recommendations, see C. Gonzalez Fischer and T. Garnett, *Plates, Pyramids, Planet – Developments in National Healthy and Sustainable Dietary Guidelines: A State of Play Assessment, Food and Agricultural Organization of the United Nations:* Rome, 2016; and P. Behrens et al., 'Evaluating the Environmental Impacts of Dietary Recommendations', *Proceedings of the National Academy of Sciences of the United States of America*, 114/51 (2017), 13412–17. doi: 10.1073/pnas.1711889114. For how taxes can encourage people to move away from meat-based diets, see M. Springmann et al., 'Health-Motivated Taxes on Red and Processed Meat: A Modelling Study on Optimal Tax Levels and Associated Health Impacts', *PLOS ONE*, 13/11 (2018). doi: 10.1371/journal. pone.0204139

69 For the controversy on the word 'burgers': Daniel Boffey, '"Veggie Discs" to Replace Veggie Burgers in EU Crackdown on Food Labels' (14 Apr. 2019), https://www. theguardian.com/food/2019/apr/04/eu-to-ban-non-meat-product-labels-veggie-burgers-and-vegan-steaks, accessed 16 Dec. 2019. For controversy on the word 'milk': Umair Irfan, '"Fake Milk": Why the Dairy Industry Is Boiling over Plant-Based Milks', *Vox* (21 Dec. 2018), https://www.vox.com/2018/8/31/17760738/almond-milk-dairy-soy-oat-labeling-fda, accessed 4 Jan. 2020.

70 FAO, *The State of World Fisheries and Aquaculture 2016: Contributing to Food Security*

and Nutrition for All, (Rome: FAO, 2016), http://www.fao.org/3/a-i5555e.pdf, accessed 16 Dec. 2019.

71   Daniel Pauly, 'Oceana Board Member Daniel Pauly: Fisheries on the Brink' [video], YouTube (uploaded 8 Sep. 2006), https://www.youtube.com/watch?v=Tf1EgeHDxpA, accessed 16 Dec. 2016.

72   Mukhisa Kituyi and Peter Thomson, '90% of Fish Stocks Are Used Up – Fisheries Subsidies Must Stop Emptying the Ocean', World Economic Forum (13 Jul. 2018), https://www.weforum.org/agenda/2018/07/fish-stocks-are-used-up-fisheries-subsidies-must-stop/, accessed 16 Dec. 2019.

73   R. Hilborn et al., 'The Environmental Cost of Animal Source Foods', Frontiers in Ecology and the Environment, 16/6 (2018), 329–35. doi: 10.1002/fee.1822

74   M.J. Behrenfeld et al., 'Climate-Driven Trends in Contemporary Ocean Productivity', Nature, 444/7120 (2007), 752–5. doi: 10.1038/nature05317

75   For the proportion of the EU budget spent on agricultural subsidies, see European Commission, 'Fact Check on the EU Budget' (2018), https://ec.europa.eu/info/strategy/eu-budget/how-it-works/fact-check_en, accessed 16 Dec. 2019. For the estimate of 50% going to the worst performers, see Greenpeace, 'Feeding the Problem – the Dangerous Intensification of Animal Farming in Europe' (12 Feb. 2018), https://www.greenpeace.org/eu-unit/issues/nature-food/1803/feeding-problem-dangerous-intensification-animal-farming/, accessed 16 Dec. 2019. What's worse, there is evidence that subsidies are being channelled to oligarchs and populists: Selam Gebrekidan, Matt Apuzzo and Benjamin Novak, 'The Money Farmers: How Oligarchs and Populists Milk the E.U. for Millions', New York Times (3 Nov. 2019), https://www.nytimes.com/2019/11/03/world/europe/eu-farm-subsidy-hungary.html, accessed 4 Jan. 2020.

76   This according to a 2016 study: K.R. Siegel et al., 'Association of Higher Consumption of Foods Derived from Subsidized Commodities with Adverse Cardiometabolic Risk among US Adults', JAMA Internal Medicine, 176/8 (2016), 1124–32. doi: 10.1001/jamainternmed.2016.2410

77   In fact, even taking into account these subsidies it can still be cheaper to eat a vegetarian diet than a meat one, but this very much depends on where you live. One Harvard University review found savings of $746 a year from going vegetarian: M.M. Flynn and A.R. Schiff, 'Economical Healthy Diets (2012): Including Lean Animal Protein Costs More than Using Extra Virgin Olive Oil', Journal of Hunger & Environmental Nutrition, 10/4 (2015), 467–82. doi: 10.1080/19320248.2015.1045675

78   See United Nations Sustainable Development sub-goal 2B, 'Goal 2: Zero Hunger', https://www.un.org/sustainabledevelopment/hunger/. For attempts to reform subsidies in the UK and EU, see European Union Court of Auditors, 'Opinion No 7/2018 Concerning Commission Proposals for Regulations Relating to the Common Agricultural Policy for the Post-2020 Period', Official Journal of the European Union, 62 (1 Feb. 2019), https://eur-lex.europa.eu/legal-content/EN/TXT/HTML/?uri=OJ:C:2019:041:FULL&from=FR, accessed 16 Dec. 2019.

79   Sarah Gordon, 'UK to Replace EU Farm Subsidies with 7-Year Transition Scheme', Financial Times (11 Sep. 2018), available to subscribers at https://www.ft.com/content/0ee3cfbe-b5e4-11e8-bbc3-ccd7de085ffe, accessed 16 Dec. 2019.

80   On the percentage of small-scale farms and families producing food worldwide, see B.E. Graeub et al., 'The State of Family Farms in the World', World Development, 87 (2016), 1–15. doi: 10.1016/j.worlddev.2015.05.012. On the percentage of seeds sold

by four companies, see OECD, *Concentration in Seed Markets: Potential Effects and Policy Responses*, 2018. doi: 10.1787/9789264308367-en

81 Felicity Lawrence, 'The Global Food Crisis: ABCD of Food – How the Multinationals Dominate Trade', *The Guardian* (2 Jun. 2011), https://www.theguardian.com/global-development/poverty-matters/2011/jun/02/abcd-food-giants-dominate-trade, accessed 16 Dec. 2019.

82 Pesticide regulation: there are many such examples, including M.D. Boone et al., 'Pesticide Regulation Amid the Influence of Industry', *BioScience*, 64/10 (2014), 917–22. doi: 10.1093/biosci/biu138. International trade policy: K. Gawande and B. Hoekman, 'Lobbying and Agricultural Trade Policy in the United States', *International Organization*, 60/3 (2006), 527–61. doi: 10.1017/S0020818306060243.

83 Jacqui Fatka, 'Farmer Share of Food Dollar Declines', *Feedstuffs* (26 Apr. 2019), https://www.feedstuffs.com/news/farmer-share-food-dollar-declines, accessed 16 Dec. 2019.

84 All data from the US Department of Agriculture: S.L. Wang, R. Nehring and R. Mosheim, 'Agricultural Productivity Growth in the United States: 1948–2015' (5 Mar. 2018), https://www.ers.usda.gov/amber-waves/2018/march/agricultural-productivity-growth-in-the-united-states-1948-2015/

85 A.P. Davis et al., 'High Extinction Risk for Wild Coffee Species and Implications for Coffee Sector Sustainability', *Science Advances*, 5/1 (2019) 1–10. doi: 10.1126/sciadv.aav3473

86 Emily Chasan and Gerson Freitas Jr, 'Climate Change Is Already Costing Meat and Air Producers a Lot', *Bloomberg* (4 Sep. 2019), https://www.bloomberg.com/news/articles/2019-09-03/climate-change-is-already-costing-meat-and-dairy-producers-a-lot, accessed 16 Dec. 2019.

87 See for example V. Varma and D.P. Bebber, 'Climate Change Impacts on Banana Yields Around the World', *Nature Climate Change*, 9/10 (2019), 752–7. doi: 10.1038/s41558-019-0559-9; and D.P. Bebber, 'Climate Change Effects on Black Sigatoka Disease of Banana', *Philosophical Transactions of The Royal Society B Biological Sciences*, 374/1775 (2019). doi: 10.1098/rstb.2018.0269

88 This won't all happen at once; the problems will arrive in different decades depending on the food type as weather systems change and temperatures rise.

89 W. Xie et al., 'Decreases in Global Beer Supply Due to Extreme Drought and Heat', *Nature Plants*, 4/11 (2018), 964–73. doi: 10.1038/s41477-018-0263-1

90 Damian Carrington, 'UK Chips an Inch Shorter After Summer Heatwave – Report', *The Guardian* (5 Feb. 2019), https://www.theguardian.com/environment/2019/feb/05/uk-chips-an-inch-shorter-after-summer-heatwave-report, accessed 16 Dec. 2019.

91 Arthur Neslen, 'Italy Sees 57% Drop in Olive Harvest as Result of Climate Change, Scientist Says', *The Guardian* (5 Mar. 2019), https://www.theguardian.com/world/2019/mar/05/italy-may-depend-on-olive-imports-from-april-scientist-says, accessed 16 Dec. 2019.

92 H. Fraga et al., 'Modelling Climate Change Impacts on Viticultural Yield, Phenology and Stress Conditions in Europe', *Global Change Biology*, 22/11 (2016), 3774–88. doi: 10.1111/gcb.13382

93 Julie Al-Zoubi, 'Italy Fears Running Out of Olive Oil by April', *Olive Oil Times* (21 Feb. 2019), https://www.oliveoiltimes.com/business/italy-fears-running-out-of-olive-oil-by-april/67010, accessed 16 Dec. 2019.

Chapter Six – Hope – Green Shoots

1   IPCC, 'Climate Change and Land: An IPCC Special Report on Climate Change, Desertification, Land Degradation, Sustainable Land Management, Food Security, and Greenhouse Gas Fluxes in Terrestrial Ecosystems' (summary for policymakers approved draft) (7 Aug. 2019). doi: 10.4337/9781784710644.

2   This is an approximate estimate from studies such as the following: T. Searchinger et al., 'Creating a Sustainable Food Future', *World Resources Institute* (2018); and M. Springmann et al., 'Options for Keeping the Food System within Environmental Limits', *Nature*, 562/7728 (2018), 519–25. doi: 10.1038/s41586-018-0594-0, along with the biologist E.O. Wilson's concept of half-earth as described in his book *Half-Earth: Our Planet's Fight for Life*, Liveright: New York, 2017.

3   This passage owes a debt to Kenneth Boulding's 'Spaceship Earth': Kenneth E. Boulding, 'The Economics of the Coming Spaceship Earth' (1966), http://arachnid. biosci.utexas.edu/courses/THOC/Readings/Boulding_SpaceshipEarth.pdf, accessed 3 Jan. 2020.

4   See for example 'EAT-Lancet Commission Summary Report', Eat-Lancet Commission, *Lancet*, 32 (2019). doi: 10.1016/S0140-6736(18)31788-4; and Searchinger et al., 'Creating a Sustainable Food Future'.

5   'Eat-Lancet Comission Summary Report', Eat-Lancet Commission.

6   Searchinger et al., 'Creating a Sustainable Food Future'.

7   Leasing land like this already happens in Ethiopia and other countries. A journalistic overview is available: Daniel A. Medina, 'In Ethiopia, Foreign Investment is a Fancy Word for Stealing Land', *Quartz* (17 Oct. 2014), https://qz.com/275489/in-ethiopia-foreign-investment-is-a-fancy-word-for-stealing-land/, accessed 16 Dec. 2019.

8   IPCC, 'Climate Change and Land'.

9   J. Poore and T. Nemecek, 'Reducing Food's Environmental Impacts Through Producers and Consumers', *Science*, 360/6392 (2018), 987–92. doi: 10.1126/science.aaq0216

10 S. Jose, D. Walter and B. Mohan Kumar, 'Ecological Considerations in Sustainable Silvopasture Design and Management', *Agroforestry Systems*, 93/1 (2019), 317–31. doi: 10.1007/s10457-016-0065-2. There are other technical options, which may include methane inhibitors in cows.

11 B.H. Meijer, 'Chemical Maker DSM Sees Strong Demand for Methane-Reducing Cow Feed Additive', *Reuters* (30 Sep. 2019), https://uk.reuters.com/article/us-climate-change-dsm/chemical-maker-dsm-sees-strong-demand-for-methane-reducing-cow-feed-additive-idUKKBN1WF1E8, accessed 16 Dec. 2019.

12 Humans use around 75% of ice-free land. 1% is used for infrastructure, 12% has been converted to crops, 37% is pasture, 22% managed forests and 28% is not intensively managed: IPCC, 'Climate Change and Land'.

13 We showed this in high-income nations by combining large amounts of spatial and food production data with trade information. See Z. Sun et al., 'Dietary Change in High-Income Nations Alone Can Lead to Substantial Double Climate Dividend', *Nature Food*, 3 (2022), 29–37. doi: 10.1038/s43016-021-00431-5.

14 T.P. Van Boeckel et al., 'Global Trends in Antimicrobial Resistance in Animals in Low- and Middle-Income Countries', *Science*, 365/6459 (2019). doi: 10.1126/science.aaw1944

15 M. Springmann et al., 'Health-Motivated Taxes on Red and Processed Meat: A Modelling Study on Optimal Tax Levels and Associated Health Impacts', *PLOS One*, 13/11 (2018). doi: https://doi.org/10.1371/journal.pone.0204139

16　W.K. Dodds et al., 'Eutrophication of U.S. Freshwaters: Analysis of Potential Economic Damages', *Environmental Science and Technology*, 43/1, 12–19. doi: 10.1021/es801217q

17　C. Nawroth et al., 'Farm Animal Cognition-Linking Behavior, Welfare and Ethics', *Front. Vet. Sci.* 6/24 (2019), 1–16. doi: 10.3389/fvets.2019.00024

18　L. Marino and K. Allen, 'The Psychology of Cows', *Animal Behavior Cognition*, 4/4 (2017), 474–98. doi: 10.26451/abc.04.04.06.2017

19　C. Brown, 'Familiarity with the Test Environment Improves Escape Responses in the Crimson Spotted Rainbowfish, *Melanotaenia duboulayi*', *Animal Cognition*, 4/2 (2001), 109–13. doi: 10.1007/s100710100105

20　On the idea of sentience being on a scale, see Frans de Waals, *Are We Smart Enough to Know How Smart Animals Are?*, W.W. Norton & Company: New York, 2016. On animals feeling stress and pain, see Aliza le Roux, 'What Animals Can Teach Us About Stress', *The Conversation* (17 Jul. 2015), https://theconversation.com/what-animals-can-teach-us-about-stress-42500, accessed 16 Dec. 2019.

21　M. Niles et al., '*Climate Change & Food Systems: Assessing Impacts and Opportunities*', Meridian Institute (2017), https://www.researchgate.net/publication/317170041_Climate_Change_Food_Systems_Assessing_Impacts_and_Opportunities, accessed 16 Dec. 2019.

22　Mike Berners-Lee, *There Is No Planet B: A Handbook for the Make or Break Years*, Cambridge University Press: Cambridge, 2019.

23　P. Behrens et al., 'Evaluating the Environmental Impacts of Dietary Recommendations', *Proceedings of the National Academy of Sciences of the United States of America*, 114/51 (2017), 13412–17. doi: 10.1073/pnas.1711889114

24　M.M. Vilmori-Andrieux, *The Vegetable Garden*, London: John Murray: 1885.

25　UK estimates: Ipsos MORI survey, as reported by The Economist, 'Interest in Veganism is Surging', *The Economist* (29 Jan. 2020), https://www.economist.com/graphic-detail/2020/01/29/interest-in-veganism-is-surging, accessed 4 Feb. 2020. US estimates: Janet Forgrieve, 'The Growing Acceptance of Veganism', *Forbes* (2 Nov. 2018), https://www.forbes.com/sites/janetforgrieve/2018/11/02/picturing-a-kindler-gentler-world-vegan-month/#1c7ff6002f2b, accessed 21 Dec. 2019.

26　E.E. Garnett et al., 'Impact of Increasing Vegetarian Availability on Meal Selection and Sales in Cafeterias', *Proceedings of the National Academy of Sciences of the United States of America*, 116/42 (2019), 20923–9. doi: 10.1073/pnas.1907207116

27　Third reduced meat: Rebecca Smithers, 'Third of Britons Have Stopped or Reduced Eating Meat – Report', *The Guardian* (1 Nov. 2018), https://www.theguardian.com/business/2018/nov/01/third-of-britons-have-stopped-or-reduced-meat-eating-vegan-vegetarian-report. Increase in plant-based milk: Zoe Wood, 'Plant-Based Milk the Choice for Almost 25% of Britons Now', *The Guardian* (19 Jul. 2019), https://www.theguardian.com/food/2019/jul/19/plant-based-milk-the-choice-for-almost-25-of-britons-now, both accessed 4 Jan. 2020.

28　See the data at the OECD website here: https://data.oecd.org/agroutput/meat-consumption.htm, accessed 4 Jan. 2020.

29　European Commission, 'EU Agricultural Outlook 2020–30: Sustainability Objectives to Impact Meat and Dairy Along the Supply Chain' (16 Dec. 2020), https://ec.europa.eu/info/news/eu-agricultural-outlook-2020-30-sustainability-objectives-impact-meat-and-dairy-along-supply-chain-2020-dec-16_en, accessed 9 Jan. 2021.

30　'New Food Balances (Preliminary Data)'. This is taken from the FAO's 'New Food

Balances (Preliminary Data), http://www.fao.org/faostat/en/#data/FBS/metadata, accessed 3 Mar. 2022.

31 S.M. Lonergan, D.G. Topel and D.N. Marple, 'Fresh and Cured Meat Processing and Preservation', *Science of Animal Growth and Meat Technology* (2nd ed.), Academic Press: London, 2019, 205–228. doi: 10.1016/b978-0-12-815277-5.00013-5

32 N. Stephens et al., 'Bringing Cultured Meat to Market: Technical, Sociopolitical, and Regulatory Challenges in Cellular Agriculture', *Trends in Food Science & Technology*, 78 (2018), 155–66 (2018). doi: 10.1016/j.tifs.2018.04.010

33 There may still be issues with public acceptance in the US and potentially in other countries: see W. Verbeke, P. Sans and E.J. Van Loo, 'Challenges and Prospects for Consumer Acceptance of Cultured Meat', *Journal of Integrative Agriculture*, 14/2 (2015), 285–94 (2015). doi: 10.1016/S2095-3119(14)60884-4

34 Food and Agriculture Organization of the United Nations, 'What Is Happening to Agrobiodiversity' in 'Building on Gender, Agrobiodiversity and Local Knowledge' (2004), http://www.fao.org/3/y5609e/y5609e02.htmhttp://www.fao.org/3/y5609e/y5609e02.htmhttp://www.fao.org/3/y5609e/y5609e02.htm, accessed 7 Jan. 2020.

35 T. S. Cox et al., 'Prospects for Developing Perennial Grain Crops', *Bioscience*, 56/8 (2006), 649–59. doi: 10.1641/0006-3568(2006)56[649:PFDPGC]2.0.CO;2

36 Gustavsson et al., 'Global Food Losses and Food Waste – Extent, Causes and Prevention' (Rome: FAO, 2011), http://www.fao.org/3/a-i2697e.pdf, accessed 16 Dec. 2019.

37 Ibid.

38 Department for Environment, Food & Rural Affairs, 'Family Food 2016/17: Expenditure' (26 Apr. 2018), https://www.gov.uk/government/publications/family-food-201617/expenditure, accessed 16 Dec. 2019.

39 Jonathan Bloom, 'Denmark Capitalizes on Culture to Stop Food Waste', *National Geographic* (26 Sep. 2016), https://www.nationalgeographic.com/people-and-culture/food/the-plate/2016/09/denmark-harnesses-its-own-culture-to-stop-food-waste/, accessed 16 Dec. 2019.

40 Eleanor Beardsley, 'French Food Waste Law Changing How Grocery Stores Approach Excess Food', *NPR* (24 Feb. 2018), https://www.npr.org/sections/thesalt/2018/02/24/586579455/french-food-waste-law-changing-how-grocery-stores-approach-excess-food, accessed 16 Dec. 2019.

41 Z. Conrad et al., 'Relationship Between Food Waste, Diet Quality, and Environmental Sustainability', *PLOS ONE*, 13/4 (2018). doi: 10.1371/journal.pone.0195405

42 D. Qi and B.E. Roe, 'Foodservice Composting Crowds Out Consumer Food Waste Reduction Behavior in a Dining Experiment', *American Journal of Agricultural Economics*, 99/5 (2017), 1159–71. doi: 10.1093/ajae/aax050

43 Gustavsson et al., 'Global Food Losses'.

44 Emmely Wildeboer and Paul Bosch, 'Why We Must Invest in Local Food Storage in Sub-Saharan Africa', *The Guardian* (15 Jan. 2015), https://www.theguardian.com/sustainable-business/2015/jan/15/invest-local-food-storage-sub-saharan-africa, accessed 16 Dec. 2019.

45 M. Sheahan and C.B. Barrett, 'Food Loss and Waste in Sub-Saharan Africa: A Critical Review', *Food Policy*, 70 (2017), 1–12. doi: 10.1016/j.foodpol.2017.03.012

46 D. Weisberger, V. Nichols and M. Liebman, 'Does Diversifying Crop Rotations Suppress Weeds? A Meta-Analysis', *PLOS ONE*, 14/7 (2019). doi: 10.1371/journal.pone.0219847

47 All data comes from the Global Yield Gap Atlas, www.yieldgap.org, accessed 1 Aug. 2019.

48 D. Tilman, 'Global Environmental Impacts of Agricultural Expansion: The Need for Sustainable and Efficient Practices', *Proceedings of the National Academy of Sciences of the United States of America*, 96/11 (1999), 5995–6000. doi: 10.1073/pnas.96.11.5995

49 S. Klasen et al., 'Economic and Ecological Trade-Offs of Agricultural Specialization at Different Spatial Scales', *Ecological Economics*, 122 (2016), 111–20. doi: 10.1016/j.ecolecon.2016.01.001

50 D. Rolnick et al., 'Tackling Climate Change with Machine Learning', *arXiv*, 1–97 (2019). doi: arXiv:1906.05433

51 There are many examples of such companies, such as Phytech Ltd, https://www.phytech.com/, accessed 16 Dec. 2019.

52 M. Jain et al., 'The Impact of Agricultural Interventions Can Be Doubled by Using Satellite Data', *Nature Sustainability*, 2/10 (2019), 931–4. doi: 10.1038/s41893-019-0396-x

53 One such example is the University of Sydney's RIPPA robot: 'University of Sydney, Rippa Robot Takes Farms Forward to the Future' (21 Oct. 2015), https://sydney.edu.au/news-opinion/news/2015/10/21/rippa-robot-takes-farms-forward-to-the-future-.html, accessed 16 Dec. 2019.

54 Michael J. Coren, 'Cheap Robots Are Coming for Our Farm Jobs by Taking the Most Brutal Tasks First', *Quartz* (10 Jul. 2016), https://qz.com/726667/cheap-robots-are-coming-for-our-farm-jobs-by-taking-the-most-brutal-tasks-first/, accessed 4 Jan. 2020.

55 J.M. Pleasant, 'The Science behind the Three Sisters Mound System', *Histories of Maize: Multidisciplinary Approaches to the Prehistory, Linguistics, Biogeography, Domestication, and Evolution of Maize* (Elsevier: Amsterdam, 2006), 529–37. doi: 10.1016/b978-012369364-8/50290-4

56 Brooker R.W., et al., 'Improving intercropping: a synthesiss of research in agronomy, plant physiology and ecology', *New Phytologist*, 206/1 (2014), 107-117 doi: 10.1111/nph.13132.

57 R.E. Black et al., 'Maternal and Child Undernutrition: Global and Regional Exposures and Health Consequences', *Lancet*, 371 (2008), 243–60. doi: 10.1016/S0140-6736(07)61690-0

58 Ibid.

59 H.C.J. Godfray et al., 'Food Security: The Challenge of Feeding 9 Billion People', *Science*, 327/5967 (2010), 812–18. doi: 10.1126/science.1185383.

60 M. Eisenhut and A.P.M. Weber, 'Improving Crop Yield', *Science*, 363/6422 (2019), 32–3 (2019). doi: 10.1126/science.aav8979

61 R.E. Blankenship et al., 'Comparing Photosynthetic and Photovoltaic Efficiencies and Recognizing the Potential for Improvement', *Science*, 332/6031 (2011), 805–9. doi: 10.1126/science.1200165

62 Veronique Greenwood, 'Taming the Groundcherry: With Crispr, a Fussy Fruit Inches Toward the Supermarket', *New York Times* (5 Oct. 2019), https://www.nytimes.com/2018/10/05/science/groundcherries-crispr-gene-editing.html, accessed 4 Jan. 2020.

63 Amy Maxmen, 'CRISPR Might Be the Banana's Only Hope Against a Deadly Fungus', *Nature* (24 Sep. 2019), https://www.nature.com/articles/d41586-019-02770-7, accessed 16 Dec. 2019.

64 Y. Eshed and Z.B. Lippman, 'Revolutions in Agriculture Chart a Course for Targeted Breeding of Old and New Crops', *Science*, 366/6466 (2019). doi: 10.1126/science.

aax0025. Cynthia Graber and Nicola Twilley, 'What's CRISPR Doing in Our Food?' [podcast] (7 Oct. 2019), Gastropod, https://gastropod.com/whats-crispr-doing-in-our-food/, accessed 21 Dec. 2019.

65 W. Hu et al., 'RNA-Directed Gene Editing Specifically Eradicates Latent and Prevents New HIV-1 Infection', Proceedings of the National Academy of Sciences of the United States of America, 111/31 (2014), 11461–6. doi: 10.1073/pnas.14051 86111.

66 World Health Organization, 'Food Safety: Frequently Asked Questions on Genetically Modified Foods' (May 2014), https://www.who.int/foodsafety/areas_work/food-technology/faq-genetically-modified-food/en/, accessed 17 Dec. 2019.

67 E.D. Perry et al., 'Genetically Engineered Crops and Pesticide Use in U.S. Maize and Soybeans', Science Advances, 2/8 (2016), 1–9. doi: 10.1126/sciadv.1600850

68 On DNA damage, see M. Kosicki, K. Tomberg and A. Bradley, 'Repair of Double-Strand Breaks Induced by CRISPR–Cas9 Leads to Large Deletions and Complex Rearrangements', Nature Biotechnology, 36 (2018), 765–71. doi: 10.1038/nbt.4192. On the creation of edited cells, see E. Haapaniemi et al., 'CRISPR-Cas9 Genome Editing Induces a p53-Mediated DNA Damage Response', Nature Medicine, 24/7 (2018), 927–30 (2018). doi: 10.1038/s41591-018-0049-z.

69 Lessley Anderson, 'Why Does Everyone Hate Monsanto?', Modern Farmer (4 Mar. 2014), https://modernfarmer.com/2014/03/monsantos-good-bad-pr-problem/, accessed 17 Dec. 2019.

70 Pamela C. Ronald & Raoul W. Adamchak, Tomorrow's Table: Organic Farming, Genetics, and the Future of Food (2nd ed.), Oxford University Press: Oxford, 2018.

71 For the Open Source Seed Initiative, see https://osseeds.org/. For the Biological Innovation for Open Society, see https://cambia.org/bios-landing/, both accessed 4 Jan. 2020.

72 Take pesticide regulation, for example: once a pesticide is banned, the second most toxic (or even a more toxic) compound is often used. See for instance E. Stokstad, 'European Union Expands Ban of Three Neonicotinoid Pesticides', Science (2018). doi: 10.1126/science.aau0152

73 See World Bank, 'Agricultural Land (% of Land Area)' (2015), https://data.worldbank. org/indicator/ag.lnd.agri.zs, accessed 17 Dec. 2019.

74 D. Centola et al., 'Experimental Evidence for Tipping Points in Social Convention', Science, 360/6393 (2018), 1116–19. doi: 10.1126/science.aas8827

75 M.P. White et al., 'Spending at Least 120 Minutes a Week in Nature is Associated with Good Health and Wellbeing', Scientific Reports, 9/1 (2019), 1–11. doi: 10.1038/ s41598-019-44097-3

76 D. Rojas-Rueda et al., 'Green Spaces and Mortality: A Systematic Review and Meta-Analysis of Cohort Studies', Lancet Planetary Health, 3/11 (2019), e469–77. doi: 10.1016/S2542-5196(19)30215-3

Chapter Seven – Pessimism – Where All Roads Meet

1 There are many different lines of evidence for such an assertion. See for instance J.A. Patz et al., 'Impact of Regional Climate Change on Human Health', Nature, 438/7066 (2005), 310–17. doi: 10.1038/nature04188. This paper describes impacts up until 2005 (after which many more impacts have been experienced, as described in the rest of this chapter). Looking forward, there are heat, nutrition, disease, conflict, and other

impacts of climate change, to list just two studies: C. Mora et al., 'Global Risk of Deadly Heat', *Nature Climate Change*, 7/7 (2017), 501–6. doi: 10.1038/nclimate3322; and M. Springmann et al., 'Global and Regional Health Effects of Future Food Production Under Climate Change: A Modelling Study', *Lancet*, 387/10031 (2016), 1937–46. doi:10.1016/S0140-6736(15)01156-3

2   If we look at permafrost thaw, for example: J. Hjort et al., 'Degrading Permafrost Puts Arctic Infrastructure at Risk by Mid-Century', *Nature Communications*, 9/1 (2018). doi: 10.1038/s41467-018-07557-4. Thaw is already destabilizing infrastructure in the Arctic, including, oddly, oil and gas infrastructure: Julian Lee, 'Why Vladimir Putin Suddenly Believes in Global Warming', Bloomberg (29 Sep. 2019), https://www-bloomberg-com.cdn.ampproject.org/c/s/www.bloomberg.com/amp/opinion/articles/2019-09-29/climate-change-russia-s-oil-and-gas-heartlands-are-under-threat, accessed 17 Dec. 2019.

3   Ecologist Chris Dickman estimated that a billion animals perished in the fires as reported in Lisa Cox, 'A billion animals: some of the species most at risk from Australia's bushfire crisis', *The Guardian*, (13 Jan. 2020), https://www.theguardian.com/australia-news/2020/jan/14/a-billion-animals-the-australian-species-most-at-risk-from-the-bushfire-crisis, accessed 28 Jan. 2020. The statistic for the burned area of forest comes from a Nature Climate Change special edition as reported in: 'Bushfires burned a fifth of Australia's forest: study', Phys.org, (24 Feb. 2020), https://phys.org/news/2020-02-bushfires-australia-forest.html, accessed 28 Feb. 2020.

4   T.L. Frölicher, M. Winton and J.L. Sarmiento, 'Continued Global Warming after CO_2 Emissions Stoppage', *Nature Climate Change*, 4/1 (2013), 1–5. doi: 10.1038/NCLIMATE2060

5   D. Archer and V. Brovkin, 'The Millennial Atmospheric Lifetime of Anthropogenic CO_2', *Climatic Change*, 90/3 (2008), 283–97. doi: 10.1007/s10584-008-9413-1

6   Timothy Morton, *Hyperobjects: Philosophy and Ecology after the End of the World*, University of Minnesota Press: Minneapolis, MN, 2013.

7   This is a rough approximation which assumes that the current climate forcing is 1.6 W per m² averaged over the Earth and compared to the Hiroshima nuclear bomb blast.

8   David Orr, *Down to the Wire: Confronting Climate Collapse*, Oxford University Press: Oxford, 2019, p. 189.

9   This will become increasingly true; a case in point is David Koch, a billionaire who fuelled and funded climate denial for years and who died, aged 79, in 2019.

10   To quote *Game of Thrones* writer George R.R. Martin, 'the people in Westeros are fighting their individual battles over power and status and wealth. And those are so distracting them that they're ignoring the threat of "winter is coming", which has the potential to destroy all of them and to destroy their world,' quoted in Jamie Sims, 'George R.R. Martin Answers Times Staffers' Burning Questions', *New York Times Style Magazine* (16 Oct. 2018), https://www.nytimes.com/2018/10/16/t-magazine/george-rr-martin-qanda-game-of-thrones.html, accessed 17 Dec. 2019.

11   R.E. Zeebe, A. Ridgwell and J.C. Zachos, 'Anthropogenic Carbon Release Rate Unprecedented During the Past 66 Million Years', *Nature Geoscience*, 9/4 (2016), 325–9. doi: 10.1038/ngeo2681

12   David Nield, 'There Were Trees at the South Pole the Last Time There Was This Much CO_2 in the Air', *Science Alert* (7 Apr. 2019), https://www.sciencealert.com/there-were-trees-at-the-south-pole-the-last-time-there-was-this-much-co2-in-the-

air, accessed 17 Dec. 2019. The article is Nield's report on a meeting of the Royal Meteorological Society in April 2019.

13 M. Zemp et al., 'Global Glacier Mass Changes and Their Contributions to Sea-Level Rise from 1961 to 2016', Nature, 568/7752 (2019), 382–6. doi: 0.1038/s41586-019-1071-0

14 Thomas Frank, 'After a $14 Billion Upgrade, New Orleans' Levees Are Sinking', Scientific American (11 Apr. 2019), https://www.scientificamerican.com/article/after-a-14-billion-upgrade-new-orleans-levees-are-sinking/, accessed 17 Dec. 2019.

15 Associated Press, 'Barry Spares New Orleans but Fuels Fears of Floods and Tornadoes', The Guardian (14 Jul. 2019), https://www.theguardian.com/us-news/2019/jul/14/barry-new-orleans-mississippi-flash-flood, accessed 17 Dec. 2019.

16 John Schwartz, 'Wallace Broecker, 87, Dies; Sounded Early Warning on Climate Change', New York Times (19 Feb. 2019), https://www.nytimes.com/2019/02/19/obituaries/wallace-broecker-dead.html, accessed 17 Dec. 2019.

17 G.R. Grant et al., 'The Amplitude and Origin of Sea-Level Variability During the Pliocene Epoch', Nature, 574/2 (2019). doi: 10.1038/s41586-019-1619-z

18 On historic computer model estimates, see Working Group 1, T.F. Stocker et al., eds, Climate Change 2013: The Physical Science Basis. Contribution of Working Group I to the Fifth Assessment Report of the Intergovernmental Panel on Climate Change, IPCC, Cambridge University Press: United Kingdom and New York, 2013. doi: 10.1017/CBO9781107415324. On the latest results, see P. Voosen, 'New Climate Models Predict a Warming Surge' Science (2019). doi: 10.1126/science.aax7217; and 'The CMIP6 Landscape', Nature Climate Change, 9/10 (2019), 727. doi: 10.1038/s41558-019-0599-1

19 Peter Dockrill, 'It's Official: Atmospheric CO_2 Just Exceeded 415ppm for the First Time in Human History', Science Alert (13 May 2019), https://www.sciencealert.com/it-s-official-atmospheric-co2-just-exceeded-415-ppm-for-first-time-in-human-history, accessed 17 Dec. 2019.

20 Earth System Research Laboratory Global Monitoring Division, 'Trends in Atmospheric Carbon Dioxide', https://www.esrl.noaa.gov/gmd/ccgg/trends/data.html, accessed 4 Jan. 2020.

21 X. Zhang and K. Caldeira, 'Time Scales and Ratios of Climate Forcing Due to Thermal Versus Carbon Dioxide Emissions from Fossil Fuels', Geophysical Research Letters, 42/11 (2015), 4548–55. doi: 10.1002/2015GL063514

22 Ken Caldeira, 'How Much Ice Is Melted by Each Carbon Dioxide Emission?', @ KENCALDEIRA [blog] (24 Mar. 2018), https://kencaldeira.wordpress.com/2018/03/24/how-much-ice-is-melted-by-each-carbon-dioxide-emission/, accessed 17 Dec. 2019.

23 Ben See, 'The planet you think you're living on no longer exists. Goodbye, Arctic Ice.' [Twitter post], 10.03 p.m., 19 Jul. 2019, https://twitter.com/climateben/status/1152323044045668352?lang=en, accessed 4 Jan. 2020.

24 Eleanor Ainge Roy, 'Stronger Storms Mean New 'Category Six' Scale May Be Needed', The Guardian (22 Feb. 2018), https://www.theguardian.com/environment/2018/feb/22/category-six-storms-cyclones, accessed 18 Dec. 2019.

25 Josh Solomon, 'Category 6? Scientists Warn Hurricanes Could Keep Getting Stronger', Tampa Bay Times (30 Nov. 2018), https://www.tampabay.com/weather/category-6-scientists-warn-hurricanes-could-keep-getting-stronger-20181130/, accessed 18 Dec. 2019.

26 M. Davis, S. Faurby and J.C. Svenning, 'Mammal Diversity Will Take Millions of Years to Recover from the Current Biodiversity Crisis', Proceedings of the National Academy

   of Sciences of the United States of America, 115/44 (2018), 11262–7. doi: 10.1073/
   pnas.1804906115

27 M.J. Henehan et al., 'Rapid Ocean Acidification and Protracted Earth System Recovery
   Followed the End-Cretaceous Chicxulub Impact', *Proceedings of the National Academy
   of Sciences of the United States of America*, 116/45 (2019), 22500–4. doi: 10.1073/
   pnas.1905989116

28 Bill McGuire, *Waking the Giant: How a Changing Climate Triggers Earthquakes, Tsunamis,
   and Volcanoes*, Oxford University Press: Oxford, 2012.

29 N. Watts et al., 'The Lancet Countdown on Health and Climate Change: From 25
   Years of Inaction to a Global Transformation for Public Health', *Lancet*, 391/10120
   (2018), 581–630 (2018). doi: 10.1016/S0140-6736(17)32464-9

30 C. Funk et al., 'Examining the Role of Unusually Warm Indo-Pacific Sea Surface
   Temperatures in Recent African Droughts', *Quarterly Journal of the Royal Meteorological
   Society*, 144/S1 (2018), 360–83. doi: 10.1002/qj.3266

31 C. Funk et al., 'Examining the Potential Contributions of Extreme 'Western
   V' Sea Surface Temperatures to the 2017 March–June East African Drought',
   Bulletin of the American Meteorological Society, 100/1 (2019), 51–6. doi: 10.1175/
   BAMS-D-18-0108.1

32 J.M. Robine et al., 'Death Toll Exceeded 70,000 in Europe During the Summer of 2003',
   Comptes Rendus Biologies, 331/2 (2008), 171–8. doi: 10.1016/j.crvi.2007.12.001; and
   attribution to anthropogenic climate change: P.A. Stott et al., 'Human Contribution
   to the European Heatwave of 2003', *Nature*, 432/7017 (2004), 610–14. doi: 10.1038/
   nature03089

33 Various studies support this conclusion: F.E.L. Otto et al., 'Reconciling Two Approaches
   to Attribution of the 2010 Russian Heat Wave', *Geophysical Research Letters*, 39/4
   (2012). doi: 10.1029/2011GL050422; B. Dong et al., 'The 2015 European Heat
   Wave', *Bulletin of the American Meteorological Society*, 97/12 (2016), 57–62. doi:
   10.1175/BAMS-D-16-0140.1; S.F. Kew et al., 'The Exceptional Summer Heat Wave
   in Southern Europe 2017', *Bulletin of the American Meteorological Society*, 100 (2019),
   51–5. doi: 10.1175/BAMS-D-18-0109.1; 'Heatwave in Northern Europe, Summer
   2018', *World Weather Attribution* (28 Jul. 2018), https://www.worldweatherattribution.
   org/attribution-of-the-2018-heat-in-northern-europe/; Geert Jan van Oldenborgh et
   al., 'Human Contribution to the Record-Breaking June 2019 Heatwave in France',
   World Weather Attribution (2 Jul. 2019), https://www.worldweatherattribution.
   org/human-contribution-to-record-breaking-june-2019-heatwave-in-france/; and
   Robert Vautard et al., 'Human Contribution to the Record-Breaking July 2019 Heat
   Wave in Western Europe', *World Weather Attribution* (2 Aug. 2019), https://www.
   worldweatherattribution.org/human-contribution-to-the-record-breaking-july-2019-
   heat-wave-in-western-europe/, accessed 18 Dec. 2019.

34 N. Vargas-Cuentas, A. Roman-Gonzalez and L.A. Muñoz, 'Use of Satellite Images for
   Droughts Studying: The Bolivian Case', conference paper, *International Astronautical
   Congress – IAC* (Sep. 2017).

35 K. Hansen, 'Bolivia's Lake Poopó Disappears', *NASA Earth Observatory* (23 Jan.
   2016), https://earthobservatory.nasa.gov/images/87363/bolivias-lake-poopo-disappears,
   accessed 18 Dec. 2019.

36 T. Kinouchi et al., 'Water Security in High Mountain Cities of the Andes Under
   a Growing Population and Climate Change: A Case Study of La Paz and El Alto,
   Bolivia', *Water Security*, 6/1 (2019). 10.1016/j.wasec.2019.100025; A. Mohammad

Abadi Kamarei and C.M. Rowe, 'Drought Components Change in Bolivia and Its Societal Impacts as a Response to Climate Change', American Geophysical Union Fall Meeting, 2018, abstract (2018); and P. Paterson, 'Calentamiento global y cambio climático en Sudamérica', Revista Política y Estrategia, 130 (2017), 153–88. doi: 10.26797/rpye.v0i130.133

37   Y. Imada et al., 'Climate Change Increased the Likelihood of the 2016 Heat Extremes in Asia', Bulletin of the American Meteorological Society, 98/12(2018), 97–101. doi: 10.1175/BAMS-D-17-0109.1

38   A.J. Hobday et al., 'A Hierarchical Approach to Defining Marine Heatwaves', Progress in Oceanography, 141 (2016), 227–38. doi: 10.1016/j.pocean.2015.12.014

39   T.P. Hughes et al., 'Global Warming and Recurrent Mass Bleaching of Corals', Nature, 543/7645 (2017), 373–7. doi: 10.1038/nature21707

40   E.C. Oliver et al., 'Anthropogenic and Natural Influences on Record 2016 Marine Heat Waves', Bulletin of the American Meteorological Society, 99/1 (2018), 44–8. doi: 10.1175/BAMS-D-17-0093.1

41   F.E. Otto et al., 'Anthropogenic Influence on the Drivers of the Western Cape Drought 2015–2017', Environmental Research Letters, 13/12 (2018). doi: 10.1088/1748-9326/aae9f9

42   B.I. Cook et al., 'Spatiotemporal Drought Variability in the Mediterranean over the Last 900 Years', Journal of Geophysical Research: Atmospheres, 121/5 (2016), 2060–74. doi: 10.1002/2015JD023929

43   On the contribution to political unrest in the region, see K. Bergaoui et al., 'The Contribution of Human-Induced Climate Change to the Drought of 2014 in the Southern Levant Region', Bulletin of the American Meteorological Society, 96/12 (2015), 66–70. doi: 10.1175/BAMS; and C.P. Kelley et al., 'Climate Change in the Fertile Crescent and Implications of the Recent Drought', Proceedings of the National Academy of Sciences, 112/11 (2015), 3241–6. doi: 10.1073/pnas.1421533112. For disputes over the extent to which it was a factor, see J. Selby et al., 'Climate Change and the Syrian Civil War Revisited', Political Geography, 60 (2017), 232–44. doi: 10.1016/j. polgeo.2017.05.007.

44   See Bergaoui et al., 'Contribution of Human-Induced Climate Change to the Drought of 2014'; Cook et al., 'Spatiotemporal Drought Variability in the Mediterranean'; and Kelley et al., 'Climate Change in the Fertile Crescent'.

45   D.A. Herrera et al., 'Exacerbation of the 2013–2016 Pan-Caribbean Drought by Anthropogenic Warming', Geophysical Research Letters, 45/19 (2018), 10619–26. doi: 10.1029/2018GL079408

46   N. Christidis, R.A. Betts and P.A. Stott, 'The Extremely Wet March of 2017 in Peru', Bulletin of the American Meteorological Society, 100 (2019), 51–5. doi: 10.1175/ BAMS-D-18-0110.1

47   S.Y. Wang et al., 'The Deadly Himalayan Snowstorm of October 2014: Synoptic Conditions and Associated Trends', Bulletin of the American Meteorological Society, 96/12 (2015), 89–94. doi: 10.1175/BAMS-D-15-00113.1

48   FAO, 'Bangladesh: Severe Floods in 2017 Affected Large Numbers of People and Caused Damage to the Agriculture Sector' (3 Oct. 2017), www.fao.org/3/a-i7876e. pdf, accessed 18 Dec. 2019.

49   R.H. Rimi et al., 'Risks of Pre-Monsoon Extreme Rainfall Events of Bangladesh: Is Anthropogenic Climate Change Playing a Role?', Bulletin of the American Meteorological Society, 100/1 (2019), 51–5. 10.1175/BAMS-D-18-0152.1

50 Richard Davies, 'China Floods – Wettest May for 40 Years, Over 50 Dead, More Rain to Come', *FloodList* (26 May 2015), http://floodlist.com/asia/china-floods-record-rain-50-dead, accessed 18 Dec. 2019.

51 C. Burke et al., 'Attribution of Extreme Rainfall in Southeast China During May 2015', *Bulletin of the American Meteorological Society*, 97/12 (2016), 92–6. doi: 10.1175/BAMS-D-16-0144.1

52 R.C. De Abreu et al., 'Contribution of Anthropogenic Climate Change to April–May 2017 Heavy Precipitation over the Uruguay River Basin', *Bulletin of the American Meteorological Society*, 100/1 (2018), 37–41. doi: 10.1175/BAMS-D-18-0102.1

53 ACAPS, 'Nigeria: Floods Update II', *Reliefweb* (2 Oct. 2018), https://reliefweb.int/sites/reliefweb.int/files/resources/20181002_acaps_briefing_note_nigeria_floods_update_ii.pdf, accessed 18 Dec. 2019.

54 H. Haider, 'Climate Change in Nigeria: Impacts and Responses', K4D Helpdesk Report 675 (10 Oct. 2016), Institute of Development Studies: Brighton, 2019, https://opendocs.ids.ac.uk/opendocs/handle/20.500.12413/14761; and C.C. Olanrewaju et al., 'Impacts of Flood Disasters in Nigeria: A Critical Evaluation of Health Implications and Management', *Jàmbá: Journal of Disaster Risk Studies*, 11/1 (2019), 557. doi: 10.4102/jamba.v11i1.557

55 National Oceanic and Atmospheric Administration (NOAA), 'Fast Facts – Hurricane Costs', https://coast.noaa.gov/states/fast-facts/hurricane-costs.html, accessed 16 Nov. 2019.

56 On rainfalls during Harvey being three times more likely, see G.J. Van Oldenborgh et al., 'Attribution of Extreme Rainfall from Hurricane Harvey, August 2017', *Environmental Research Letters*, 12/12 (2017). doi: 10.1088/1748-9326/aa9ef2. On the increases in rainfall during Katrina, Irma and Maria, see C.M. Patricola and M.F. Wehner, 'Anthropogenic Influences on Major Tropical Cyclone Events', *Nature*, 563/7731 (2018), 339–46. doi: 10.1038/s41586-018-0673-2

57 Committee on Extreme Weather Events and Climate Change Attribution, *Attribution of Extreme Weather Events in the Context of Climate Change*, National Academies Press: Washington, DC, 2016. doi: 10.17226/21852

58 Helena Smith, Sam Jones and Martin Farrer, 'Greece Wildfires: Scores Dead as Holiday Resort Devastated', *The Guardian* (24 Jul. 2018), https://www.theguardian.com/world/2018/jul/23/greeks-urged-to-leave-homes-as-wildfires-spread-near-athens, accessed 19 Dec. 2019.

59 Sheri Fink, 'The Deadly Choices at Memorial', *New York Times Magazine* (25 Aug. 2009), https://www.nytimes.com/2009/08/30/magazine/30doctors.html, accessed 19 Dec. 2019.

60 Kelley et al., 'Climate Change in the Fertile Crescent'. There is controversy, though, about the level to which it contributed: J. Selby et al., 'Climate Change and the Syrian Civil War Revisited: A Rejoinder', *Political Geography*, 60 (2017), 253–255. doi: 10.1016/j.polgeo.2017.08.001

61 Caroline Criado-Perez, *Invisible Women: Exposing Data Bias in a World Designed for Men*, Penguin: London, 2019.

62 K. Pistone, I. Eisenman and V. Ramanathan, 'Radiative Heating of an Ice-Free Arctic Ocean', *Geophysical Research Letters*, 46/13 (2019), 7474–7480. doi: 10.1029/2019GL082914

63 IPCC. Projections of Future Changes in Climate – AR4 WGI Summary for Policymakers. *Clim. Chang. Work. Gr. I Phys. Sci. Basis* (2007).

64   Stocker et al., eds, *Climate Change 2013*.

65   J.A. Screen and C. Deser, 'Pacific Ocean Variability Influences the Time of Emergence of a Seasonally Ice-Free Arctic Ocean', *Geophysical Research Letters*, 46/4 (2019), 2222–31. doi: 10.1029/2018GL081393

66   D.A. Sutherland et al., 'Direct Observations of Submarine Melt and Subsurface Geometry at a Tidewater Glacier', *Science*, 365/6451 (2019), 369–74. doi: 10.1126/science.aax3528

67   J. Mouginot et al., 'Forty-Six Years of Greenland Ice Sheet Mass Balance from 1972 to 2018', *Proceedings of the National Academy of Sciences of the United States of America*, 116/19 (2019), 9239–44. doi: 10.1073/pnas.1904242116

68   See Andrew Freedman, 'Greenland is on Track to Lose Most Ice on Record this Year and Has Already Shed 250 Billion Tons', *Washington Post* (8 Aug. 2019), https://www.washingtonpost.com/weather/2019/08/08/greenland-is-track-record-melt-year-having-already-lost-billion-tons-ice/; and Nerilie Abram, 'Time Will Tell if this Is a Record Summer for Greenland Ice Melt, but the Pattern over the Past 20 Years Is Clear', *The Conversation* (2 Jul. 2019), https://theconversation.com/time-will-tell-if-this-is-a-record-summer-for-greenland-ice-melt-but-the-pattern-over-the-past-20-years-is-clear-119307, both accessed 19 Dec. 2019.

69   For the fish mass figure, see R.W. Wilson et al., 'Contribution of Fish to the Marine Inorganic Carbon Cycle', *Science*, 323/5912 (2009), 359–62. doi: 10.1126/science.1157972. For the annual weight of food produced by humans, see the FAO's facts and figures here: FAO, 'SAVE FOOD: Global Initiative on Food Loss and Waste Reduction': http://www.fao.org/save-food/resources/keyfindings/en/, accessed 19 Dec. 2019.

70   Thomas Mote, 'Greenland's Ice Wasn't Supposed to Melt Like Last Week Until 2070', *The Hill* (4 Aug. 2018), https://thehill.com/opinion/energy-environment/456112-greenlands-ice-sheet-wasnt-expected-to-melt-like-this-until-2070, accessed 19 Dec. 2019.

71   L.M. Farquharson et al., 'Climate Change Drives Widespread and Rapid Thermokarst Development in Very Cold Permafrost in the Canadian High Arctic', *Geophysical Research Letters*, 46/12 (2019), 6681–9. doi: 10.1029/2019GL082187, as reported by Nadia Suleman, 'Canada's Permafrost Is Thawing 70 Years Earlier than Expected, Study Shows. Scientists Are "Quite Surprised"', *TIME* (19 Jun. 2019), https://time.com/5610084/canadas-permafrost-thawing-surprising/, accessed 19 Dec. 2019.

72   C.D. Koven et al., 'A Simplified, Data-Constrained Approach to Estimate the Permafrost Carbon–Climate Feedback', *Philosophical Transactions of the Royal Society A: Mathematical, Physical and Engineering Sciences*, 373/2054 (2015). doi: 10.1098/rsta.2014.0423

73   Alec Luhn, 'Thawing Siberian Permafrost Soil Risks Rise of Anthrax and Prehistoric Diseases', *The Telegraph* (14 Apr. 2019), https://www.telegraph.co.uk/global-health/climate-and-people/thawing-siberian-permafrost-soil-risks-rise-anthrax-prehistoric/, accessed 19 Dec. 2019.

74   X.J. Walker et al., 'Increasing Wildfires Threaten Historic Carbon Sink of Boreal Forest Soils', *Nature*, 572/7770 (2014), 520–3. doi: 10.1038/s41586-019-1474-y

75   S. Veraverbeke et al., 'Lightning as a Major Driver of Recent Large Fire Years in North American Boreal Forests', *Nature Climate Change*, 7/7 (2017), 529–34. doi: 10.1038/nclimate3329

76   W.A. Kurz et al., 'Mountain Pine Beetle and Forest Carbon Feedback to Climate Change', *Nature*, 452/7190 (2008), 987–90. doi: 10.1038/nature06777

77   J. Hansen and L. Nazarenko, 'Soot Climate Forcing via Snow and Ice Albedos', *Proceedings of the National Academy of Sciences of the United States of America*, 101/2 (2004), 423–8. doi: 10.1073/pnas.2237157100

78   C.L. Parkinson, 'A 40-y Record Reveals Gradual Antarctic Sea Ice Increases Followed by Decreases at Rates Far Exceeding the Rates Seen in the Arctic', *Proceedings of the National Academy of Sciences of the United States of America*, 116/29 (2019), 14414–23. doi: 10.1073/pnas.1906556116

79   H.D. Pritchard et al., 'Antarctic Ice-Sheet Loss Driven by Basal Melting of Ice Shelves', *Nature*, 484/7395 (2012), 502–5. doi: 10.1038/nature10968

80   E. Rignot and S.S. Jacobs, 'Rapid Bottom Melting Widespread Near Antarctic Ice Sheet Grounding Lines', *Science*, 296/5575 (2002), 2020–3. doi: 10.1126/science.1070942

81   E. Rignot et al., 'Four Decades of Antarctic Ice Sheet Mass Balance from 1979–2017', *Proceedings of the National Academy of Sciences of the United States of America*, 116/4 (2019), 1095–1103. doi: 10.1073/pnas.1812883116

82   Nerilie Abram, Matthew England and Matt King, 'Arctic Ice Loss Is Worrying, but the Giant Stirring in the South Could Be Even Worse', *The Conversation* (11 Jul. 2019), https://theconversation.com/arctic-ice-loss-is-worrying-but-the-giant-stirring-in-the-south-could-be-even-worse-119822, accessed 19 Dec. 2019.

83   Jeff Goodell, '"The Fuse Has Been Blown", and the Doomsday Glacier Is Coming for Us All', *Rolling Stone* (29 Dec. 2021), https://www.rollingstone.com/politics/politics-features/doomsday-glacier-thwaites-antarctica-climate-crisis-1273841/, accessed 9 Jan. 2022.

84   S. Dangendorf et al., 'Persistent Acceleration in Global Sea-Level Rise Since the 1960s', *Nature Climate Change*, 9/9 (2010), 705–10. doi: 10.1038/s41558-019-0531-8

85   For the worst-case estimate, see Adam Vaughan, 'Sea Level Rise Could Hit 2 Metres by 2100 – Much Worse than Feared', *New Scientist* (20 May 2019), https://www.newscientist.com/article/2203700-sea-level-rise-could-hit-2-metres-by-2100-much-worse-than-feared/, accessed 19 Dec. 2019. For the population figures of who will be affected by this rise, see J.L. Bamber et al., 'Ice Sheet Contributions to Future Sea-Level Rise from Structured Expert Judgment', *Proceedings of the National Academy of Sciences of the United States of America*, 166 (2019), 11195–200. For the consensus view, see J.A. Church et al., 'Sea Level Change', in Working Group 1, Stocker et al., eds, *Climate Change 2013*, 895–900. doi: 10.1007/978-1-4020-4399-4_309

86   G.R. Grant et al., 'Amplitude and Origin of Sea-Level Variability During the Pliocene Epoch', *Nature*, 574/7777 (2019), 237-241. doi: 10.1038/s41586-019-1619-z.

87   D.J.R. Thornalley et al., 'Anomalously Weak Labrador Sea Convection and Atlantic Overturning During the Past 150 Years', *Nature*, 556/7700 (2018), 227–30. doi: 10.1038/s41586-018-0007-4

88   This is a big statement, but comes from the evidence that climate sensitivity may be higher than thought, along with a committed warming of perhaps 1.5 degrees Celsius by 2100 and a review of tipping points, as summarized in W. Steffen et al., 'Trajectories of the Earth System in the Anthropocene', *Proceedings of the National Academy of Sciences of the United States of America*, 115/33 (2018), 8252–9. doi/10.1073/pnas.1810141115

89   K. Brysse et al., 'Climate Change Prediction : Erring on the Side of Least Drama.' *Global Environmental Change*, 23/1 (2013), 327–37. doi: 10.1016/j.gloenvcha.2012.10.008

90   Nafeez Ahmed, 'IPCC Reports "Diluted" Under "Political Pressure" to Protect Fossil Fuel Interests', *The Guardian* (15 May 2014), https://www.theguardian.com/

environment/earth-insight/2014/may/15/ipcc-un-climate-reports-diluted-protect-fossil-fuel-interests, accessed 19 Dec. 2019.

91 The same sorts of processes happen in the thermokarst ponds in regions of permafrost: K. Walter Anthony et al., '21st-Century Modeled Permafrost Carbon Emissions Accelerated by Abrupt Thaw Beneath Lakes', *Nature Communications*, 9/1 (2018), 3262. doi: 10.1038/s41467-018-05738-9

92 IPCC, 'Climate Change and Land: An IPCC Special Report on Climate Change, Desertification, Land Degradation, Sustainable Land Management, Food Security, and Greenhouse Gas Fluxes in Terrestrial Ecosystems' (summary for policymakers approved draft) (7 Aug. 2019). doi: 10.4337/9781784710644

93 S. Díaz et al., 'Summary for Policymakers of the Global Assessment Report on Biodiversity and Ecosystem Services', *IPBES* (2019), 1–39.

94 C. Mora et al., 'Broad Threat to Humanity from Cumulative Climate Hazards Intensified by Greenhouse Gas Emissions', *Nature Climate Change*, 8/12 (2018), 1062–71. doi: 10.1038/s41558-018-0315-6

95 D.K. Ray et al., 'Climate Variation Explains a Third of Global Crop Yield Variability', *Nature Communications*, 6 (2015), 1–9. doi: 10.1038/ncomms6989

96 D.K. Ray et al., 'Climate Change Has Likely Already Affected Global Food Production', *PLOS ONE*, 14/5 (2019), 1–18. doi: 10.1371/journal.pone.0217148

97 For example see the attribution studies shown here: Roz Pidcock, Rosamund Pearce and Robert McSweeny, 'Attributing Extreme Weather to Climate Change', *Carbon Brief* (11 Mar. 2019), https://www.carbonbrief.org/mapped-how-climate-change-affects-extreme-weather-around-the-world, accessed 19 Dec. 2019.

98 See for instance the reporting on grain supplies: Eric Schroeder, 'Bangladesh Rice Output Forecast Lower', *World Grain* (5 Aug. 2019), https://www.world-grain.com/articles/12419-bangladesh-rice-output-forecast-lower; *World Grain*, 'Weather Weighs on ADM Earnings' (2 Aug. 2019), https://www.world-grain.com/articles/12412-weather-weighs-on-adm-earnings; Arvin Donley, 'E.U. Wheat Production Forecast Lower', *World Grain* (1 Aug. 2019), https://www.world-grain.com/articles/12404-eu-wheat-production-forecast-lower; Ron Sterk, 'U.S. Organic Hard Red Winter Wheat Down Sharply in May–June' (31 Jul. 2019), https://www.world-grain.com/articles/12393-us-organic-hard-red-winter-wheat-down-sharply-in-may-june; Susan Reidy, 'Heatwave Field Fires Halt Harvest in France', *World Grain* (26 Jul. 2019), https://www.world-grain.com/articles/12382-heatwave-field-fires-halt-harvest-in-france; and *World Grain*, 'IGC Revises Global Grain Output Lower' (26 Jul. 2019), https://www.world-grain.com/articles/12381-igc-revises-global-grain-output-lower, accessed 19 Dec. 2019.

99 B.G. Meerburg et al., 'Do Nonlinear Temperature Effects Indicate Severe Damages to US Crop Yields Under Climate Change?', *Proceedings of the National Academy of Sciences of the United States of America*, 106/43 (2009), 15594–8. doi: 10.1073/pnas.0910618106

100 J. Jägermeyer et al, 'Climate Impacts on Global Agriculture Emerge Earlier in New Generation of Climate and Crop Models', *Nature Food*, 2 (2021), 873–85, doi: 10.1038/s43016-021-00400-y

101 IPCC, 'Climate Change and Land'.

102 F. Gaupp et al., Increasing Risks of Multiple Breadbasket Failure Under 1.5 and 2 °C Global Warming', *Agricultural Systems*, 175 (2019), 34–45. doi: 10.1016/j.agsy.2019.05.010

103 F. Gaupp et al., 'Changing Risks of Simultaneous Global Breadbasket Failure', *Nature Climate Change*, 568/7752 (2019), 382–6. doi: 10.1038/s41558-019-0600-z.

104 The water irrigation figures are based on Tariq Khorkhar, 'Globally, 70% of Freshwater is Used for Agriculture', *World Bank* blog (22 Mar. 2017), https://blogs.worldbank.org/opendata/chart-globally-70-freshwater-used-agriculture. The water stress statement is based on World Resources Institute data described in Somini Sengupta and Weiyi Cai, 'A Quarter of Humanity Faces Looming Water Crises', *New York Times* (6 Aug. 2019), https://www.nytimes.com/interactive/2019/08/06/climate/world-water-stress.html, both accessed 19 Dec. 2019.

105 This was first noted by John Holdren, who was senior adviser to Barack Obama on science and technology. See James Kanter and Andrew C. Revkin, 'World Scientists Near Consensus on Warming', *New York Times* (30 Jan. 2007), https://www.nytimes.com/2007/01/30/world/30climate.html, accessed 9 Jan. 2022.

106 Damian Carrington, '"Reality Check": Global CO2 Emissions Shooting Back to Record Levels', *The Guardian* (4 Nov. 2021), https://www.theguardian.com/environment/2021/nov/04/reality-check-global-co2-emissions-shooting-back-to-record-levels, accessed 9 Jan. 2022.

107 IPCC, 'Climate Change and Land'.

108 A.E. Raftery et al., 'Less than 2 °C Warming by 2100 Unlikely', *Nature Climate Change*, 7/9 (2017), 637–41. doi: 10.1038/nclimate3352

109 UN Environment Programme, *Emissions Gap Report 2017* (2017), https://www.unenvironment.org/resources/emissions-gap-report-2017, accessed 19 Dec. 2019.

110 This is calculated using the Global Carbon Atlas using consumption-based accounting and adding aviation and shipping: http://www.globalcarbonatlas.org/en/content/welcome-carbon-atlas, accessed 19 Dec. 2019. The UK government figures, not including emissions from imported goods, aviation and shipping are from Department for Business, Energy and Industrial Strategy, *2018 UK Greenhouse Gas Emissions, Provisional Figures* (2019), https://assets.publishing.service.gov.uk/government/uploads/system/uploads/attachment_data/file/790626/2018-provisional-emissions-statistics-report.pdf, accessed 19 Dec. 2019. See also Office for National Statistics, 'Net Zero and the Different Official Measures of the UK's Greenhouse Gas Emissions', (24 Jul. 2019), https://www.ons.gov.uk/economy/environmentalaccounts/articles/netzeroandthedifferentofficialmeasuresoftheuksgreenhousegasemissions/2019-07-24, accessed 9 Jan. 2022.

111 On oil platforms, see BBC News, 'Oil and Gas UK Welcomes Budget Tax Move' (22 Nov. 2017), https://www.bbc.com/news/uk-scotland-scotland-business-42085447. On cuts to solar subsidies, see Jillian Ambrose, 'Home Solar Panel Installations Fall by 94% as Subsidies Cut', *The Guardian* (5 Jun. 2019), https://www.theguardian.com/environment/2019/jun/05/home-solar-panel-installations-fall-by-94-as-subsidies-cut, both accessed 19 Dec. 2019.

112 Without having to rely on dangerous, untested techno-fixes.

113 Jennifer Hansler, 'Pompeo: Melting Sea Ice Presents "New Opportunities for Trade"', *CNN* (7 May 2019) https://edition.cnn.com/2019/05/06/politics/pompeo-sea-ice-arctic-council/index.html, accessed 19 Dec. 2019.

114 Thomas L. Friedman, 'Going Cheney on Climate', *New York Times* (8 Dec. 2019), https://www.nytimes.com/2009/12/09/opinion/09friedman.html, accessed 13 Jan. 2019.

115 V.P. Tynkkynen and N. Tynkkynen, 'Climate Denial Revisited: (Re)contextualising

Russian Public Discourse on Climate Change during Putin 2.0', *Europe-Asia Studies*, 70/7 (2018), 1103–20. doi: 10.1080/09668136.2018.1472218

116 For fossil fuel reserves see 'After the Ice: Mineral Riches of the Arctic', 911 Metallurgist, https://www.911metallurgist.com/blog/mineral-riches-of-the-arctic, accessed 3 Jan. 2020. Arctic sea lanes as preferable to southern lanes like the Panama Canal: Jugal K. Patel and Henry Fountain, 'As Arctic Ice Vanishes, New Shipping Routes Open', *New York Times* (3 May 2017), https://www.nytimes.com/interactive/2017/05/03/science/earth/arctic-shipping.html, accessed 10 Dec. 2019.

117 Less than one third of the vegetation clearing for the 2019 programme was completed by October: J.D. Morris, 'PG&E Is Less than One-Third Done with Its 2019 Tree-Trimming Work', *San Francisco Chronicle* (1 Oct. 2019), https://www.sfchronicle.com/business/article/PG-E-is-less-than-one-third-done-with-its-2019-14483596.php. For a description of emergency planning mismanagement, see John Myers and Taryn Luna, 'California Utilities – Not Lawmakers – Are Calling the Shots on Power Outages to Prevent Wildfires', *Los Angeles Times* (Oct. 2019), https://www.latimes.com/california/story/2019-10-26/california-utilities-politicians-wildfire-power-outage-analysis, both accessed 3 Jan. 2010.

118 Jeff Daniels, 'California's Largest Utility Warns Tens of Thousands of People Could See Power Shut off Due to Extreme Fire Risk', *CNBC* (7 Jun. 2019), https://www.cnbc.com/2019/06/07/californias-largest-utility-warns-tens-of-thousands-of-people-could-see-power-shut-off-due-to-extreme-fire-risk.html, accessed 19 Dec. 2019.

119 Marjolijn Haasnoot, 'Deltares Brengt Mogelijke Gevolgen van Versnelde Zeespiegelstijging voor Nederland in Kaart' [Deltares (Deltares Knowledge Institute) Maps Out the Possible Consequences of Accelerated Sea Rise for the Netherlands] (18 Sep. 2018), https://www.deltares.nl/nl/nieuws/deltares-brengt-mogelijke-gevolgen-van-versnelde-zeespiegelstijging-voor-nederland-kaart/, accessed 19 Dec. 2019.

120 Rolf Schuttenhelm, 'In Face of Rising Sea Levels the Netherlands "Must Consider Controlled Withdrawal"', *Vrij Nederland* (9 Feb. 2019), https://www.vn.nl/rising-sea-levels-netherlands/, accessed 19 Dec. 2019.

121 This apparently happened during the California shut-off, when a man was unable to reach his battery-powered oxygen respirator in time. See Scott McDonald, 'California Man Dependent on Oxygen Died 12 Minutes After PG&E Power Shutoff', *Newsweek* (11 Oct. 2019), https://www.newsweek.com/california-man-dependent-oxygen-died-12-minutes-after-pge-power-shutoff-1464766, accessed 19 Dec. 2019.

122 On connections between climate and other issues, see G.J. Abel et al., 'Climate, Conflict and Forced Migration', *Global Environmental Change*, 54 (2019), 239–49. doi: 10.1016/j.gloenvcha.2018.12.003. On the belief that these connections will deepen over time, see K.J. Mach et al., 'Climate as a Risk Factor for Armed Conflict', *Nature*, 57/7764 (2019). doi: 10.1038/s41586-019-1300-6

123 It should be noted that even more suffering occurred in European refugee housing, with women feeling the brunt most, as documented in Perez, *Invisible Women*.

124 For instance see Jessica Brown, 'Last Chance Tourism: Is This Trend Just Causing More Damage?', *The Independent* (7 Jun. 2018), https://www.independent.co.uk/news/long_reads/last-chance-tourism-travel-great-barrier-reef-amazon-machu-picchu-a8363466.html, accessed 3 Jan. 2020.

125 Jamie Smyth, 'Cost Overruns near $50bn as Australia's LNG Boom Falters', *Financial Times* (30 Oct. 2016), available to subscribers at https://www.ft.com/content/29667e96-9f15-11e6-891e-abe238dee8e2, accessed 19 Dec. 2019.

126 For an excellent academic overview, see J. Timmons Roberts et al., 'Rebooting a Failed Promise of Climate Finance', *Nature Climate Change*, 11 (2021), 180–2, doi: 10.1038/s41558-021-00990-2. For a detailed report, see Oxfam International, 'Climate Finance Shadow Report 2020: Assessing Progress Towards the $100 Billion Commitment', Oct. 2020, doi:10.21201/2020.6621, oxfamilibrary.openrepository.com/bitstream/handle/10546/621066/bp-climate-finance-shadow-report-2020-201020-en.pdf, accessed 9 Jan. 2022.

127 Quote Investigator, 'What Has Posterity Ever Done for Us?', https://quoteinvestigator.com/2018/05/09/posterity-ever/, accessed 4 Jan. 2020.

128 B.J. van Ruijven, E. De Cian and Ian Sue Wing, 'Amplification of Future Energy Demand Growth Due to Climate Change', *Nature Communications*, 10/1 (2009), 1–12. doi: 10.1038/s41467-019-10399-3

129 BusinessGreen staff, 'Survey: A Third of Public Suffering from "Eco Anxiety"', *BusinessGreen* (4 Jul. 2019), https://www.businessgreen.com/bg/news/3078422/survey-a-third-of-public-suffering-from-eco-anxiety, accessed 19 Dec. 2019.

130 Vikram Dodd and Matthew Taylor, 'Police Call for Tougher Sentences to Deter Extinction Rebellion', *The Guardian* (18 Jul. 2019), https://www.theguardian.com/environment/2019/jul/18/police-call-for-tougher-sentences-to-deter-extinction-rebellion, accessed 19 Dec. 2019.

131 On temperatures making it impossible to work, see J.G. Zivin and M. Neidell, 'Temperature and the Allocation of Time: Implications for Climate Change', *Journal of Labor Economics*, 32/1 (2014), 1–26. doi: 10.1086/671766. On the link between temperatures and suicide rates, see M. Burke et al., 'Higher Temperatures Increase Suicide Rates in the United States and Mexico', *Nature Climate Change*, 8/8 (2018), 723–9. doi: 10.1038/s41558-018-0222-x

132 Lyme disease: K.L. Ebi et al., 'Detecting and Attributing Health Burdens to Climate Change', *Environmental Health Perspectives*, 125/8, 1–8 (2017). doi: 10.1289/EHP1509. Malaria: P.V.V. Le et al., 'Predicting the Direct and Indirect Impacts of Climate Change on Malaria in Coastal Kenya', *PLOS ONE*, 14/2 (2019), 1–18. doi: 10.1371/journal.pone.0211258; and A.S. Siraj et al., 'Altitudinal Changes in Malaria Incidence in Highlands of Ethiopia and Colombia', *Science*, 343/6175 (2014), 1154–8. doi: 10.1126/science.1244325. For remaining diseases see X. Wu et al., 'Impact of Climate Change on Human Infectious Diseases: Empirical Evidence and Human Adaptation', *Environment International*, 86 (2016). doi: 10.1016/j.envint.2015.09.007. Readers might be interested to know that tick-borne encephalitis has now been detected in the UK (the risk is very low but is linked to warming): Michelle Roberts, 'Brain Illness Spread by Ticks Has Reached UK', BBC News (29 Oct. 2019), https://www.bbc.com/news/health-50206382, accessed 4 Jan. 2020.

133 Ibid.

134 C. Mora et al., 'Broad Threat to Humanity from Cumulative Climate Hazards Intensified by Greenhouse Gas Emissions', *Nature Climate Change*, 8/12 (2018), 1062–71. doi: 10.1038/s41558-018-0315-6

135 For example: IPCC, 'Land Is a Critical Resource, IPCC Report Says' (8 Aug. 2019), https://www.ipcc.ch/2019/08/08/land-is-a-critical-resource_srccl/. From the main report: IPCC, 'Climate Change and Land'. Further reporting is available here: Stephen Leahy, 'World Food Crisis Looms If Carbon Emissions Go Unchecked, UN Says', *National Geographic* (8 Aug. 2018), https://www.nationalgeographic.com/environment/2019/08/ipcc-un-food-security/, accessed 19 Dec. 2019.

136 Frank Laczko and Christine Aghazarm, eds, *Migration, Environment and Climate Change: Assessing the Evidence*, International Organization for Migration: Bangkok, 2009.

137 *Economic Times*, 'Bangladesh to Erect Barbed Wire Fence on Border with India' (12 Jul. 2018) https://economictimes.indiatimes.com/news/defence/bangladesh-to-erect-barbed-wire-fence-on-border-with-india/articleshow/54508281.cms, accessed 20 Dec. 2019.

138 Lauren Markham, 'How Climate Change Is Pushing Central American Migrants to the US', *The Guardian* (6 Apr. 2019), https://www.theguardian.com/commentisfree/2019/apr/06/us-mexico-immigration-climate-change-migration, accessed 20 Dec. 2019.

139 Thucydides, *History of the Peloponnesian War* (trans. By R. Crawley), Dent: London, 1914, pp. 394-5.

140 See for instance https://survivalcondo.com/, accessed 20 Dec. 2019.

141 James Lovelock, *The Vanishing Face of Gaia*, Basic Books: New York, 2017, p. 25.

Chapter Eight – Hope – Making Up for Lost Time

1 As the IPCC puts it, 'Every action matters. Every bit of warming matters. Every year matters. Every choice matters.' [Twitter post], 9.37 a.m., 11 Dec. 2018, https://twitter.com/ipcc_ch/status/1072425174820114432?lang=en, accessed 20 Dec. 2019.

2 For the figure since the pre-industrial age: UK Met Office, 'What Is Climate Change?' (2019), https://www.metoffice.gov.uk/weather/learn-about/climate-and-climate-change/climate-change/index, accessed 20 Dec. 2019. On being committed to further increases this century, see T. Mauritsen and R. Pincus, 'Committed Warming Inferred from Observations'. *Nature Climate Change*, 7 (2017), 652–5. doi: 10.1038/nclimate3357

3 Chloé Farand, 'Four Countries Have Declared Climate Emergencies, yet Give Billions to Fossil Fuels', *Climate Home News* (24 Jun. 2019), https://www.climatechangenews.com/2019/06/24/four-countries-declared-climate-emergencies-give-billions-fossil-fuels/, accessed 20 Dec. 2019.

4 James Kunstler, *The Long Emergency: Surviving the Converging Catastrophes of the Twenty-First Century*, Grove Atlantic: New York, 2006.

5 For this reason it may actually be better to use concentrations in the atmosphere rather than carbon budgets, which will continue to shift based on the estimates of warming from climate models – which are getting better at incorporating feedbacks. Budgets can also be extended by assuming more negative emissions in the future, as we'll discuss later in the chapter.

6 Joseph Stiglitz, 'The Climate Crisis Is Our Third World War. It Needs a Bold Response', *The Guardian* (4 Jun. 2019), https://www.theguardian.com/commentisfree/2019/jun/04/climate-change-world-war-iii-green-new-deal, accessed 20 Dec. 2019.

7 K.J. Mach et al., 'Climate as a Risk Factor for Armed Conflict'. *Nature*, 57/7764 (2019). doi: 10.1038/s41586-019-1300-6

8 This process wasn't perfect; the highway system is of course one infrastructural problems we now face when trying to decarbonize.

9 Matthew Yglesias, 'Elizabeth Warren's Latest Big Idea Is "Economic Patriotism"', *Vox* (4 Jun. 2019), https://www.vox.com/2019/6/4/18650850/elizabeth-warren-economic-patriotism-green-marshall-plan, accessed 20 Dec. 2019.

10 Joana Setzer and Rebecca Byrnes, 'Global Trends in Climate Change Litigation: 2019 Snapshot', Grantham Research Institute on Climate Change and the Environment

(4 Jul. 2019), http://www.lse.ac.uk/GranthamInstitute/publication/global-trends-in-climate-change-litigation-2019-snapshot/, accessed 20 Dec. 2019.

11 Arthur Neslen, 'Dutch Appeals Court Upholds Landmark Climate Change Ruling', *The Guardian* (9 Oct. 2018), https://www.theguardian.com/environment/2018/oct/09/dutch-appeals-court-upholds-landmark-climate-change-ruling, accessed 20 Dec. 2019.

12 Ibid.

13 Dennis van Berkel, 'Climate Case Heard Before Supreme Court on 24 May 2019', *Urgenda* (May 2019), https://www.urgenda.nl/en/themas/climate-case/, accessed 20 Dec. 2019.

14 In total, around 50% of all emissions have been emitted since 1990, after Exxon and Shell knew about climate change in the 1980s. For more details see Karen Savage, 'New Research Tying 20 Companies to One-Third of Global Emissions Aids Liability Argument', *Climate Liability News* (9 Oct. 2019), https://www.climateliabilitynews.org/2019/10/09/global-emissions-carbon-majors-richard-heede/, accessed 21 Dec. 2019.

15 R. Doll and A.B. Hill, 'Smoking and Carcinoma of the Lung; Preliminary Report', *British Medical Journal*, 2/4682 (1950), 739–48. doi: 10.1136/bmj.2.4682.739

16 W.J. Jones and G.A. Silvestri, 'The Master Settlement Agreement and Its Impact on Tobacco Use 10 Years Later: Lessons for Physicians About Health Policy Making', *Chest*, 137/3 (2010), 692–700. doi: 10.1378/chest.09-0982

17 Almost half a million people die each year from cigarette smoking in the US alone: Centers for Disease Control and Prevention, 'Smoking & Tobacco Use, Fast Facts' (n.d.), https://www.cdc.gov/tobacco/data_statistics/fact_sheets/fast_facts/index.htm, accessed 21 Dec. 2019.

18 Naomi Oreskes and Erik M. Conway, *Merchants of Doubt: How a Handful of Scientists Obscured the Truth on Issues from Tobacco Smoke to Global Warming*, Bloomsbury Publishing: London, 2012.

19 Leslie Hook, 'Oil Majors Gear up for Wave of Climate Change Liability Lawsuits', *Financial Times* (8 Jun. 2019), available to subscribers at https://www.ft.com/content/d5fbeae4-869c-11e9-97ea-05ac2431f453, accessed 21 Dec. 2019.

20 W. Cornwall, 'A New "Blob" Menaces Pacific Ecosystems', *Science*, 365/6459 (2019), 1233. doi: 10.1126/science.365.6459.1233

21 The compelling story of how this has unfolded was broadcast in a podcast for *Drilled*: Drilled, 'S2 E5: David vs. Goliath' (13 May 2019), https://drilled.libsyn.com/s2e5-david-vs-goliath, accessed 21 Dec. 2019.

22 David Hasemyer, 'Massachusetts Sues Exxon over Climate Change, Accusing the Oil Giant of Fraud', *Inside Climate News* (25 Oct. 2019), https://insideclimatenews.org/news/24102019/massachusetts-sues-exxon-climate-change-investor-fraud-misleading-advertising-healey, accessed 21 Dec. 2019.

23 Nicholas Kusnetz, 'Exxon and Oil Sands Go on Trial in New York Climate Fraud Case', *Inside Climate News* (17 Oct. 2019), https://insideclimatenews.org/news/16102019/exxon-oil-sands-trial-climate-change-fraud-new-york-rex-tillerson, accessed 21 Dec. 2019.

24 We could hypothesize that if we had more of an overlap between generations we might be doing better at recognizing the threat.

25 'Greta Thunberg: The Rebellion Has Begun', *Medium* (31 Oct. 2018), https://medium.com/wedonthavetime/the-rebellion-has-begun-d1bffe31d3b5, accessed 21 Dec. 2019.

26 M.C. Monroe, 'Children Teach Their Parents', *Nature Climate Change*, 9 (2019), 435–46. doi: 10.1038/s41558-019-0478-9

27 BBC News, 'UK Government Declares Climate Emergency' (1 May 2019), https://www.bbc.com/news/uk-politics-48126677, accessed 21 Dec. 2019.

28 Green Party dominance in EU Parliament: Emma Graham-Harrison, 'A Quiet Revolution Sweeps Europe as Greens Become a Political Force', *The Guardian* (2 Jun. 2019), https://www.theguardian.com/politics/2019/jun/02/european-parliament-election-green-parties-success; Greens getting second highest vote in German EU elections: Jon Stone, 'European Election Results: Green Parties Surge as "Green Wave" Hits EU', *The Independent* (26 May 2019), https://www.independent.co.uk/news/world/europe/european-election-result-greens-green-wave-spd-germany-a8931351.html, accessed 21 Dec. 2019.

29 Douglas Broom, 'Sweden Has Invented a Word to Encourage People Not to Fly. And It's Working', *World Economic Forum* (5 Jun. 2019), https://www.weforum.org/agenda/2019/06/sweden-has-invented-a-word-to-encourage-people-not-to-fly-and-it-s-working/, accessed 21 Dec. 2019.

30 David Reid, 'Boeing and Airbus to See Reduced Jet Demand as Climate Awareness Grows, UBS Says', *CNBC* (30 Sep. 2019), https://www.cnbc.com/2019/09/30/boeing-and-airbus-to-see-reduced-plane-demand-as-climate-awareness-grows.html, accessed 21 Dec. 2019.

31 For example see Eurostat, 'Railway Passenger Transport Statistics – Quarterly and Annual Data' (Oct. 2019), https://ec.europa.eu/eurostat/statistics-explained/index.php/Railway_passenger_transport_statistics_-_quarterly_and_annual_data; and David Burroughs, 'European Passenger Rail Market Growing, Says EU Report', *International Railway Journal* (7 Feb. 2019), https://www.railjournal.com/regions/europe/european-passenger-rail-market-growing-says-eu-report/, accessed 21 Dec. 2019.

32 Tess Riley, 'Just 100 Companies Responsible for 71% of Global Emissions, Study Says', *The Guardian* (10 Jul. 2017), https://www.theguardian.com/sustainable-business/2017/jul/10/100-fossil-fuel-companies-investors-responsible-71-global-emissions-cdp-study-climate-change, accessed 21 Dec. 2019.

33 Matt Bors, 'Mister Gotcha', *The Nib* (13 Sep. 2016), https://thenib.com/mister-gotcha, accessed 21 Dec. 2019.

34 For just two examples of many studies which reach the same conclusion, see G.T. Kraft-Todd et al., 'Credibility-Enhancing Displays Promote the Provision of Non-Normative Public Goods', *Nature, 563*/7730 (2018), 245–8. doi: 10.1038/s41586-018-0647-4; and J.M. Jachimowicz et al., 'The Critical Role of Second-Order Normative Beliefs in Predicting Energy Conservation', *Nature Human Behaviour, 2* (2018), 757–64. doi: 10.1038/s41562-018-0434-0, along with many other studies. For a journalistic overview with more examples, see Leor Hackel and Gregg Sparkman, 'Reducing Your Carbon Footprint Still Matters: In Fact, Getting Politicians and Industry to Address Climate Change May Start at Home', *Slate* (26 Oct. 2018), https://slate.com/technology/2018/10/carbon-footprint-climate-change-personal-action-collective-action.html, accessed 21 Dec. 2019.

35 C. Roberts and F.W. Geels, 'Conditions for Politically Accelerated Transitions: Historical Institutionalism, the Multi-Level Perspective, and Two Historical Case Studies in Transport and Agriculture', *Technological Forecasting and Social Change*, vol. 140(C) (2019), 221–40. doi: 10.1016/j.techfore.2018.11.019

36 Olaf Storbeck, 'Munich Re Boss Wants Higher Carbon Emission Costs', *Financial Times* (9 Jun. 2019), available to subscribers at https://www.ft.com/content/f68d336e-892e-11e9-97ea-05ac2431f453, accessed 21 Dec. 2019.

37 Arthur Neslen, 'Climate Change Could Make Insurance Too Expensive for Most People – Report', *The Guardian* (21 Mar. 2019), https://www.theguardian.com/environment/2019/mar/21/climate-change-could-make-insurance-too-expensive-for-ordinary-people-report, accessed 21 Dec. 2019.

38 Storbeck, 'Munich Re Boss'.

39 'No coal, no caveats' policy: James Kyne and Leslie Hook, 'Development Bank Halts Coal Financing to Combat Climate Change', *Financial Times* (12 Dec. 2018), available to subscribers at https://www.ft.com/content/7d0814f0-fd6f-11e8-ac00-57a2a826423e; EBRD quitting all fossil fuels: Scott Carpenter, 'The European Investment Bank Has Quit Fossil Fuels – but Borrowers Still Have Their Pick of Lenders', *Forbes* (18 Nov. 2019), https://www.forbes.com/sites/scottcarpenter/2019/11/18/the-european-investment-bank-quits-fossil-fuels-but-borrowers-still-have-their-pick-of-lenders/#3b28d4217349, accessed 21 Dec. 2019.

40 Jonathan Watts, 'China Pledge to Stop Funding Coal Projects "Buys Time for Emissions Target"', *The Guardian* (22 Sep. 2021), https://www.theguardian.com/world/2021/sep/22/china-pledge-to-stop-funding-coal-projects-buys-time-for-emissions-target, accessed 9 Jan. 2022.

41 S. Battiston et al., 'A Climate Stress-Test of the Financial System', *Nature Climate Change*, 7 (2017), 283–28. doi: 10.2139/ssrn.2726076

42 BBC News, 'Climate Change: Central Banks Warn of Financial Risks in Open Letter' (17 Apr. 2019), https://www.bbc.com/news/world-47965284, accessed 21 Dec. 2019.

43 J. Meckling, T. Sterner and G. Wagner, 'Policy Sequencing Toward Decarbonization', *Nature Energy*, 2/12 (2017), 1–5. doi: 10.1038/s41560-017-0025-8

44 Mathew Carr and Lisa Pham, 'Bank of England Climate Tests Weigh Disorder in Stock Market', *Bloomberg* (2 Aug. 2019), https://www.bloomberg.com/news/articles/2019-08-02/u-k-insurers-to-weigh-stock-market-disorder-caused-by-climate, accessed 21 Dec. 2019.

45 David Fickling, 'The World's Last Coal Plant Will Soon Be Built', Bloomberg (15 May 2019), https://www.bloomberg.com/opinion/articles/2019-05-15/coal-s-end-foreshadowed-in-iea-s-plant-investment-report, accessed 21 Dec. 2019.

46 Dropping from S&P 500: Kevin Crowley and Brandon Kochkodin, 'Exxon Poised to Drop from S&P 500's Top 10 for the First Time Ever', Bloomberg (30 Aug. 2019), https://www.bloomberg.com/news/articles/2019-08-30/exxon-poised-to-drop-from-s-p-500-s-top-10-for-first-time-ever, accessed 21 Dec. 2019. Negative outlook: Avi Salzman, 'Moody's Downgrades Exxon's Credit Outlook on Cash-Flow Worries', Barron's (22 Nov. 2019), https://www.barrons.com/articles/moodys-downgrades-exxons-credit-outlook-on-cash-flow-worries-51574442361, accessed 4 Jan. 2020.

47 Harry Dempsey and David Sheppard, 'Redburn Says Big Oil No Longer a "Buy" as Peak Oil Demand Looms', *Financial Times* (5 Sep. 2019), available to subscribers at https://www.ft.com/content/b214d780-cffd-11e9-99a4-b5ded7a7fe3f, accessed 21 Dec. 2019.

48 Mark Lewis, 'Wells, Wires, and Wheels – EROCI and the Tough Road Ahead for Oil', BNP Paribas (2 Aug. 2019), https://investors-corner.bnpparibas-am.com/investment-themes/sri/petrol-eroci-petroleum-age/, accessed 21 Dec. 2019.

49 Amy Harder, 'Wall Street Is Starting to Care About Climate Change', *AXIOS* (26 Jun. 2017), https://www.axios.com/wall-street-is-starting-to-care-about-climate-change-1513303205-f97cf14c-c921-4ad0-b37a-12832acea4fb.html, accessed 21 Dec. 2019.

50 There are many research articles showing that natural settings are good for creativity, stress, happiness, and more. For more see Jill Suttie, 'How Nature Can Make You Kinder, Happier, and More Creative', Greater Good, Berkeley University (2. Mar. 2016), https://greatergood.berkeley.edu/article/item/how_nature_makes_you_kinder_happier_more_creative, accessed 21 Dec. 2019.

51 Jonathan Watts, 'Deforestation of Brazilian Amazon Surges to Record High', *The Guardian* (4 Jun. 2019), https://www.theguardian.com/world/2019/jun/04/deforestation-of-brazilian-amazon-surges-to-record-high-bolsonaro, accessed 21 Dec. 2019.

52 Fabiano Maisonnave, 'Amazon's Indigenous Warriors Take on Invading Loggers and Ranchers', *The Guardian* (29 Aug. 2019), https://www.theguardian.com/world/2019/aug/29/xikrin-people-fight-back-against-amazon-land-grabbing, accessed 21 Dec. 2019.

53 B.W. Griscom et al., 'Natural Climate Solutions', *Proceedings of the National Academy of Sciences of the United States of America,* 114/44 (2017), 11645–50. doi: 10.1073/pnas.1710465114

54 Johnny Wood, 'Europe Bucks Global Deforestation Trend', World Economic Forum (25 Jul. 2019), https://www.weforum.org/agenda/2019/07/forest-europe-environment/, accessed 3 Jan. 2020.

55 F. Lu et al., 'Effects of National Ecological Restoration Projects on Carbon Sequestration in China from 2001 to 2010', *Proceedings of the National Academy of Sciences of the United States of America,* 115/16 (2018), 4039–44. doi: 10.1073/pnas.1700294115

56 X. Chu et al., 'Assessment on Forest Carbon Sequestration in the Three-North Shelterbelt Program Region, China', *Journal of Cleaner Production,* 215 (2019), 382–9. doi: 10.1016/j.jclepro.2018.12.296

57 P. Smith, 'Soil Carbon Sequestration and Biochar as Negative Emission Technologies', *Global Change Biology,* 22/3 (2016), 1315–24. doi: 10.1111/gcb.13178

58 Increasing water constraints: M. Zastrow, 'China's Tree-Planting Drive Could Falter in a Warming World', *Nature,* 573/7775 (2019). doi: 10.1038/d41586-019-02789-w. Increased wildfires and forests turning into sources of greenhouse gases: A. Baccini et al., 'Tropical Forests Are a Net Carbon Source Based on Aboveground Measurements of Gain and Loss', *Science,* 358/6360 (2017), 230–4. doi: 10.1126/science.aam5962

59 One could argue that solar radiation management and more interventionist geoengineering schemes are also *dei ex machina*. But for reasons discussed later, they are far more problematic.

60 David Roberts, 'Sucking Carbon out of the Air Won't Solve Climate Change: But It Might Fill in a Few Key Pieces of the Clean Energy Puzzle', *Vox* (16 Jul. 2018), https://www.vox.com/energy-and-environment/2018/6/14/17445622/direct-air-capture-air-to-fuels-carbon-dioxide-engineering, accessed 21 Dec. 2019.

61 V. Masson-Delmotte et al., eds, 'Summary for Policymakers', in *Global Warming of 1.5°C. An IPCC Special Report on the Impacts of Global Warming of 1.5°C Above Pre-Industrial Levels and Related Global Greenhouse Gas Emission Pathways, in the Context of Strengthening the Global Response to the Threat of Climate Change, Sustainable Development, and Efforts to Eradicate Poverty,* IPCC, World Meteorological Organization: Geneva, 2018. doi: 10.1017/CBO9781107415324

62 C. Le Quéré, 'Temporary Reduction in Daily Global CO2 Emissions During the COVID-19 Forced Confinement', *Nature Climate Change,* 10 (2020), 647–53, doi: 10.1038/s41558-020-0797-x

63   G.P. Peters et al., 'Rapid Growth in CO2 Emissions after the 2008–2009 Global Financial Crisis', *Nature Climate Change*, 2/1 (2011), 2–4. doi: 10.1038/nclimate1332

64   F. Creutzig et al., 'The Mutual Dependence of Negative Emission Technologies and Energy Systems', *Energy & Environmental Science*, 12/6 (2019), 1805–17. doi: 10.1039/C8EE03682A

65   J. van Zalk and P. Behrens, 'The Spatial Extent of Renewable and Non-Renewable Power Generation: A Review and Meta-Analysis of Power Densities and Their Application in the U.S.' *Energy Policy*, 123 (2018), 83–91. doi: 10.1016/j.enpol.2018.08.023

66   Y. Jin et al., 'Water Use of Electricity Technologies : A Global Meta-Analysis', *Renewable and Sustainable Energy Reviews*, 115 (2019), 109391. doi: 10.1016/j.rser.2019.109391

67   D.W. Keith et al., 'A Process for Capturing Carbon Dioxide from the Atmosphere', *Joule*, 2/8 (2018), 1573–94. doi: 10.1016/j.joule.2018.05.006, as reported here: R. Service, 'Cost Plunges for Capturing Carbon Dioxide from the Air', *Science* (7 Jun. 2018). doi: 10.1126/science.aau4107

68   M. Fasihi, O. Efimova and C. Breyer, 'Techno-Economic Assessment of CO_2 Direct Air Capture Plants', *Journal of Cleaner Production*, 224 (2019), 957–80. doi: 10.1016/j.jclepro.2019.03.086. It's hard to say whether this is an over- or underestimate. Given the experience with the cost of solar, we might assume that DAC could get even cheaper than this, but the technology is obviously very different, there are other competitors, and we simply cannot be sure.

69   F. Creutzig et al., 'The Mutual Dependence of Negative Emission Technologies and Energy Systems', *Energy & Environmental Science*, 12 (2019), 1805–17. doi: 10.1039/C8EE03682A

70   Akshat Rathi and David Yanofsky, 'What Saudi Arabia's 200 GW Solar Power Plant Would Look Like – if Placed in Your Neighborhood', *Quartz* (1 Apr. 2018), https://qz.com/1240862/what-saudi-arabias-200-gw-solar-power-plant-would-look-like-from-space/, accessed 4 Jan. 2020.

71   R.F. Service, 'Renewable Bonds', *Science*, 365/6459 (2019), 1236–9. doi: 10.1126/science.365.6459.1236

72   Roberts, 'Sucking Carbon out of the Air Won't Solve Climate Change'.

73   Martin Gilbert, *Winston S. Churchill: The Challenge of War, 1914-1916 (Volume III)*, Rosetta Books: New York, 2015, p. 10.

74   D. Centola et al., 'Experimental Evidence for Tipping Points in Social Convention', *Science*, 360/6393 (2018), 1116–9. doi: 10.1126/science.aas8827. For reporting on this see: University of Pennsylvania, 'Research Finds Tipping Point for Large-scale Social Change' (7 Jun. 2018), https://www.asc.upenn.edu/news-events/news/research-finds-tipping-point-large-scale-social-change, accessed 21 Dec. 2019.

75   For more see Erica Chenoweth, 'It May Only Take 3.5% of the Population to Topple a Dictator – with Civil Disobedience', *The Guardian* (1 Feb. 2017), https://www.theguardian.com/commentisfree/2017/feb/01/worried-american-democracy-study-activist-techniques, accessed 21 Dec. 2019; and see also Chenoweth's book, *Why Civil Resistance works: The Strategic Logic of Nonviolent Conflict*, MIT Press: Cambridge, MA, 2011. Some of the messages have been questioned by others, such as A. Malm in *How to Blow Up a Pipeline: Learning to Fight in a World on Fire*, Verso: London, 2021.

76   Matthew Taylor, Jonathan Watts and John Bartlett, 'Climate Crisis: 6 Million People Join Latest Wave of Global Protests', *The Guardian* (27 Sep. 2019), https://www.

theguardian.com/environment/2019/sep/27/climate-crisis-6-million-people-join-latest-wave-of-worldwide-protests, accessed 21 Dec. 2019.

77   Whack-a-mole is a fairground game where you hit moles with a mallet as they appear from holes in a table. As each mole is hit, others pop up in different orders, meaning you need to hit increasing numbers of moles in a constant effort to keep up.

Chapter Nine – Pessimism – Counting the Costs

1   J. Rockström et al., 'A Safe Operating Space for Humanity', *Nature*, 461 (2009), 472–5. doi: 10.1038/461472a

2   Although the offending chemicals have been phased out, progress has been slower than most people believe, and it will take several more decades before the ozone hole approaches full recovery.

3   W. Steffen et al., 'Planetary Boundaries: Guiding Human Development on a Changing Planet', *Science*, 347/6223 (2015), 1217. doi: 10.1126/science.1259855

4   Michael Kelly, 'The 1992 Campaign: The Democrats – Clinton and Bush Compete to Be Champion of Change; Democrat Fights Perceptions of Bush Gain', *New York Times* (31 Oct. 1992), https://www.nytimes.com/1992/10/31/us/1992-campaign-democrats-clinton-bush-compete-be-champion-change-democrat-fights.html, accessed 4 Feb. 2020.

5   Nelson Gaylord, *Beyond Earth Day: Fulfilling the Promise*, University of Wisconsin Press: Madison, 2002.

6   Rutger Hoekstra, *Replacing GDP by 2030: Towards a Common Language for the Well-Being and Sustainability Community*, Cambridge University Press: Cambridge, 2019.

7   Danny Dorling and Stuart Gietel-Basten, 'Life Expectancy in Britain Has Fallen so Much That a Million Years of Life Could Disappear by 2058 – Why?', *The Conversation* (29 Nov. 2017), https://theconversation.com/life-expectancy-in-britain-has-fallen-so-much-that-a-million-years-of-life-could-disappear-by-2058-why-88063, accessed 21 Dec. 2019.

8   Jamie Ducharme, 'More Millennials are Dying "Deaths of Despair" as Overdose and Suicide Rates Climb', *Time* (13 Jun. 2019), https://time.com/5606411/millennials-deaths-of-despair/, accessed 21 Dec. 2019.

9   Hoekstra, *Replacing GDP by 2030*.

10   These consumers don't need to be individuals. Usually this final consumption expenditure is split into households, governments and non-profits.

11   Robert F. Kennedy, 'Remarks at the University of Kansas', John F. Kennedy Library (Mar. 1968), https://www.jfklibrary.org/learn/about-jfk/the-kennedy-family/robert-f-kennedy/robert-f-kennedy-speeches/remarks-at-the-university-of-kansas-march-18-1968, accessed 3 Jan. 2019. The quote is worth repeating in its entirety: 'Too much and for too long, we seemed to have surrendered personal excellence and community values in the mere accumulation of material things. Our Gross National Product... if we judge the United States of America by that... counts air pollution and cigarette advertising, and ambulances to clear our highways of carnage. It counts special locks for our doors and the jails for the people who break them. It counts the destruction of the redwood and the loss of our natural wonder in chaotic sprawl. It counts napalm and counts nuclear warheads and armored cars for the police to fight the riots in our cities. It counts Whitman's rifle and Speck's knife, and the television programs which glorify violence in order to sell toys to our children. Yet the gross

national product does not allow for the health of our children, the quality of their education or the joy of their play. It does not include the beauty of our poetry or the strength of our marriages, the intelligence of our public debate or the integrity of our public officials. It measures neither our wit nor our courage, neither our wisdom nor our learning, neither our compassion nor our devotion to our country, it measures everything in short, except that which makes life worthwhile. And it can tell us everything about America except why we are proud that we are Americans.'

12 For Canada and the UK respectively; other countries came in between: P. Van De Ven et al., 'Including Unpaid Household Activities; an Estimate of Its Impact on Macro-Economic Indicators in the G7 Economies and the Way Forward', *OECD Working Paper No. 91* (25 Jul. 2018), 1–31.

13 On the rise in cases of rickets, see M. Goldacre, N. Hall and D.G.R. Yeates, 'Hospitalisation for Children with Rickets in England: A Historical Perspective', *Lancet*, 383/9917 (2014), 597–8. doi: 10.1016/S0140-6736(14)60211-7. On the increase in food bank numbers, see Sean Coughlan, 'Food Bank Supplies Help Record Numbers', *BBC News* (25 Apr. 2019), https://www.bbc.com/news/education-48037122, accessed 21 Dec. 2019.

14 Incarceration rates: for example, in the US incarceration rates are 655 people per 100,000 and the highest in the world, compared to between 51 in Finland and 140 in the UK. These data are from the World Prison Brief online database, at https://www.prisonstudies.org/. Maternal deaths: in Europe there are around two to ten deaths per 10,000 births; in the US this is seventeen deaths per 10,000 births. See https://data.unicef.org/topic/maternal-health/maternal-mortality/, accessed 21 Dec. 2019. Social mobility: Daily Chart, 'Americans Overstate Social Mobility in Their Country', *The Economist* (14 Feb. 2018), https://www.economist.com/graphic-detail/2018/02/14/americans-overestimate-social-mobility-in-their-country, accessed 3 Jan. 2020.

15 Spiegel Staff, 'Economists Search for a New Definition of Well-Being', *Spiegel Online* (22 Sep. 2009), https://www.spiegel.de/international/business/beyond-gdp-economists-search-for-new-definition-of-well-being-a-650532.html, accessed 21 Dec. 2019.

16 Robert D. Hershey Jr., 'Environmental Factors to Be Calculated in G.D.P.', *New York Times* (20 Apr. 1994), https://www.nytimes.com/1994/04/20/business/environmental-factors-to-be-calculated-in-gdp.html, accessed 21 Dec. 2019.

17 Hoekstra, *Replacing GDP by 2030*, p. 4.

18 See for example the United Nations Sustainable Development Goal Tracker, 'Promote Inclusive and Sustainable Economic Growth, Employment and Decent Work for All', 2019: https://sdg-tracker.org/economic-growth, accessed 21 Dec. 2019.

19 Moody's for instance mention GDP when making adjustments: Sam Ro, 'Moody's Slashes 6 Euro Countries, Cuts Outlook on Aaa Rated France, Austria, and UK', *Business Insider* (13 Feb. 2012), https://www.businessinsider.com/moodys-downgrades-europe-2012-2?international=true&r=US&IR=T. For further information see Will Smale, 'What Are Credit Ratings Agencies?', BBC News (25 Jun. 2016), https://www.bbc.com/news/business-36629099, accessed 21 Dec. 2019.

20 Nordhaus as quoted in L. Roberts, 'Academy Panel Split on Greenhouse Adaptation', *Science*, 253/5025 (1991), 1206. doi: 10.1126/science.253.5025.1206. Similar thinking is expounded in the 1991 paper: W.D. Nordhaus, 'To Slow or Not to Slow: The Economics of The Greenhouse Effect', *Economic Journal*, 101/407 (1991), 920–37. doi: 10.2307/2233864

21 Dirk Philipsen, *The Little Big Number: How GDP Came to Rule the World and What to Do About It*, Princeton University Press: Princeton, NJ, 2015, p. 171.

22 Chris Mooney and Andrew Freedman, 'The World Needs a Massive Carbon Tax in Just 10 Years to Limit Climate Change, IMF Says', *Washington Post* (10 Oct. 2019), https://www.washingtonpost.com/climate-environment/2019/10/10/world-needs-massive-carbon-tax-just-years-limit-climate-change-imf-says/, accessed 3 Jan. 2020.

23 $33 trillion in 1995 dollars, roughly $60 trillion today: R. Costanza et al., 'The Value of the World's Ecosystem Services and Natural Capital', *Nature*, 387/15 (1997), 253–60. 10.1016/S0921-8009(98)00020-2

24 Thomas Homer-Dixon, *Carbon Shift*, Random House: Canada, 2009, p. 171.

25 For instance, in 2019 the UK Office of National Statistics estimated the nation's natural capital assets to be just £951 billion. We should assume that these policy reports, since they don't highlight some of the important caveats we have, have an impact on the way policymakers think about nature and its value. See: 'UK Natural Capital Accounts: 2019', Office for National Statistics (18 Oct. 2019), https://www.ons.gov.uk/economy/environmentalaccounts/bulletins/uknaturalcapitalaccounts/2019, accessed 22 Dec. 2019.

26 *The Lancet*, 'The Lancet Commission on Pollution and Health' (19 Oct. 2017), https://gahp.net/the-lancet-report-2/, accessed 22 Dec. 2019.

27 A. Tukker et al., 'Exiopol – Development and Illustrative Analyses of a Detailed Global Mr Ee Sut/Iot', *Economic Systems Research*, 25/1 (2013), 50–70. doi: 10.1080/09535314.2012.761952

28 If total warming reaches two degrees Celsius rather than 1.5, as it is likely to; GDP reductions are estimated very conservatively at around 25% by 2100 without even including the ecological damage. See: M. Burke, W.M. Davis and N.S. Diffenbaugh, 'Large Potential Reduction in Economic Damages Under UN Mitigation Targets', *Nature*, 557/7706 (2018), 549–53. doi: 10.1038/s41586-018-0071-9

29 For instance: K.E. Pickett and R.G. Wilkinson, 'Income Inequality and Health: A Causal Review', *Social Science & Medicine*, 128 (2015), 316–26. doi: 10.1016/j.socscimed.2014.12.031. This section is heavily informed by Kate Pickett and Richard G. Wilkinson, *The Inner Level: How More Equal Societies Reduce Stress, Restore Sanity and Improve Everyone's Well-Being*, Penguin: London, 2018.

30 The Equality Trust, 'Inequality Costs UK £39 Billion per Year' (14 Mar. 2014), https://www.equalitytrust.org.uk/news/inequality-costs-uk-%C2%A339-billion-year, accessed 22 Dec. 2019.

31 Biodiversity loss: G.M. Mikkelson, A. Gonzalez and G.D. Peterson, 'Economic Inequality Predicts Biodiversity Loss', *PLOS One*, 2/5 (2007), 3–7. doi: 10.1371/journal.pone.0000444; and T.G. Holland, G.D. Peterson and A. Gonzalez. A Cross-National Analysis of How Economic Inequality Predicts Biodiversity Loss. *Conservation Biology* 23, 1304–1313 (2009). Consumption of unhealthy goods: L. Walasek and G.D.A. Brown, 'Income Inequality, Income, and Internet Searches for Status Goods: A Cross-National Study of the Association Between Inequality and Well-Being', *Social Indicators Research*, 129/3 (2015), 1001–14. doi: 10.1007/s11205-015-1158-4; and idem, 'Income Inequality and Status Seeking: Searching for Positional Goods in Unequal U.S. States', *Psychological Science*, 26/4 (2015), 527–33. doi: 10.1177/0956797614567511. Environmentally damaging goods: N. Stotesbury and D. Dorling, 'Understanding Income Inequality and Its Implications: Why Better Statistics Are Needed', *Statistics Views* (21 Oct. 2015), https://www.statisticsviews.com/

details/feature/8493411/Understanding-Income-Inequality-and-its-Implications-Why-Better-Statistics-Are-N.html, accessed 22 Dec. 2019. Emissions and air pollution: A. Drabo, 'Impact of Income Inequality on Health: Does Environment Quality Matter?', *Environment and Planning A: Economy and Space*, 43/1 (2011), 146–65. doi: 10.1068/a43307; L. Cushing et al., 'The Haves, the Have-Nots, and the Health of Everyone: The Relationship Between Social Inequality and Environmental Quality', *Annual Review of Public Health*, 36 (2015), 193–209. doi: 10.1146/annurev-publhealth-031914-122646; and A.K. Jorgenson et al., 'Income Inequality and Residential Carbon Emissions in the United States: A Preliminary Analysis', *Human Ecology Review*, 22/1 (2015), 93–105. doi: 10.22459/HER.22.01.2015.06

32   H. Dittmar et al., 'The Relationship Between Materialism and Personal Well-Being: A Meta-Analysis', *Journal of Personality and Social Psychology*, 107/5 (2014), 879–924. doi: 10.1037/a0037409

33   Joe Pinsker, 'The Reason Many Ultrarich People Aren't Satisfied with Their Wealth', *The Atlantic* (4 Dec. 2018), https://www.theatlantic.com/family/archive/2018/12/rich-people-happy-money/577231/, accessed 22 Dec. 2019.

34   This indicator starts with standard GDP statistics and adds beneficial activities like unpaid caring and volunteering. It then subtracts environmental and social damages that are straightforward to measure, such as the cost of incarceration and cleaning water of chemical pollution (although it doesn't include the full spectrum of damages).

35   Overall, they found that GPI per capita doesn't seem to increase beyond a GDP per capita of around $7,000.

36   On well-being having peaked in the late 1970s, see I. Kubiszewski et al., 'Beyond GDP: Measuring and Achieving Global Genuine Progress', *Ecological Economics*, 93 (2013), 57–68. doi: 10.1016/j.ecolecon.2013.04.019. On GDP growth during the same period: other researchers using different approaches broadly agree with these results (the dates of peak well-being may change, but generally lie in between 1970 and 2000). See for instance E. Neumayer, *Weak Versus Strong Sustainability: Exploring the Limits of Two Opposing Paradigms* (4th ed.), Edward Elgar Publishing: Cheltenham, UK and Northampton, MA, USA, 2003.

37   Helen Roxburgh, 'How Clean Indoor Air Is Becoming China's Latest Luxury Must-Have', *The Guardian* (27 Mar. 2018), https://www.theguardian.com/cities/2018/mar/27/china-clean-air-indoor-quality-shanghai-cordis-hongqiao-filters, accessed 22 Dec. 2019.

38   R.A. Easterlin et al., 'The Happiness–Income Paradox Revisited', *Proceedings of the National Academy of Sciences of the United States of America*, 107/52 (2010), 22463–8. doi: 10.1073/pnas.1015962107

39   Our relative place in society as more important than absolute income: Z. Yu and L. Chen, 'Income and Well-Being: Relative Income and Absolute Income Weaken Negative Emotion, but Only Relative Income Improves Positive Emotion', *Frontiers in Psychology*, 7/257 (2016), 1–6. doi: 10.3389/fpsyg.2016.02012; and R. Ball and K. Chernova, 'Absolute Income, Relative Income, and Happiness', *Social Indicators Research*, 88/3 (2008), 497–529. doi: 10.1007/s11205-007-9217-0. GDP growth as not indicative of civic or environmental well-being: See Pickett and Wilkinson, *The Inner Level*.

40   C. Le Quéré et al., 'Drivers of Declining CO_2 Emissions in 18 Developed Economies', *Nature Climate Change*, 9/3 (2019), 213–17. doi: 10.1038/s41558-019-0419-7

41   T.O. Wiedmann et al., 'The Material Footprint of Nations', *Proceedings of the National Academy of Sciences of the United States of America*, 112/20 (2015), 9–10. doi: 10.1073/pnas.1220362110

42   This effect as happening within countries: J.D. Ward et al., 'Is Decoupling GDP
     Growth from Environmental Impact Possible?', PLOS ONE, 11/10 (2016), 1–14.
     doi: 10.1371/journal.pone.0164733. Richer Chinese provinces importing materials
     from poorer regions: J. Meng, P. Behrens et al., 'Provincial and Sector-Level Material
     Footprints in China', Proceedings of the National Academy of Sciences of the United States
     of America, 116/52 (2019), 1–6. doi: 10.1073/pnas.1903028116
43   E. Van der Voet et al., 'Environmental Implications of Future Demand Scenarios for
     Metals: Methodology and Application to the Case of Seven Major Metals', Journal
     of Industrial Ecology, 23/1 (2019), 141–55. doi: 10.1111/jiec.12722
44   William Stanley Jevons, VII: The Coal Question (2nd ed.), London: Macmillan and
     Company, 1866, OCLC 464772008, retrieved 21 Jul. 2008.
45   There are many studies giving between 5% and 30%, for instance 5–15% here:
     B.A. Thomas and I.L. Azevedo, 'Estimating Direct and Indirect Rebound Effects
     for U.S. Households with Input–Output Analysis, Part 1: Theoretical Framework',
     Ecological Economics, 86 (2013), 199–210. doi: 10.1016/j.ecolecon.2012.12.003;
     and 26% here: T. Barker, P. Ekins and T. Foxon, 'The Macro-Economic Rebound
     Effect and the UK Economy', Energy Policy, 35/10 (2007), 4935–46. doi: 10.1016/j.
     enpol.2007.04.009
46   S. Bruns, A. Moneta and D.I. Stern, 'Macroeconomic Time-Series Evidence that
     Energy Efficiency Improvements Do not Save Energy', SSRN Electronic Journal (2019).
     doi: 10.2139/ssrn.3336357
47   E.J. Mishan, The Costs of Economic Growth, Praeger Publishers: New York, 1967.
48   M. Büchs and M. Koch, 'Challenges for the Degrowth Transition: The Debate About
     Wellbeing', Futures, 105 (2019), 155–65. doi: 10.1016/j.futures.2018.09.002
49   There is another large discussion about the ethical frameworks used in economics.
     Utility is related to the utilitarian view of economics, which is dominant in most
     liberal economies.
50   Glenn Curtis, Russia: A Country Study, Federal Research Division, Library of Congress:
     Washington DC, 1996.
51   T. Forster et al., 'How Structural Adjustment Programs Affect Inequality: A
     Disaggregated Analysis of IMF Conditionality, 1980–2014', Social Science Research, 80
     (2019), 83–113. doi: 10.1016/j.ssresearch.2019.01.001. Important exceptions include
     South Korea, Singapore and China, as argued by Ha-Joon Chang in Economics: The
     User's Guide, Pelican Books: Louisiana, 2014.
52   Lucy Siegle, 'This Much I Know: Amory Lovins', The Guardian (23 Mar. 2008),
     https://www.theguardian.com/environment/2008/mar/23/ethicalliving.lifeandhealth4,
     accessed 22 Dec. 2019.
53   Phillip Inman, 'Study: Big Corporations Dominate List of World's Top Economic
     Entities', The Guardian (12 Sep. 2016), https://www.theguardian.com/business/2016/
     sep/12/global-justice-now-study-multinational-businesses-walmart-apple-shell,
     accessed 22 Dec. 2019.
54   Emily Stewart, 'America's Monopoly Problem, in One Chart', Vox (26 Nov. 2018),
     https://www.vox.com/2018/11/26/18112651/monopoly-open-markets-institute-report-
     concentration, accessed 22 Dec. 2019.
55   Avner Mendelson, 'Survival Strategy: Cut the Number of Banks in Half', American
     Banker (30 Jan. 2018), https://www.americanbanker.com/opinion/survival-strategy-
     cut-the-number-of-banks-in-half, accessed 22 Dec. 2019.
56   See for instance A.H.M. Noman, S.G. Chan and C.R. Isa, 'Does Competition Improve

Financial Stability of the Banking Sector in ASEAN Countries? An Empirical Analysis'. *PLOS ONE*, 12/5 (2017), 1–27. doi: 10.1371/journal.pone.0176546

57 J.G. Matsusaka, 'When Do Legislators Follow Constituent Opinion? Evidence from Matched Roll Call and Referendum Votes', *Chicago Booth Stigler Centre Working Papers* (2017), 1–50.

58 Lord Acton, 'Letter to Archbishop Model Creighton', Hanover College (5 Apr. 1887), https://history.hanover.edu/courses/excerpts/165acton.html, accessed 22 Dec. 2019.

59 Facebook: IBIS World, 'Social Networking Sites Industry in the US – Market Research Report' (15 Dec. 2018), https://www.ibisworld.com/industry-trends/specialized-market-research-reports/technology/electronic-online-entertainment/social-networking-sites.html. Google: J. Clement, 'Worldwide Desktop Market Share of Leading Search Engines from January 2010 to July 2019', Statista (3 Dec. 2019), https://www.statista.com/statistics/216573/worldwide-market-share-of-search-engines/. Amazon: Open Markets Institute, 'E-commerce Industry' (2018), https://concentrationcrisis.openmarketsinstitute.org/industry/e-commerce/, accessed 22 Dec. 2019.

60 See for instance W. McDowall et al., 'The Development of Wind Power in China, Europe and the USA: How Have Policies and Innovation System Activities Co-Evolved?', *Technology Analysis & Strategic Management*, 25 (2013), 163–85. doi: 10.1080/09537325.2012.759204, with impacts described here: P. Behrens et al., 'Environmental, Economic, and Social Impacts of Feed-in Tariffs: A Portuguese Perspective 2000–2010', *Applied Energy*, 173 (2016), 309–19 (2016). doi: 10.1016/j.apenergy.2016.04.044

61 J. Peters, 'Labour Market Deregulation and the Decline of Labour Power in North America and Western Europe', *Policy and Society*, 27/1 (2008), 83–98. doi: 10.1016/j.polsoc.2008.07.007

62 For instance: BBC News, 'JD Sports and Asos Warehouses like "Dark Satanic Mills"' (7 May 2019), https://www.bbc.co.uk/news/uk-england-48186516; and Sean Farrell, 'Unions Lobby Investors to Press Amazon over UK Working Conditions', *The Guardian* (20 May 2019), https://www.theguardian.com/technology/2019/may/20/unions-lobby-investors-to-press-amazon-over-uk-working-conditions, accessed 22 Dec. 2019.

63 Sendhil Mullainathan and Eldar Shafir, *Scarcity: Why Having Too Little Means So Much*, Macmillan: New York, 2003.

64 Economic Policy Institute, 'The Productivity–Pay Gap' (Jul. 2019), https://www.epi.org/productivity-pay-gap/, accessed 22 Dec. 2019.

65 Lawrence Mishel and Julia Wolfe, 'CEO Compensation Has Grown 940% Since 1978', *Economic Policy Institute* (14 Aug. 2019), https://www.epi.org/publication/ceo-compensation-2018/, accessed 22 Dec. 2019.

66 Branko Milanovic, *Capitalism, Alone: The Future of the System that Rules the World*, Belknap Press: Cambridge, MA, 2019.

67 Thomas Piketty, *Capital in the Twenty-First Century*, Harvard University Press: Cambridge, MA, 2013.

68 Neil M. Barofsky, 'Where the Bailout Went Wrong', *New York Times* (29 Mar. 2011), https://www.nytimes.com/2011/03/30/opinion/30barofsky.html, accessed 22 Dec. 2019.

69 Wolfgang Streeck, *How Will Capitalism End? Essays on a Failing System*, Verso: London, 2017, p. 41.

70 This figure uses GDP, since this is how statistics are gathered. See Laura Wood, 'Analyzing the Advertising Industry in the United States 2017 – Research and Markets', *BusinessWire* (8 Aug. 2017), https://www.businesswire.com/news/

home/20170808005969/en/Analyzing-Advertising-Industry-United-States-2017--, accessed 22 Dec. 2019.

71   This is a concern often raised by writers such as Yuval Noah Harari and Evgeny Morozov.

72   BBC News, 'Russia "Meddled in All Big Social Media" Around US Election' (17 Dec. 2018), https://www.bbc.com/news/technology-46590890, accessed 4 Jan. 2020.

73   G. Ainslie and V. Haendcl, 'The Motives of the Will', in E. Gottheil et al., eds, *Etiology Aspects of Alcohol and Drug Abuse*, Charles C. Thomas: Springfield, IL, 1983, 119–140.

74   M. Budolfson et al., 'The Comparative Importance for Optimal Climate Policy of Discounting, Inequalities and Catastrophes', *Climate Change*, 145/3–4 (2017), 481–94. doi: 10.1007/s10584-017-2094-x

75   The Stern report still underestimated how bad the situation was, assuming a 0.95% growth in global emissions in the years between 2000 and 2005, whereas the truth was around 2.4%, as described here: K. Anderson and J. Jewell, 'Debating the Bedrock of Climate-Change Mitigation Scenarios', *Nature*, 573/7774 (2019), 348–9. doi: 10.1038/d41586-019-02744-9

76   Simon Dietz, *A Long-Run Target for Climate Policy: the Stern Review and Its Critics, Supporting Research for the UK Committee on Climate Change's Inaugural Report on Building a Low-Carbon Economy – the UK's Contribution to Tackling Climate Change* (2 May 2008), http://personal.lse.ac.uk/dietzs/A%20long-run%20target%20for%20 climate%20policy%20-%20the%20Stern%20Review%20and%20its%20critics.pdf, accessed 7 Jan. 2020.

77   Idem, 'Economics: Current Climate Models Are Grossly Misleading', *Nature*, 530/7591 (2016), 407–9 (2016). doi: 10.1038/530407a

78   M. Fleurbaey and S. Zuber, 'Climate Policies Deserve a Negative Discount Rate', *Chicago Journal of International Law*, 13/2 (2013), 1–33.

79   J. Emmerling et al., 'The Role of the Discount Rate for Emission Pathways and Negative Emissions', *Environmental Research Letters*, 14/10 (2019). doi: 10.1088/1748-9326/ab3cc9

80   Mark Lynas, *Six Degrees: Our Future on a Hotter Planet*, National Geographic: Washington DC, 2008.

81   M.L. Weitzman, 'On Modeling and Interpreting the Economics of Catastrophic Climate Change', *Review of Economics and Statistics*, 91/1 (2009), 1–19. doi: 10.1162/ rest.91.1.1

82   R.S. Pindyck, 'Climate Change Policy: What Do the Models Tell Us?', *Journal of Economic Literature*, 51/3 (2013), 860–72. doi: 10.1257/jel.51.3.860

83   'Optimal' warming: W. Nordhaus, 'Projections and Uncertainties About Climate Change in an Era of Minimal Climate Policies', *American Economic Journal: Economic Policy*, 10/3 (2018), 333–60. doi: 10.1257/pol.20170046. We have already seen how the chance of a multi-breadbasket failure increases non-linearly from 1.5 to two degrees Celsius. A further 1.5 degrees Celsius is difficult to imagine.

84   K. Anderson and G. Peters, 'The Trouble with Negative Emissions', *Science*, 354/6309 (2016), 182–3. doi: 10.1126/science.aah4567

85   David Roberts, 'Pulling CO_2 out of the Air and Using It Could Be a Trillion-Dollar Business: Meet "Carbon Capture and Utilisation", Which Puts CO2 to Work Making Valuable Products', *Vox* (22 Nov. 2019), https://www.vox.com/energy-and-environment/2019/9/4/20829431/climate-change-carbon-capture-utilization-sequestration-ccu-ccs, accessed 22 Dec. 2019.

86 C. Mora et al., 'Broad Threat to Humanity from Cumulative Climate Hazards Intensified by Greenhouse Gas Emissions', *Nature Climate Change*, 8/12 (2018), 1062–71. doi: 10.1038/s41558-018-0315-6

87 G.E.P. Box, 'Science and Statistics', *Journal of the American Statistical Association*, 71/356 (1976), 791–9. doi: 10.2307/2286841

88 HM Government, *Leading on Clean Growth: Government Response to the Committee on Climate Change 2019 Progress Report to Parliament – Reducing UK Emissions*, HM Government: London, 2019, https://www.gov.uk/government/publications/committee-on-climate-changes-2019-progress-reports-government-responses, accessed 22 Dec. 2019.

89 For more on this see Dani Rodrik's book, *Economics Rules: The Rights and Wrongs of the Dismal Science*, W.W. Norton & Company: New York, 2015.

90 E. Gómez-Baggethun and J.M. Naredo, 'In Search of Lost Time: The Rise and Fall of Limits to Growth in International Sustainability Policy', *Sustainability Science*, 10/3 (2015), 385–95. doi: 10.1007/s11625-015-0308-6

91 The original report was in the 1970s: D.H.M. Meadows et al., *The Limits to Growth: A Report for the Club of Rome's Project on the Predicament of Mankind* (1972). doi: 10.1111/j.1752-1688.1972.tb05230.x , but updates since then find we are tracking the forecast in some areas: U. Bardi, 'Limits to Growth', in *International Encyclopedia of the Social & Behavioral Sciences* (2015), 138–43. doi: 10.1016/B978-0-08-097086-8.91047-X

92 Thirty countries for microcomponents, seventy Apple stores worldwide: Magdalena Petrova, 'We Traced What It Takes to Make an iPhone, from Its Initial Design to the Components and Raw Materials Needed to Make It a Reality', CNBC (14 Dec. 2018), https://www.cnbc.com/2018/12/13/inside-apple-iphone-where-parts-and-materials-come-from.html, accessed 4 Jan. 2020.

93 James S. Henry, 'Taxing Tax Havens: How to Respond to the Panama Papers', *Foreign Affairs* (12 Apr. 2016), https://www.foreignaffairs.com/articles/panama/2016-04-12/taxing-tax-havens, accessed 22 Dec. 2019.

94 For more see Mariana Mazzucato, *The Entrepreneurial State: Debunking Public vs. Private Sector Myths*, Anthem Press: London, 2013.

95 As argued ibid.

96 Heather Stewart, 'This Is How We Let the Credit Crunch Happen, Ma'am ...', *The Guardian* (26 Jul. 2009), https://www.theguardian.com/uk/2009/jul/26/monarchy-credit-crunch, accessed 22 Dec. 2019.

Chapter Ten – Hope – Valuing the Future

1 'Too cheap to meter': 'Abundant Power from Atom Seen: It Will Be Too Cheap for Our Children to Meter, Strauss Tells Science Writers', *New York Times* (17 Sep. 1954), https://www.nytimes.com/1954/09/17/archives/abundant-power-from-atom-seen-it-will-be-too-cheap-for-our-children.html. 'Not the slightest indication': Alexander Moszkowski, *Einstein, Einblicke in seine Gedankenwelt; gemeinverständliche Betrachtungen über die Relativitätstheorie und ein neues Weltsystem, entwickelt aus Gesprächen mit Einstein*, Hoffmann und Campe: Hamburg, 1921.

2 J. Bessen, 'More Machines, Better Machines... or Better Workers?' *Journal of Economic History*, 72/1 (2012), 44–74. doi: 10.1017/s0022050711002439

3 For more see Richard Baldwin, *The Globotics Upheaval: Globalisation, Robotics and the Future of Work*, Oxford University Press: Oxford, 2019.

4   E. J. Hobsbawm, 'The Machine Breakers', *Past and Present*, 1 (1952), 57–70, doi:10.1093/past/1.1.57

5   M. Hermanussen, 'Stature of Early Europeans', *Hormones*, 2/3 (2003), 175–8. doi: 10.14310/horm.2002.1199

6   Karl Polanyi, *The Great Transformation*, Farrar & Rinehart: New York, 1944.

7   Branko Milanovic, *Global Inequality: A New Approach for the Age of Globalization*, Harvard University Press: Cambridge, MA, 2016.

8   Joseph A. Schumpeter, *Capitalism, Socialism and Democracy*, Harper Perennial: New York, 1950, pp. 31–2.

9   This splitting of benign and malign impacts on inequality is also taken from Milanovic, *Global Inequality*.

10 As many economists have argued, this tends to make a nonsense of Ricardo's theory of competitive advantage.

11 C. Lakner and B. Milanovic, 'Global Income Distribution: From the Fall of the Berlin Wall to the Great Recession', *World Bank Economic Review*, 30/2 (2016), 203–32. doi: 10.1596/1813-9450-6719

12 See for example Nick Routley, 'Visualizing the Trillion-Fold Increase in Computing Power', *Visual Capitalist* (4 Nov. 2017), https://www.visualcapitalist.com/visualizing-trillion-fold-increase-computing-power/, accessed 22 Dec. 2019.

13 M.R. Gillings, M. Hilbert and D.J. Kemp, 'Information in the Biosphere: Biological and Digital Worlds', *Trends in Ecology & Evolution*, 31/3 (2016), 180–9. doi: 10.1016/j.tree.2015.12.013

14 K.D. Tsavdaridis et al., 'Application of Structural Topology Optimisation in Aluminium Cross-Sectional Design', *Thin-Walled Structures*, 139 (2019), 372–8. doi: 10.1016/j.tws.2019.02.038

15 Many commentators are worried about the millions of truck drivers who might be displaced by automated vehicles – but this is looking in the wrong place; the messy real world is still a challenge to machine learning, but the controlled environment of a warehouse isn't. James Vincent, 'Amazon's Latest Warehouse Machine Demonstrates the Slow Drip of Automation', *The Verge* (13 May 2019), https://www.theverge.com/2019/5/13/18617465/amazon-automation-warehouse-workers-replaced-technology-jobs, accessed 22 Dec. 2019.

16 There are still huge challenges with automated vehicles, as AIs can be easy to fool – but increased computing speed and data will continue to help: D. Heaven, 'Why Deep-Learning AIs Are so Easy to Fool', *Nature*, 574/7777 (2019), 163–6. doi: 10.1038/d41586-019-03013-5

17 C.B. Frey and M. Osborne, 'The Future of Employment: How Susceptible Are Jobs to Computerisation?', *Oxford Martin School* (2013). An updated version of this paper was published in *Technological Forecasting and Social Change*, 114 (2017), 254–80. doi:10.1016/j.techfore.2016.08.019.

18 Bank of England: Larry Elliot, 'Robots Threaten 15m UK Jobs, Says Bank of England's Chief Economist', *The Guardian* (12 Nov. 2015), https://www.theguardian.com/business/2015/nov/12/robots-threaten-low-paid-jobs-says-bank-of-england-chief-economist; EU report: European Commission, 'AI The Future of Work? Work of the Future!' (3 May 2019), https://ec.europa.eu/digital-single-market/en/news/future-work-work-future. The report to the US Congress found that workers on the lowest wages face an 83% chance of having their work lost to robots within the next few decades: J. Furman et al., 'Artificial Intelligence, Automation, and the Economy', Executive Office

of the President (Dec. 2016), https://obamawhitehouse.archives.gov/blog/2016/12/20/ artificial-intelligence-automation-and-economy, accessed 5 Jan. 2020.

19   As economist Richard Baldwin points out, it would be impossible for global trade flows to double in a little over a year, since we'd have to build many more ports and transportation links. Yet the trade in world information has doubled every two years for decades now: Eshe Nelson, 'Globots and Telemigrants: The New Languages of the Future of Work', *Quartz* (14 Jun. 2019), https://qz.com/1642691/richard-baldwin-on-the-inhumanely-fast-next-phase-of-globalization/, accessed 22 Dec. 2019.

20   Will Dahlgreen, '37% of British Workers Think Their Jobs Are Meaningless', *YouGov* (12 Aug. 2015), https://yougov.co.uk/topics/lifestyle/articles-reports/2015/08/12/british-jobs-meaningless, accessed 22 Dec. 2019.

21   For a rough accounting see David Graeber's *Bullshit Jobs: A Theory*, Simon & Schuster: New York, 2018.

22   See Jim Harter, 'Employee Engagement on the Rise in the U.S.', *Gallup* (26 Aug. 2018), https://news.gallup.com/poll/241649/employee-engagement-rise.aspx; and Tomas Chamorro-Premuzic, 'Workplace Disengagement Is a "Worldwide Epidemic"', *Management Today* (18 Dec. 2014), https://www.managementtoday.co.uk/workplace-disengagement-worldwide-epidemic/article/1301164, accessed 22 Dec. 2019.

23   Steve Crabtree, 'Worldwide, 13% of Employees Are Engaged at Work', *Gallup* (8 Oct. 2013), https://news.gallup.com/poll/165269/worldwide-employees-engaged-work.aspx, accessed 22 Dec. 2019.

24   John Maynard Keynes, 'Economic Possibilities for Our Grandchildren' (1930), in *Essays in Persuasion*, W.W. Norton & Company: New York, 1963, p. 5.

25   In fact, they found a productivity increase of 40%: Sophie Jackman, 'Microsoft's Japanese Division Switched to a 4-Day Workweek – Then Productivity Skyrocketed', *TIME* (4 Nov. 2019), https://time.com/5717401/microsoft-4-day-workweek/, accessed 22 Dec. 2019.

26   Ben Laker and Thomas Roulet, 'Will the 4-Day Workweek Take Hold in Europe?', *Harvard Business Review* (5 Aug. 2019), https://hbr.org/2019/08/will-the-4-day-workweek-take-hold-in-europe, accessed 22 Dec. 2019.

27   J. Nässén and J. Larsson, 'Would Shorter Working Time Reduce Greenhouse Gas Emissions? An Analysis of Time Use and Consumption in Swedish Households', *Environment and Planning C: Government and Policy*, 33/4 (2015), 726–45. doi: 10.1068/c12239

28   K.W. Knight, E.A. Rosa and J.B. Schor, 'Could Working Less Reduce Pressures on the Environment? A Cross-National Panel Analysis of OECD Countries, 1970–2007', *Global Environmental Change*, 23/4 (2013), 691–700. doi: 10.1016/j.gloenvcha.2013.02.017

29   Repetitive 'thinking tasks': Jared Spataro, 'Bringing AI to Excel – 4 New Features Announced Today at Ignite', Microsoft Blogs (24 Sep. 2018), https://www.microsoft.com/en-us/microsoft-365/blog/2018/09/24/bringing-ai-to-excel-4-new-features-announced-today-at-ignite/. Pattern-recognition tasks: H.A. Haenssle et al., 'Man Against Machine: Diagnostic Performance of a Deep Learning Convolutional Neural Network for Dermoscopic Melanoma Recognition in Comparison to 58 Dermatologists', *Annals of Oncology*, 29/8 (2018), 1836–42. doi: 10.1093/annonc/mdy166. Breast cancer screening: Ian Sample, 'AI System Outperforms Experts in Spotting Breast Cancer', *The Guardian* (1 Jan. 2020), https://www.theguardian.com/society/2020/jan/01/ai-system-outperforms-experts-in-spotting-breast-cancer, accessed 3 Jan. 2020.

30 Sendhil Mullainathan and Eldar Shafir, *Scarcity: Why Having Too Little Means So Much*, Macmillan: New York, 2003.

31 'Daniel Kahneman', *Desert Island Discs*, BBC (16 Aug. 2013), https://www.bbc.co.uk/programmes/b038112v, accessed 22 Dec. 2019.

32 Elizabeth Dunn and Michael Norton, *Happy Money: The Science of Happier Spending*, Simon & Schuster: New York, 2014.

33 R. Romaniuc and C. Bazart, 'Intrinsic and Extrinsic Motivation', in Jürgen G. Backhaus, ed., *Encyclopedia of Law and Economics*, Springer: New York, 2015, 1–4. doi: 10.1007/978-1-4614-7883-6_270-1, p. 699.

34 D. Kahneman et al., 'A Survey Method for Characterizing Daily Life Experience: The Day Reconstruction Method', *Science*, 306/5702 (2004), 1776–80 (2004). doi: 10.1126/science.1103572

35 Ellie Mae O'Hagan, 'Love the Idea of a Universal Basic Income? Be Careful What You Wish For', *The Guardian* (23 Jun. 2017), https://www.theguardian.com/commentisfree/2017/jun/23/universal-basic-income-ubi-welfare-state, accessed 22 Dec. 2019.

36 Joi Ito, 'The Paradox of Universal Basic Income', *Wired* (29 Mar. 2018), https://www.wired.com/story/the-paradox-of-universal-basic-income/, accessed 22 Dec. 2019.

37 In a letter to King Frederick William of Prussia in 1767, as cited in Jonathan Kimmelman, *Gene Transfer and the Ethics of First-in-Human Research*, Cambridge University Press: Cambridge, 2019, p. 6.

38 Despite these recent improvements, many experts now see a carbon tax as a simpler, more reliable approach which doesn't need as much technocratic control.

39 D. Klenert et al., 'Making Carbon Pricing Work for Citizens', *Nature Climate Change*, 8/8 (2018), 669–77. doi: 10.1038/s41558-018-0201-2

40 European Commission, 'State of the Energy Union 2021: Renewables Overtake Fossil Fuels As the EU's Main Power Source', (26 Oct. 2021), https://ec.europa.eu/commission/presscorner/detail/en/IP_21_5554, accessed 9 Jan 2022.

41 William Nordhaus, who we heard about in Chapter Nine, thinks this price should be closer to $300: Chris Mooney and Andrew Freeman, 'The World Needs a Massive Carbon Tax in Just 10 Years to Limit Climate Change, IMF Says', *Washington Post* (10 Oct. 2019), https://www.washingtonpost.com/climate-environment/2019/10/10/world-needs-massive-carbon-tax-just-years-limit-climate-change-imf-says/, accessed 22 Dec. 2019. Other research suggests that the price should ramp up very quickly in early years before tapering off, given how quick decarbonization has to be now: K.D. Daniel, R.B. Litterman and G. Wagner, 'Declining CO_2 Price Paths', *Proceedings of the National Academy of Sciences of the United States of America*, 116/42 (2019), 20886–91. doi: 10.1073/pnas.1817444116

42 Elke Asen, 'Carbon Taxes in Europe', *Tax Foundation* (14 Nov. 2019), https://taxfoundation.org/carbon-taxes-in-europe-2019/, accessed 22 Dec. 2019.

43 Sam Fleming and Chris Giles, 'EU Risks Trade Fight over Carbon Border Tax Plans', *Financial Times* (15 Oct. 2019), available to subscribers at https://www.ft.com/content/154368c8-ef55-11e9-ad1e-4367d8281195, accessed 3 Jan. 2020.

44 J. Heitzig and U. Kornek, 'Bottom-up Linking of Carbon Markets under Far-Sighted Cap Coordination and Reversibility', *Nature Climate Change*, 8/3 (2018), 204–9. doi: 10.1038/s41558-018-0079-z.

45 Economists such as Joe Stiglitz, Rutger Hoekstra and Kate Raworth, among many others.

46 One example is Jack Philips, 'Principles of National Capital Accounting', Office for National Statistics (24 Feb. 2017), https://www.ons.gov.uk/economy/environmentalaccounts/methodologies/principlesofnaturalcapitalaccounting, accessed 22 Dec. 2019, although there are many more. Iceland prioritizing well-being: BBC News, 'Iceland Puts Well-Being Ahead of GDP in Budget' (3 Dec. 2019), https://www.bbc.com/news/world-europe-50650155, accessed 3 Jan. 2020.

47 Rutger Hoekstra, *Replacing GDP by 2030: Towards a Common Language for the Well-Being and Sustainability Community*, Cambridge University Press: Cambridge, 2019, p. 238.

48 It probably isn't too great a leap. For instance, economists today often use analogies from the forces of nature – think 'innovation ecosystems', 'rivers of credit', 'riding out the storm' or 'torrents of debt'. Equally, natural scientists sometimes talk in the language of economists when they discuss 'natural capital' or 'ecosystem services'.

49 Dani Rodrik, *Economics Rules: The Rights and Wrongs of the Dismal Science*, W.W. Norton & Company: New York, 2015, p. 116.

50 James Lovelock, *Novacene: The Coming Age of Hyperintelligence*, Allen Lane: London, 2019.

51 Daron Acemoglu and James A. Robinson, *Why Nations Fail: The Origins of Power, Prosperity, and Poverty*, Currency: New York, 2013.

52 P. H. Hauser, D. G. Rand, A. Peysakhovich and M.A. Nowak, 'Cooperating with the Future', *Nature*, 511 (2014), 220–23. https://doi.org/10.1038/nature13530.

53 D. Kruse, 'Does Employee Ownership Improve Performance?', *IZA World of Labor*, 311 (2016). doi: 10.15185/izawol.311; P.S. Martins, 'Dispersion in Wage Premiums and Firm Performance', *Economics Letters*, 101, 63–65 (2008); and J. Lampel, A. Bhalla and P. Jha, 'Model Growth: Do Employee-Owned Businesses Deliver Sustainable Performance?' (2010), City University London, https://openaccess.city.ac.uk/id/eprint/16278/, accessed 22 Dec. 2019.

54 Andrew Winston, 'Is the Business Roundtable Statement Just Empty Rhetoric?', *Harvard Business Review* (30 Aug. 2019), https://hbr.org/2019/08/is-the-business-roundtable-statement-just-empty-rhetoric, accessed 22 Dec. 2019.

55 Lydia DePillis, 'Shareholder Activism Is on the Rise, but Companies Are Fighting Back', CNN (31 Jan. 2019), https://edition.cnn.com/2019/01/30/investing/activist-shareholders/index.html, accessed 22 Dec. 2019.

56 Kelly Gilblom, 'Big Money Starts to Dump Stocks That Pose Climate Risks', *Bloomberg* (7 Aug. 2019), https://www.bloomberg.com/news/articles/2019-08-07/big-money-starts-to-dump-stocks-that-pose-climate-risks, accessed 22 Dec. 2019.

57 J.C.J.M. van den Bergh, 'A Third Option for Climate Policy Within Potential Limits to Growth', *Nature Climate Change*, 7/2 (2017), 107–12. doi: 10.1038/nclimate3113

58 Adam Smith, *An Inquiry into the Nature and Causes of the Wealth of Nations*, University of Chicago: Chicago, 1977, p. 138.

59 C.L. Spash, 'The Economics of Climate Change Impacts à la Stern: Novel and Nuanced or Rhetorically Restricted?', *Ecological Economics*, 63/4 (2007), 706–713. doi: 10.1016/j.ecolecon.2007.05.017

60 J.M. Keynes, *The General Theory of Employment, Interest, and Money*, Springer Nature: Switzerland, 2018. doi: 10.1007/978-3-319-70344-2, *Preface vii.*

61 A.H. Maslow, 'A Theory of Human Motivation', *Psychological Review*, 50/4 (1943), 370–96. doi: 10.1037/h0054346

62   For example Kahneman et al., 'A Survey Method for Characterizing Daily Life Experience'.

Epilogue – Pessimism – Are We Almost at the End?

1   To our knowledge, the first time the question was asked was in: K. Tsiolkovsky, 'The Planets are Occupied by Living Beings' (1933), Archives of the Tsiolkovsky State Museum of the History of Cosmonautics, Kaluga, Russia. The story of Fermi's walk is told in Eric M. Jones, '"Where Is Everybody?" An Account of Fermi's Question' (1985), Los Alamos National Laboratory, document identifier: LA-10311-MS/UC-34B, http://www.fas.org/sgp/othergov/doe/lanl/la-10311-ms.pdf, accessed 3 Jan. 2020.

2   E.A. Petigura, A.W. Howard and G.W. Marcy, 'Prevalence of Earth-Size Planets Orbiting Sun-Like Stars', *Proceedings of the National Academy of Sciences of the United States of America*, 110/48 (2013), 19273–8. doi: 10.1073/pnas.1319909110. And there may be life that we don't know of yet on planets we previously thought were unsuitable for life.

3   Robin Hanson, 'The Great Filter – Are We Almost Past It?' (15 Sep. 1998), http://mason.gmu.edu/~rhanson/greatfilter.html, accessed 3 Jan. 2020.

4   Nick Bostrom, *Superintelligence: Paths, Dangers, Strategies*, Oxford University Press: Oxford, 2014.

5   Yale Climate Connections, 'Why Climate Change Is a "Threat Multiplier"' (20 Jun. 2019), https://www.yaleclimateconnections.org/2019/06/why-climate-change-is-a-threat-multiplier/, accessed 4 Jan. 2020.

6   Such as those in Cape Town, Chennai and Karachi.

7   Thomas Homer-Dixon, *The Upside of Down: Catastrophe, Creativity, and the Renewal of Civilization*, Island Press: Washington, DC, 2006.

8   Ted Macdonald, 'How Broadcast TV Networks Covered Climate Change in 2020', Media Matters for America (3 Oct. 2021), https://www.mediamatters.org/broadcast-networks/how-broadcast-tv-networks-covered-climate-change-2020, accessed 9 Jan. 2022. The four networks included were ABC, CBS, NBC and Fox.

9   For reporting see Damian Carrington, *The Guardian* (15 Sep. 2021), https://www.theguardian.com/environment/2021/sep/15/cake-mentioned-10-times-more-than-climate-change-on-uk-tv-report, accessed 9 Jan. 2022.

10   For just one of many such articles, see the reporting, image and lack of climate change discussion here: Victoria Pease, 'Scotland Enjoys Warmest September Day for More Than a Century', *STV* (8 Sep. 2021), https://news.stv.tv/scotland/scotland-could-see-warmest-september-day-in-a-century, accessed 9 Jan. 2022.

11   Stories are several times more memorable than facts: G.H. Bower and Michal C. Clark, 'Narrative Stories as Mediators for Serial Learning', *Psychonomic Science*, 14/4 (1969), 181–2. doi: https://doi.org/10.3758/BF03332778.

12   John Gibbons, 'Climate Change Reporting Should Be Obligatory', *Irish Times* (26 Apr. 2019), https://www.irishtimes.com/opinion/climate-change-reporting-should-be-obligatory-1.3871618, accessed 22 Dec. 2019.

13   D. Pauly, 'Anecdotes and the Shifting Baseline Syndrome of Fisheries', *Trends in Ecology & Evolution*, 10/10, 430 (1995). doi: 10.1016/S0169-5347(00)89171-5

14   F.C. Moore et al., 'Rapidly Declining Remarkability of Temperature Anomalies May Obscure Public Perception of Climate Change', *Proceedings of the National Academy*

of Sciences of the United States of America, 116/11 (2019), 4905–10. doi: 10.1073/pnas.1816541116

15   Richard B. Rood, 'Let's Call It: 30 Years of Above Average Temperatures Means the Climate Has Changed', *The Conversation* (26 Feb. 2015), https://theconversation.com/lets-call-it-30-years-of-above-average-temperatures-means-the-climate-has-changed-36175, accessed 22 Dec. 2019.

16   D.M. Kahan, 'Ideology, Motivated Reasoning, and Cognitive Reflection: An Experimental Study', *SSRN Electronic Journal*, 8/4 (2013), 407–24. doi: 10.2139/ssrn.2182588

17   There are many such studies on this, for example C. Drummond and B. Fischhoff, 'Individuals with Greater Science Literacy and Education Have More Polarized Beliefs on Controversial Science Topics', *Proceedings of the National Academy of Sciences of the United States of America*, 114/36 (2017), 9587–92. doi: 10.1073/pnas.1704882114

18   Max Planck, *Scientific Autobiography and Other Papers*, Williams & Norgate Ltd: London, 1950, p. 33.

19   Elisa Shearer and Katerina Eva Matsa, 'News Use Across Social Media Platforms 2018', *Pew Research Center Journalism & Media* (10 Sep. 2018), https://www.journalism.org/2018/09/10/news-use-across-social-media-platforms-2018/, accessed 22 Dec. 2019.

20   BBC News, 'Ofcom Reports More People Using Social Media for News' (24 Jul. 2019), https://www.bbc.co.uk/news/entertainment-arts-49098430, accessed 22 Dec. 2019.

21   Shearer and Matsa, 'News Use Across Social Media Platforms 2018'.

22   For a recent study on the efficacy of lies and repetition see D. Cash et al., 'The Effect of Statement Type and Repetition on Deception Detection', *Cognitive Research: Principles and Implications*, 4/1 (2019), doi: 10.1186/s41235-019-0194-z

23   J. Allgaier, 'Science and Environmental Communication on YouTube: Strategically Distorted Communications in Online Videos on Climate Change and Climate Engineering', *Frontiers in Communications*, 4/36 (2019), 1–15. doi: 10.3389/fcomm.2019.00036

24   Stephanie Kirchgaessner, 'The Obscure Law That Explains Why Google Backs Climate Deniers', *The Guardian* (11 Oct. 2019), https://www.theguardian.com/environment/2019/oct/11/obscure-law-google-climate-deniers-section-230, accessed 22 Dec. 2019.

25   There are many such critiques of this behaviour, for instance: Mehreen Kahn, 'Facebook Rules on Political Advertising Criticised by EU Parties', *Financial Times* (28 Mar. 2019), available to subscribers at https://www.ft.com/content/0dab95ba-5156-11e9-b401-8d9ef1626294, or Julia Carrie Wong, 'Google Latest Tech Giant to Crack down on Political Ads as Pressure on Facebook Grows', *The Guardian* (21 Nov. 2019), https://www.theguardian.com/technology/2019/nov/20/google-political-ad-policy-facebook-twitter. Regulators looking at behaviour: Jack Nicas, Karen Weise and Mike Isaac, 'How Each Big Tech Company May Be Targeted by Regulators', *New York Times* (8 Sep. 2019), https://www.nytimes.com/2019/09/08/technology/antitrust-amazon-apple-facebook-google.html, all accessed 3 Jan. 2020.

26   Geoffrey West, *Scale: The Universal Laws of Life and Death in Organisms, Cities and Companies*, Penguin Press, 2017, p. 414.

27   From the first forest clearings by humans more than 10,000 years ago, to the Industrial Revolution and today.

28   This story is told in Clive Hamilton, *Earthmasters: The Dawn of the Age of Climate Engineering*, Yale University Press: New Haven, CT, 2014, p. 76.

29   P. Irvine et al., 'Halving Warming with Idealized Solar Geoengineering Moderates Key Climate Hazards', *Nature Climate Change*, 9/4 (2019), 295–9. doi: 10.1038/s41558-019-0398-8

30   Working Group 1, T.F. Stocker et al., eds, *Climate Change 2013: The Physical Science Basis. Contribution of Working Group I to the Fifth Assessment Report of the Intergovernmental Panel on Climate Change*, IPCC, Cambridge University Press: United Kingdom and New York, 2013. doi: 10.1017/CBO9781107415324

31   D. Shindell and C.J. Smith, 'Climate and Air-Quality Benefits of a Realistic Phase-out of Fossil Fuels', *Nature*, 573/7774 (2019), 408–11. doi: 10.1038/s41586-019-1554-z

32   M.J. Henehan et al., 'Rapid Ocean Acidification and Protracted Earth System Recovery Followed the End-Cretaceous Chicxulub Impact', *Proceedings of the National Academy of Sciences of the United States of America*, 116/45 (2019), 22500–4. doi: 10.1073/pnas.1905989116. Researcher Michael Henehan was quoted as saying: 'If 0.25 was enough to precipitate a mass extinction, we should be worried.' The pH of the ocean may drop by 0.4 units by the end of the century with continuing carbon emissions (and 0.15 units if temperature rise is limited to two degrees Celsius). For the reporting on this see Damian Carrington, 'Ocean Acidification Can Cause Mass Extinctions, Fossils Reveal', *The Guardian* (21 Oct. 2019), https://www.theguardian.com/environment/2019/oct/21/ocean-acidification-can-cause-mass-extinctions-fossils-reveal, accessed 22 Dec. 2019.

33   T.M. Lenton, 'Can Emergency Geoengineering Really Prevent Climate Tipping Points?', in J.J. Blackstock and S. Low, eds, *Geoengineering Our Climate? Ethics, Politics and Governance*, CRC Press: Boca Raton, FL, 2018.

34   Daniel Levitt et al., 'Deadly Weather: The Human Cost of 2018's Climate Disasters – Visual Guide', *The Guardian* (21 Dec. 2018), https://www.theguardian.com/environment/ng-interactive/2018/dec/21/deadly-weather-the-human-cost-of-2018s-climate-disasters-visual-guide, accessed 22 Dec. 2019.

35   Oliver Morton gives a good outline about how this could happen without conflict in his book *The Planet Remade*. Essentially, poorer states could start out secretly on a small scale (Morton calls the group 'The Concert'), and things could move on from there in a more tension-diffusing manner. See more in Oliver Morton, *The Planet Remade: How Geoengineering Could Change the World*, Princeton University Press: Princeton, NJ, 2016.

36   A.C. Lin, 'Does Geoengineering Present a Moral Hazard?', *Ecology Law Quarterly*, 40/3 (2013), 673–712. doi: 10.15779/Z38JP1J

37   For an example see F. Biermann, 'It Is Dangerous to Normalize Solar Geoengineering Research', *Nature*, 30 (2021), 595, doi: 10.1038/d41586-021-01724-2

38   Kravitz, B. et al. 'Geoengineering: Whiter skies?', *Geophysical Research Letters*, 39/L11801 (2012), doi: 10.1029/2012GL051652

39   The cartoon can be seen here: Dania El Akkawi, 'Climate Change and COVID-19: There Is More Than One Curve to Flatten', *The Cairo Review of Global Affairs* (28 Jun. 2020), https://www.thecairoreview.com/midan/climate-change-and-covid-19-there-is-more-than-one-curve-to-flatten/, accessed 16 Jan. 2022.

40   See many media articles; for example, cleaner air in the US: Kate Abnett, 'Coronavirus Lockdowns Give Europe's Cities Cleaner Air, *Reuters* (29 Mar. 2020), https://www.reuters.com/article/us-health-coronavirus-air-pollution-idUSKBN21G0XA, accessed 16 Jan. 2022; cleaner water in India: Press Trust of India, 'Lockdown Leads to Cleaner Water Bodies; Minister Says Time to "Introspect", Make India Water-Secured',

Hindu Times (22 May 2020), https://www.thehindu.com/news/national/lockdown-leads-to-cleaner-water-bodies-minister-says-time-to-introspect-make-india-water-secured/article31650233.ece, accessed 16 Jan. 2022; cleaner water in Italy: Catherine Clifford, 'The Water in Venice, Italy's Canals Is Running Clear Mid the COVID-19 Lockdown – Take a Look', *CNBC* (18 Mar. 2020), https://www.cnbc.com/2020/03/18/photos-water-in-venice-italys-canals-clear-amid-covid-19-lockdown.html, accessed 16 Jan. 2022; and lower noise: Johnathan Amos, 'Coronavirus Lockdowns Reduced Human "Rumble"', BBC News (24 Jul. 2022), https://www.bbc.com/news/science-environment-53518751, accessed 16 Jan. 2022.

41   For example, see the Organization of Economic Development, 'Making the Green Recovery Work for Jobs, Income and Growth', *OECD Policy Responses to Coronavirus* (6 Oct. 2020), https://www.oecd.org/coronavirus/policy-responses/making-the-green-recovery-work-for-jobs-income-and-growth-a505f3e7/, accessed 16 Jan. 2022; and Mark Zandi, 'Biden's Build Back Better Plan Will Improve Nearly Every Community in America', *CNN Business* (11 May 2021), https://edition.cnn.com/2021/05/11/perspectives/biden-build-back-better/index.html, accessed 16 Jan. 2022.

42   Deborah Gleeson, 'Wealthy Nations Starved the Developing World of Vaccines. Omicron Shows the Cost of This Greed', *The Conversation* (30 Nov. 2021), https://theconversation.com/wealthy-nations-starved-the-developing-world-of-vaccines-omicron-shows-the-cost-of-this-greed-172763, accessed 16 Jan. 2022.

43   Jeff Tollefson, 'COVID Curbed Carbon Emissions in 2020 – but Not by Much', Nature News (15 Jan. 2021), https://www.nature.com/articles/d41586-021-00090-3, accessed 16 Jan. 2022. But emission reductions lie around 5–6%. The best budget available is probably from the GCP at Global Carbon Project, 'Global Carbon Project 2021 (Version 1.0)' (2021), doi:10..18160/gcp-2021 The figure of 6.2% includes fossil fuel emission declines of 5.4% and around 0.8% from land use changes.

44   See Miria Pigato, 'Green or Brown: The COVID-19 Crisis and the Road to Recovery', *World Bank Blogs* (15 Nov. 2021), https://blogs.worldbank.org/developmenttalk/green-or-brown-covid-19-crisis-and-road-recovery, accessed 16 Jan. 2022 and The International Energy Agency, 'Tracking Sustainable Recoveries', *IEA* (Oct. 2021), https://www.iea.org/reports/sustainable-recovery-tracker/tracking-sustainable-recoveries, accessed 16 Jan. 2022.

45   UNAIDS, 'A Dose of Reality: How Rich Countries and Pharmaceutical Corporations Are Breaking Their Vaccine Promises', UNAIDS (21 Oct. 2021), https://www.unaids.org/en/resources/presscentre/featurestories/2021/october/20211021_dose-of-reality, accessed 16 Jan. 2022; and Mohit Mookim, 'The World Loses Under Bill Gates' Vaccine Colonialism', *Wired* (19 May 2021), https://www.wired.com/story/opinion-the-world-loses-under-bill-gates-vaccine-colonialism/, accessed 16 Jan. 2022.

46   Jeremy Mark and Vasuki Shastry, 'Let Them Eat Communiqués: Rich Countries' Pandemic Inaction, *New Atlanticist* (20 Oct. 2021), https://www.atlanticcouncil.org/blogs/new-atlanticist/let-them-eat-communiques-rich-countries-pandemic-inaction/, accessed 16 Jan. 2022.

47   See Justin Worland, 'Did We Just Blow Our Last, Best Chance to Tackle Climate Change?', *Time* (30 Dec. 2021), https://time.com/6130470/climate-change-2021-build-back-better/, accessed 16 Jan. 2020; and Ariel Wittenberg, 'Pandemic Economic Recovery Could Worsen Climate Change Health Impacts', *Scientific American* (21 Oct. 2021), https://www.scientificamerican.com/article/pandemic-economic-recovery-could-worsen-climate-change-health-impacts/, accessed 16 Jan. 2022.

48   This phrase has been attributed to several people, including Fredric Jameson and Slavoj Žižek.

49   Guardian Sport and agencies, 'Schalke Chairman Under Pressure to Resign over Racist Comments', *The Guardian* (5 Aug. 2019), https://www.theguardian.com/football/2019/aug/05/schalke-chairman-under-pressure-to-resign-over-racist-comments, accessed 22 Dec. 2019.

50   Sam Adler-Bell, 'Why White Supremacists are Hooked on Green Living', *The New Republic* (24 Sep. 2019), https://newrepublic.com/article/154971/rise-ecofascism-history-white-nationalism-environmental-preservation-immigration, accessed 22 Dec. 2019.

51   Adam K. Raymond, 'The World's 500 Richest People Increased Their Wealth by $1.2 Trillion in 2019', *New York Magazine* (27 Dec. 2019), http://nymag.com/intelligencer/2019/12/worlds-500-richest-upped-wealth-by-usd1-2-trillion-in-2019.html, accessed 3 Jan. 2020.

52   Brenna Hughes Neghaiwi and Simon Jessop, 'In 2020 the Ultra-Rich Got Richer. Now They're Bracing for the Backlash', *Reuters* (25 Mar. 2021), https://www.reuters.com/article/us-wealth-billionaires-outlook-insight-idUSKBN2BH0J7, accessed 9 Jan. 2022.

53   N. Butt et al., 'The Supply Chain of Violence', *Nature Sustainability*, 2 (2019), 742–7. doi: 10.1038/s41893-019-0349-4

54   International leaders offered a pitiful carrot of $20 million and no stick to stop deforestation activities: Julian Borger and Jonathan Watts, 'G7 Leaders Agree Plan to Help Amazon Countries Fight Wildfires', *The Guardian* (26 Aug. 2019), https://www.theguardian.com/world/2019/aug/26/g7-leaders-agree-plan-to-help-amazon-countries-fight-wildfires, accessed 3 Jan. 2020.

55   Martin Gilbert, *Winston S. Churchill: The Challenge of War, 1914–1916 (Volume III)*, Rosetta Books: New York, 2015, p. 10.

56   Edward L. Bernays, *Propaganda*, H. Liveright: New York, 1928, p. 135.

57   Wikipedia, 'List of Cognitive Biases', https://en.wikipedia.org/wiki/List_of_cognitive_biases, accessed 22 Dec. 2019.

58   N.D. Weinstein and W.M. Klein, 'Unrealistic Optimism: Present and Future', *Journal of Social and Clinical Psychology*, 15/1 (1996), 1–8. doi: 10.1521/jscp.1996.15.1.1

Epilogue – Hope – The Grass Is Greener

1   This categorization of human history was wonderfully outlined in: Simon L. Lewis and Mark A. Maslin, *Human Planet: How We Created the Anthropocene*, Pelican: London, 2018.

2   J. Mokyr, 'The Intellectual Origins of Modern Economic Growth', *Journal of Economic History*, 65/2 (2005), 285–351.

3   We are a long way from being able to do this; see for example attempts such as Biosphere 2 in the US to the Eden Project in the UK.

4   James Lovelock also makes this argument in his book *Novacene: The Coming Age of Hyperintelligence*, Allen Lane: London, 2019.

5   For instance, see the stories in Mark O'Connell's *Notes from an Apocalypse: A Personal Journey to the End of the World and Back*, Doubleday: New York, 2020. At one event super-rich clients reportedly asked questions such as 'How do I maintain authority over my security force after the [collapse] event?', https://www.salon.com/2018/07/16/we-asked-psychologists-why-so-many-rich-people-think-the-apocalypse-is-coming/, accessed 4 Jan. 2020.

6   G.W.F. Hegel, *Philosophy of Right* (1820), trans. S.W. Dyde, G. Bell: London, 1896, preface.

7   Among the many stories was Melissa Fisher's, which is described in heart-wrenching prose here: Melissa Fisher, 'The Covid Supplement Lifted Me out of Poverty. Then It Was Cut and My Life Went Back to the Way It Was', *The Guardian* (29 Dec. 2021), https://www.theguardian.com/commentisfree/2021/dec/29/the-covid-supplement-lifted-me-out-of-poverty-then-it-was-cut-and-my-life-went-back-to-the-way-it-was, accessed 16 Jan. 2022.

8   X. Zhang, X. Chen and X. Zhang, 'The Impact of Exposure to Air Pollution on Cognitive Performance', *Proceedings of the National Academy of Sciences of the United States of America*, 115/37 (2018), 9193–7. doi: 10.1073/pnas.1809474115

9   This is more generally called the Overton window, and describes the breadth of ideas which are tolerated in public discussion – from Unthinkable and Radical to Sensible and Popular. The Overton window moves over time; for example, African American rights in the US were unthinkable until they became policy over the course of several decades in the twentieth century.

10   Adam Vaughan, 'Most People in the UK Back Limits on Flying to Tackle Climate Change', *New Scientist* (18 Sep. 2019), https://www.newscientist.com/article/2216743-most-people-in-the-uk-back-limits-on-flying-to-tackle-climate-change/, accessed 22 Dec. 2019.

11   Martin Samuel, 'Lewis Hamilton's "Celebrity Green" Shtick Is so Hard to Swallow. At a Time of Climate Crisis F1 Is a Very Expensive Luxury', *Mail Online* (28 Oct. 2019), https://www.dailymail.co.uk/sport/formulaone/article-7623067/MARTIN-SAMUEL-COLUMN-Lewis-Hamiltons-celebrity-green-shtick-hard-swallow.html, accessed 22 Dec. 2019.

12   Civil servants: Richard Partington, 'Wellbeing Should Replace Growth as "Main Aim of UK Spending"', *The Guardian* (24 May 2019), https://www.theguardian.com/politics/2019/may/24/wellbeing-should-replace-growth-as-main-aim-of-uk-spending. Politicians: Christine Emba, 'What Nation Isn't Obsessed with Ensuring Economic Growth? New Zealand, Apparently', *Washington Post* (15 Jun. 2019), https://www.washingtonpost.com/opinions/what-nation-isnt-obsessed-with-ensuring-economic-growth-new-zealand-apparently/2019/06/14/f2aeabb8-8ee4-11e9-b08e-cfd89bd36d4e_story.html and Allegra Stratton, 'David Cameron Aims to Make Happiness the New GDP', *The Guardian* (14 Nov. 2010), https://www.theguardian.com/politics/2010/nov/14/david-cameron-wellbeing-inquiry. Business leaders: Patrick Jenkins, 'City Fund Managers Call for Rethink of Capitalism', *Financial Times* (14 Nov. 2019), available to subscribers at https://www.ft.com/content/1999422c-057a-11ea-9afa-d9e2401fa7ca, all accessed 4 Jan. 2020.

13   Michael Kerr, 'I'm a Travel Writer, but I'm Not Going to Fly Any More', *The Telegraph* (3 Oct. 2019), https://www.telegraph.co.uk/travel/comment/travel-writer-giving-up-flying/, accessed 22 Dec. 2019.

14   Nathaniel Geiger, 'There's Evidence That Climate Activism Could Be Swaying Public Opinion in the US', *The Conversation* (21 Sep. 2019), https://theconversation.com/theres-evidence-that-climate-activism-could-be-swaying-public-opinion-in-the-us-123740, accessed 22 Dec. 2019; and J.K. Swim, N. Geiger and M.L. Lengieza, 'Climate Change Marches as Motivators for Bystander Collective Action', *Frontiers in Communication*, 4/4 (2019). doi: 10.3389/fcomm.2019.00004

15   Jonathan Tirone and Boris Groendahl, 'Greta Effect Shakes up Austrian Politics

in Signal for Europe', *Bloomberg* (29 Sep. 2019), https://www.bloomberg.com/news/articles/2019-09-29/greta-effect-shakes-up-austrian-politics-in-signal-for-europe

16 Lisa Friedman, Maggie Astor and Linda Qiu, 'CNN Climate Town Hall: Here's What You Need to Know', *New York Times* (6 Sep. 2019), https://www.nytimes.com/live/2019/democrats-climate-town-hall, accessed 22 Dec. 2019.

17 G. Albrecht et al., 'Solastalgia: The Distress Caused by Environmental Change', *Australasian Psychiatry*, 15 (2007), 95–8. doi: 10.1080/10398560701701288

18 D. Eisenman et al., 'An Ecosystems and Vulnerable Populations Perspective on Solastalgia and Psychological Distress After a Wildfire', *EcoHealth*, 12/4 (2015), 602–10. doi: 10.1007/s10393-015-1052-1

19 *The Lancet*'s inclusion of solastalgia: S. Whitmee et al., 'Safeguarding Human Health in the Anthropocene Epoch: Report of the Rockefeller Foundation-Lancet Commission on Planetary Health', *The Lancet*, 386/10007 (2015), 1973–2028. doi: 10.1016/S0140-6736(15)60901-1. On various countries declaring mental-health crises, see Fiona Charlson, 'The Rise of "Eco-Anxiety": Climate Change Affects Our Mental Health, Too', *The Conversation* (17 Sep. 2019), https://theconversation.com/the-rise-of-eco-anxiety-climate-change-affects-our-mental-health-too-123002, accessed 22 Dec. 2019.

20 Ipsos MORI, 'Concern About Climate Change Reaches Record Levels with Half Now "Very Concerned"'(Aug. 2019), https://www.ipsos.com/ipsos-mori/en-uk/concern-about-climate-change-reaches-record-levels-half-now-very-concerned, accessed 3 Jan. 2020.

21 David Orr, *Down to the Wire: Confronting Climate Collapse*, Oxford University Press: Oxford, 2019, p. 172.

22 Ibid., p. 173.

23 Simon L. Lewis and Mark A. Maslin, *Human Planet: How We Created the Anthropocene*, Pelican: London, 2018.

ACKNOWLEDGEMENTS

This book owes a debt of gratitude to so many researchers who have done the difficult work of collecting data, building models and communicating results. It's tough to keep up with the rapid pace of developments and not a week goes by without many important papers that deserve to be read. I try to read as many as I can but with such voluminous output the role of journalists and science communicators is becoming more important than ever. Journalists like David Roberts, Simon Evans and Brad Plumer help us all to join the dots and elaborate major debates. Another important source of aggregated information is the many wonderful podcasts now available. I am addicted to The Energy Transition Show, Columbia Energy Exchange, Political Climate, Drilled, Outrage and Optimism, Resources Radio, Energy Policy Now, Gastropod, Climate One, and the Energy Gang, to name just a few. Thank you to the many journalists, presenters, guests and scientists who make these shows possible. If you are interested in further reading, a selected list of books, journal articles and podcasts used in the preparation of this book is available at www.drpaulbehrens.com.

I am deeply grateful to friends and colleagues who have given their expert suggestions on different sections of the book. My special thanks to Thijs Bosker, Nick Golledge, Rutger Hoekstra, René Kleijn, Arjan de Koning, José Mollogon, Meredith Niles, João Dias Rodrigues, Nadia Soudzilovskaia, Craig Stevens, Martina Vijver and David Zetland. Of course, the errors that remain are all mine. I'd like to thank Anya Al-Salem, Niko Kelnreiter and Nyasha Grecu for their keen eyes and research skills. Thank you also to Glen Peters, Ester van de Voet and Chris Nelder for their help with several specific details in the book. A huge thanks to James Shapiro, not only for being among the first readers of the final manuscript, but also for his support, guidance and friendship.

I am very fortunate to work with many wonderful colleagues who are too numerous to mention at two Leiden University institutes: Leiden University College (LUC) and the Institute of Environmental Sciences (CML). I am truly grateful to Judi Mesman and Arnold Tukker, both of whom have been unfailingly supportive during the writing of this book. The friendship of Ed Frettingham, Kai Hebel, Steve Becker, Isabella Chen, Fransijn Bulhof, Kim and Regan Watts kept me sane during the process of writing this book. I have also benefitted from two writing residencies at Faber Olot and Faber Andorra.

Enormous thanks to Susie Nicklin for her belief in the book, to Alex Spears, Juliet Garcia and Sarah Terry for their efforts in corralling the book to publication, and to Will Atkins for his keen eye. Thank you to Michael Salu who designed such an incredible eye-catching cover. My deepest thanks to Robert Caskie for his hard work in making this book a reality and to Liza de Block for her support and enthusiasm.

Thank you to Liz, Ken and Libby for their incredible support over the years. Thanks for your keen eye for graphical design, Libby. Finally, words can't do justice to the help, love and support I received from Caoilinn during the writing of this book. Without her this book simply wouldn't have been possible.

ABOUT THE AUTHOR

PAUL BEHRENS is Associate Professor in Environmental Change at Leiden University. His research and writing has appeared in leading scientific journals and media outlets such as the New York Times, Scientific American, Nature Food, Nature Sustainability, PNAS, Nature Energy, TEDx, Politico, Thomson Reuters and the BBC.